ISNM
International Series of Numerical Mathematics
Vol. 145

Modern Developments in Multivariate Approximation

5th International Conference, Witten-Bommerholz (Germany), September 2002

Werner Haussmann
Kurt Jetter
Manfred Reimer
Joachim Stöckler
Editors

Birkhäuser Verlag
Basel · Boston · Berlin

Editors:

Werner Haussmann
Mathematisches Institut
Gerhard-Mercator-Universität Duisburg
47048 Duisburg
Germany
haussmann@math.uni-duisburg.de

Kurt Jetter
Institut für Angewandte Mathematik und Statistik
Universität Hohenheim
70593 Stuttgart
Germany
jetter@uni-hohenheim.de

Manfred Reimer
Fachbereich Mathematik
Universität Dortmund
44221 Dortmund
Germany
manfred.reimer@math.uni-dortmund.de

Joachim Stöckler
Fachbereich Mathematik
Universität Dortmund
44221 Dortmund
Germany
joachim.stoeckler@math.uni-dortmund.de

2000 Mathematics Subject Classification 41-06; 41A63, 49K20, 65D30

A CIP catalogue record for this book is available from the Library of Congress, Washington D.C., USA

Bibliographic information published by Die Deutsche Bibliothek
Die Deutsche Bibliothek lists this publication in the Deutsche Nationalbibliografie; detailed bibliographic data is availa
in the Internet at <http://dnb.ddb.de>.

ISBN 3-7643-2195-4 Birkhäuser Verlag, Basel – Boston – Berlin

Member of the BertelsmannSpringer Publishing Group
Printed on acid-free paper produced of chlorine-free pulp. TCF ∞
Printed in Germany
ISBN 3-7643-2195-4

9 8 7 6 5 4 3 2 1

www.birkhause

Preface

These Proceedings contain the main topics and talks presented at the Fifth International Conference on Multivariate Approximation. The meeting took place during the week of September 22–27, 2002 at the well–known "Haus Bommerholz", the guest–house of the University of Dortmund. It hosted 49 participants from 10 countries, and the program included 11 invited one–hour lectures, 21 contributed talks, and a problem session. The articles collected here are carefully peer–refereed and suitably edited for publication, and they reveal the state of the art in the field.

Following the tradition of this series of conferences, the meeting was aimed at advancing selected topics of Multivariate Approximation Theory. These include refinable function vectors, vector subdivision, shift–invariant systems, elliptic control problems, discrepancy of point sets, approximation and differential equations, radial basic function approximation, N–term approximation in L_p, and one–sided L_1–approximation, among others.

One of our distinguished main lecturers was Prof. Will A. Light, University of Leicester. His lecture on "Error Estimates for Radial Basic Function Approximation" was a highlight of the conference. Nobody in the audience could know that this was to be one of his last conference lectures. Less than three months later we received the almost unbelievable message of his sudden death. We are mourning a mathematician of high scientific merit, an inspiring lecturer, a fine colleague and friend.

The editors would like to express their thanks to all who have given their support to the conference, and to the preparation of this book. In particular, our thanks go to the Deutsche Forschungsgemeinschaft for their funding of the conference, to the University of Dortmund and the staff of Haus Bommerholz for their help in running the conference, and to the University of Dortmund for the financial support of this proceedings volume. We are grateful to all the authors and referees who helped us maintain a high scientific standard. In addition, we acknowledge the constant help of Dr. Hermann Hoch (Duisburg) during the production of this proceedings volume. Last but not least, we would like to thank the publisher for including this book in the *International Series of Numerical Mathematics*.

June 2003

Werner Haußmann Kurt Jetter

Manfred Reimer Joachim Stöckler

Acknowledgment

We are deeply indebted to the

Deutsche Forschungsgemeinschaft (DFG)

whose generous funding made this conference possible.

The production of this volume was sponsored by the

Universität Dortmund

The Editors

Table of Contents

Participants

Elena E. Berdysheva, *Institut für Angewandte Mathematik und Statistik, Universität Hohenheim, D–70593 Stuttgart, Germany*
email: berdyshe@uni-hohenheim.de

Laura Beutel, *Institut für Angewandte Mathematik, Universität Dortmund, D–44221 Dortmund, Germany*
email: laura.beutel@math.uni-dortmund.de

Kai Bittner, *Abteilung Numerik, Fakultät für Mathematik und Wirtschaftswissenschaften, Universität Ulm, D–89069 Ulm, Germany*
email: bittner@mathematik.uni-ulm.de

Heribert Blum, *Institut für Angewandte Mathematik, Universität Dortmund, D–44221 Dortmund, Germany*
email: heribert.blum@mathematik.uni-dortmund.de

Dietrich Braess, *Institut für Mathematik, Ruhr-Universität Bochum, D–44780 Bochum, Germany*
email: braess@num.ruhr-uni-bochum.de

Johann Brauchart, *Institut für Mathematik A, Technische Universität Graz, Steyrergasse 30, A-8010 Graz, Austria*
email: brauchart@finanz.math.tu-graz.ac.at

Maria Charina, *Institut für Angewandte Mathematik, Universität Dortmund, D–44221 Dortmund, Germany*
email: maria.charina@math.uni-dortmund.de

Charles K. Chui, *Department of Mathematics and Computer Science, University of Missouri - St. Louis, 8001 Natural Bridge Road, St. Louis, MO 63121-4499, USA*, and *Statistics Department, Stanford University, Stanford, CA 94305, USA*
email: chui@math.umsl.edu, cchui@stat.stanford.edu

Wolfgang Dahmen, *Institut für Geometrie und Praktische Mathematik, RWTH Aachen, D–52056 Aachen, Germany*
email: dahmen@igpm.rwth-aachen.de

Ronald A. DeVore, *Industrial Mathematics Institute, University of South Carolina, Columbia, SC 29208, USA*
email: devore@math.sc.edu

Dimiter P. Dryanov, *Département de Mathématiques et de Statistique, Université de Montréal, C.P. 6128, Montréal, Québec H3C 3J7, Canada*
email: dryanovd@dms.umontreal.ca

Nira Dyn, *School of Mathematical Sciences, Tel Aviv University, Ramat Aviv, 69978 Tel Aviv, Israel*
email: niradyn@math.tau.ac.il

Karol Dziedziul, *Faculty of Applied Mathematics, Technical University of Gdansk, ul. Narutowicza 11/12, 80-952 Gdansk, Poland*
email: kdz@mifgate.pg.gda.pl

Paul Gauthier, *Département de Mathématiques et de Statistique, Université de Montréal, C.P. 6128, Montréal, Québec H3C 3J7, Canada*
email: gauthier@dms.umontreal.ca

Heiner Gonska, *Institut für Mathematik,*
Gerhard–Mercator–Universität Duisburg, D–47048 Duisburg, Germany
email: gonska@math.uni-duisburg.de

Peter J. Grabner, *Institut für Mathematik A,*
Technische Universität Graz, Steyrergasse 30, A–8010 Graz, Austria
email: peter.grabner@tu-graz.at

Werner Haußmann, *Institut für Mathematik,*
Gerhard–Mercator–Universität Duisburg, D–47048 Duisburg, Germany
email: haussmann@math.uni-duisburg.de

Armin Iske, *Zentrum Mathematik,*
Technische Universität München, D–80290 München, Germany
email: iske@ma.tum.de

Kurt Jetter, *Institut für Angewandte Mathematik und Statistik,*
Universität Hohenheim, D–70593 Stuttgart, Germany
email: kjetter@uni-hohenheim.de

Lavi Karp, *Department of Applied Mathematics,*
ORT Braude College, P.O. Box 78, 21982 Karmiel, Israel
email: karp@techunix.technion.ac.il

John Klinkhammer, *Institut für Mathematik,*
Gerhard–Mercator–Universität Duisburg, D–47048 Duisburg, Germany
email: klinkhammer@math.uni-duisburg.de

Angela Kunoth, *Institut für Angewandte Mathematik,*
Universität Bonn, D–53115 Bonn, Germany
email: kunoth@uni-bonn.de

Mingjun Lai, *Department of Mathematics,*

University of Georgia, Athens, GA 30602-7403, USA

email: mjlai@math.uga.edu

Kyoung Yong Lee, *Department of Applied Mathematics, KAIST,*

Gusong Dong 373-1, Yusong gu, Taejon 305-701, South Korea

email: kylee@math.kaist.ac.kr

Alain Le Méhauté, *Département de Mathématiques, Laboratoire Jean Leray,*

UMR 6629, Université de Nantes, 44322 Nantes Cédex 3, France

email: alm@math.univ-nantes.fr

Will A. Light († December 8, 2002)

Department of Mathematics and Computer Science,

University of Leicester, Leicester LE1 7RH, UK

Detlef Mache, *Applied Mathematics, TFH Georg Agricola Bochum,*

Herner Str. 45, D–44787 Bochum, Germany

email: mache@tfh-bochum.de

Ulrike Maier, *Fraunhofer Institute for Algorithms and*

Scientific Computing (SCAI), Schloss Birlinghofen,

D–53754 St. Augustin, Germany

email: ulrike.maier@scai.fhg.de

H. Michael Möller, *Institut für Angewandte Mathematik,*

Universität Dortmund, D–44221 Dortmund, Germany

email: moeller@math.uni-dortmund.de

Manfred W. Müller, *Institut für Angewandte Mathematik, Universität Dortmund, D–44221 Dortmund, Germany*
email: mueller@mathematik.uni-dortmund.de

Günther Nürnberger, *Fakultät für Mathematik und Informatik, Universität Mannheim, D–68131 Mannheim, Germany*
email: nuern@euklid.math.uni-mannheim.de

Gerlind Plonka, *Institut für Mathematik, Gerhard–Mercator–Universität Duisburg, D–47048 Duisburg, Germany*
email: plonka@math.uni-duisburg.de

Jürgen Prestin, *Institut für Mathematik, Medizinische Universität zu Lübeck, D–23560 Lübeck, Germany*
email: prestin@math.uni-luebeck.de

Manfred Reimer, *Institut für Angewandte Mathematik, Universität Dortmund, D–44221 Dortmund, Germany*
email: reimer@math.uni-dortmund.de

Sherman D. Riemenschneider, *Department of Mathematics, West Virginia University, Morgantown, WV 26506–6310, USA*
email: sherm@math.wvu.edu

Amos Ron, *Department of Computer Sciences, University of Wisconsin, 1210 W. Dayton St., Madison , WI 53706–1685, USA*
email: amos@cs.wisc.edu

Paul Sablonnière, *Centre de Mathématiques, INSA des Rennes, 20 Av. des Buttes de Coësmes, CS 14315, 35043 Rennes Cédex, France*
email: paul.sablonniere@insa-rennes.fr

Tomas Sauer, *Mathematisches Institut,*
Justus–Liebig–Universität Giessen, D–35392 Giessen, Germany
email: tomas.sauer@math.uni-giessen.de

Michael Scharfschwerdt, *Institut für Angewandte Mathematik,*
Universität Dortmund, D–44221 Dortmund
email: michael.scharfschwerdt@math.uni-dortmund.de

Karl Scherer, *Institut für Angewandte Mathematik,*
Universität Bonn, D–53115 Bonn, Germany
email: scherer@iam.uni-bonn.de

Joachim Stöckler, *Institut für Angewandte Mathematik,*
Universität Dortmund, D–44221 Dortmund, Germany
email: joachim.stoeckler@math.uni-dortmund.de

Manfred Tasche, *Fachbereich Mathematik,*
Universität Rostock, D–18051 Rostock, Germany
email: manfred.tasche@mathematik.uni-rostock.de

Vladimir N. Temlyakov, *Department of Mathematics,*
University of South Carolina, Columbia, SC 29208, USA
email: temlyak@math.sc.edu

Robert F. Tichy, *Institut für Mathematik A,*
Technische Universität Graz, Steyrergasse 30, A–8010 Graz, Austria
email: tichy@tugraz.at

Holger Wendland, *Institut für Numerische und Angewandte Mathematik,*
Georg–August–Universität Göttingen, D–37083 Göttingen, Germany
email: wendland@math.uni-goettingen.de

Qiang Wu, *Department of Mathematics, City University of Hong Kong, 83 Tat Chee Avenue, Kowloon, Hong Kong*
email: wu.qiang@student.cityu.edu.hk

Frank Zeilfelder, *Fakultät für Mathematik und Informatik, Universität Mannheim, D–68131 Mannheim, Germany*
email: zeilfeld@euklid.math.uni-mannheim.de

Ding-Xuan Zhou, *Department of Mathematics, City University of Hong Kong, 83 Tat Chee Avenue, Kowloon, Hong Kong*
email: mazhou@math.cityu.edu.hk

Georg Zimmermann, *Institut für Angewandte Mathematik und Statistik, Universität Hohenheim, D–70593 Stuttgart, Germany*
email: gzim@uni-hohenheim.de

Scientific Program of the Conference

Monday, September 23:

Opening: Chair: *J. Stöckler*

9:00 – 9:30	Prof. Dr. E. Becker Rektor of the University of Dortmund

1st Morning Session: Chair: *W. Haußmann*

9:30 – 10:30	*P. M. Gauthier*: Polynomial Approximation of Fundamental Solutions

2nd Morning Session: Chair: *P. Sablonnière*

11:00 – 11:30	*L. Karp*: On the Analytic Continuation of the Cauchy Transform
11:30 – 12:00	*A. Iske*: On the Local Approximation Order of Scattered Data Interpolation by Polyharmonic Splines
12:00 – 12:30	*A. Le Méhauté*: On Almost Interpolation with Radial Basis Functions

Afternoon Session: Chair: *K. Jetter*

15:30 – 16:30	*C. K. Chui*: Multivariate Refinable Function Vectors with Balanced Duals
16:30 – 17:00	*G. Zimmermann*: Three Scale Versus Matrix Refinement Equations
17:00 – 17:30	*S. D. Riemenschneider*: $\sqrt{3}$–Subdivision Schemes: Maximal Sum Rule Orders
17:30 – 18:00	*K. Dziedziul*: Shift Invariant Operators

Tuesday, September 24:

1st Morning Session: Chair: *H. M. Möller*

9:00 – 10:00	*V. N. Temlyakov*: Some Open Problems in Multivariate Approximation
10:00 – 10:30	*U. Maier*: Common Zeros of Orthogonal Polynomials on the Threedimensional Sphere

2nd Morning Session: Chair: *P. J. Grabner*

11:00 – 12:00	*R. F. Tichy*: Probabilistic Results on the Discrepancy of Point Sets
12:00 – 12:30	*J. S. Brauchart*: About Generalised Energy Functionals, Equidistributed Point Sets and Invariance Principles

Afternoon Session: Chair: *G. Nürnberger*

15:30 – 16.30	*D.–X. Zhou*: Approximation Theory Problems Arising from Learning Theory
16:30 – 17:00	*P. Sablonnière*: On Some Multivariate Quadratic Spline Quasi–Interpolants on Bounded Domains
17:00 – 17:30	*F. Zeilfelder*: Lagrange Interpolation by Bivariate Splines with Optimal Approximation Order
17:30 – 18:00	*H. Wendland*: Reconstruction of Functions from Large Data Sets

Wednesday, September 25:

1st Morning Session: Chair: *N. Dyn*

9:00 – 10:00	*W. Light*: Error Estimates for Radial Basic Function Approximation
10:00 – 10:30	*E. Berdysheva*: The Turán Problem for a Class of Polytopes

2nd Morning Session: Chair: *C. K. Chui*

11:00 – 11:30	*J. Prestin*: Localization of Functions on the d–variate Sphere and Torus
11:30 – 12:15	*M. Reimer*: Approximation of Density Functions and Reconstruction of the Approximants

Thursday, September 26:

1st Morning Session: Chair: *D.–X. Zhou*

9:00 – 10:00	*R. DeVore*: Best Basis for N–Term Approximation in L_p
10:00 – 10:30	*J. Klinkhammer*: Best One–Sided L^1–Approximation, by $B^{m,1}$–Blending Functions

2nd Morning Session: Chair: *J. Prestin*

11:00 – 12:00	*D. P. Dryanov*: Transfinite Interpolation by Blending Functions, Best One–Sided L^1–Approximation, and Cubature Formulae
12:00 – 12:30	*M. Lai*: Splines for Numerical Solution of PDE's

Afternoon Session: Chair: *S. D. Riemenschneider*

15:30 – 16:30	*A. Ron*: On Generalized Shift–Invariant Systems (GSI)
16:30 – 17:00	*N. Dyn*: Exponential Reproducing Subdivision Schemes and Spectral Interpolation
17:00 – 17:30	*M. Tasche*: Split–Radix Algorithms for Discrete Trigonometric Transforms
17:30 – 18:00	*G. Plonka*: Reversible Integer DCT Algorithms

Evening Session: Chair: *M. Reimer*

20:00 – 21:00	Problem Session

Friday, September 27:

1st Morning Session: Chair: *D. Braess*

9:00 – 10:00	*A. Kunoth*: Wavelet Methods for Linear–Quadratic Elliptic Control Problems
10:00 – 10:30	*K. Jetter*: An Identity for Multivariate Bernstein Polynomials

2nd Morning Session: Chair: *K. Jetter*

11:00 – 11:30	*M. Charina–Kehrein*: Convergence of Multivariate Non–Stationary Vector Subdivision Schemes
11:30 – 12:20	*T. Sauer*: Vector Subdivision in Several Variables and Factorization
12:20	Closing of the Conference

International Series of Numerical Mathematics
Vol. 145, ©2003 Birkhäuser Verlag Basel/Switzerland

Multivariate Approximation Methods for the Pricing of Catastrophe–Linked Bonds*

Hansjörg Albrecher, Jürgen Hartinger, and Robert F. Tichy

Dedicated to Manfred Reimer on the occasion of his 70th birthday

Abstract

In this paper we develop Quasi–Monte Carlo techniques for the evaluation of high–dimensional integrals that occur in financial applications, namely in the pricing of default–risky catastrophe–linked bonds in a model including stochastic interest rates, basis risk and default risk. It is shown that these techniques clearly outperform classical Monte Carlo integration in terms of efficiency. The methods are based on number–theoretic low–discrepancy sequences such as Halton, Sobol and Faure sequences.

1 Introduction

Quasi–Monte Carlo integration has turned out to be a powerful tool for integrals occurring in various branches of applied mathematics (see e.g. [1,2,24]). In this paper we investigate the efficiency of Quasi–Monte Carlo techniques for pricing default–risky catastrophe–linked bonds in a model including stochastic interest rates, basis risk and default risk. In Section 2 we describe the Quasi–Monte Carlo (QMC) method in general. In Section 3 we develop a suitable contingent claim model for these bonds. Section 4 discusses implementation issues and gives numerical illustrations of the superiority of QMC methods over classical Monte Carlo. Furthermore the sensitivity of CAT bond prices with respect to various model parameters is investigated.

2 Multivariate Integration Using QMC Sequences

After suitable transformation of the integration domain, an s–dimensional integral can be written in the form

$$I(f) = \int_{U^s} f(\mathbf{x})\, d\mathbf{x}, \tag{2.1}$$

*This research was supported by the Austrian Science Foundation Project S–8308–MAT

where f is a function defined on the s–dimensional unit cube $U^s = [0,1]^s$. The basic idea of Monte Carlo integration is now to choose N integration points $\mathbf{x}_1, \ldots, \mathbf{x}_n$ randomly in U^s and to approximate the integral $I(f)$ by the arithmetic mean

$$I_N(f) = \frac{1}{N} \sum_{n=1}^{N} f(\mathbf{x}_n).$$

Chebyshev's inequality gives a probabilistic bound on the integration error of Monte Carlo integration

$$P\left(\left| \int_{U^s} f(\mathbf{x})\, d\mathbf{x} - \frac{1}{N} \sum_{n=1}^{N} f(\mathbf{x}_n) \right| \geq \sqrt{\frac{\operatorname{Var}(f(\mathbf{x}))}{\epsilon N}} \right) \leq \epsilon, \tag{2.2}$$

where $\operatorname{Var}(f(\mathbf{x}))$ is the variance of $f(\xi)$, and the probable error thus turns out to be proportional to $1/\sqrt{N}$ and is independent of the dimension s (see e.g. [15]). A crucial aspect, however, is the availability of an efficient pseudo–random number generator.

The use of deterministic uniformly distributed point sequences (instead of pseudo–random sequences) has proven to be an efficient alternative to classical Monte Carlo, especially for dimensions $s < 30$. Such nonrandom sequences are called Quasi–Monte Carlo sequences. A well–known measure for the uniformness of the distribution of a sequence $\{\mathbf{y}_n\}_{1 \leq n \leq N}$ in U^s is the star–discrepancy

$$D_N^*(\mathbf{y}_n) = \sup_{I \in J_0^s} \left| \frac{A(\mathbf{y}_n; I)}{N} - \lambda_s(I) \right| ,$$

where J_0^s is the set of all intervals of the form $[0, \vec{\eta}) = [0, \eta_1) \times [0, \eta_2) \times \ldots \times [0, \eta_s)$ with $0 \leq \eta_i < 1$, $i = 1, \ldots, s$ and $A(\mathbf{y}_n; I)$ is the number of points of the sequence $\{\mathbf{y}_n\}_{1 \leq n \leq N}$ that lie in I. $\lambda_s(I)$ denotes the s–dimensional Lebesgue–measure of I.

The notion of discrepancy can be used for obtaining an upper bound for the error of Quasi–Monte Carlo integration, since by the Koksma–Hlawka inequality

$$\left| \frac{1}{N} \sum_{n=1}^{N} f(\mathbf{y}_n) - \int_{[0,1)^s} f(u) du \right| \leq V(f) D_N^* (\mathbf{y}_1, \ldots, \mathbf{y}_N), \tag{2.3}$$

where $V(f)$ is the total variation of f in the sense of Hardy and Krause (see e.g. Drmota and Tichy [10]). This error bound is deterministic (in contrast to the probabilistic error bound (2.2)). For upper bounds of the integration error in the case of unbounded variation of f, we refer to [27].

The Koksma–Hlawka inequality can be viewed as an error bound in the sense of worst case analysis. In order to establish an error bound in the sense of average case analysis we have to introduce the L^p–discrepancy for $p \geq 1$ defined by

$$D_n^{(p)}(\mathbf{y}_n) = \left(\int_{U^s} \left| \frac{A(\mathbf{y}_n; [0, \overrightarrow{\eta}))}{N} - \eta_1 \cdots \eta_s \right|^p d\eta_1 \cdots d\eta_s \right)^{1/p} .$$

The star–discrepancy corresponds to $p = \infty$. Furthermore, the L^p–discrepancy and the star–discrepancy are related by the following inequality ($C(s,p)$ denoting an explicitly known constant depending only on s and p):

$$D_N^{(p)} \leq D_N^* \leq C(s,p) \left(D_N^{(p)} \right)^{\frac{p}{s+p}} ,$$

which is due to Niederreiter, Tichy and Turnwald [23]. The central problem of average case analysis of numerical integration is to determine the minimal cost for approximation of $I(f)$ with error $\leq \epsilon$ for a given class $\mathcal{F}$ of functions, where $\mathcal{F}$ is equipped with a probability measure. A very convenient choice for $\mathcal{F}$ is the class of continuous functions equipped with the Wiener measure. The average complexity of numerical integration can be expressed as the expectation with respect to Wiener measure μ_w of the number of function evaluations and the number of operations necessary for the computation of $I_N(f)$. Woźniakowski [28] computed the average case error in terms of the L^2–discrepancy:

$$D_N^{(2)}(\mathbf{1} - \mathbf{y}_n) = \left(\int_{\mathcal{F}} \left(I(f) - I_N(f) \right)^2 d\mu_w \right)^{1/2} , \tag{2.4}$$

where $\mathbf{1} = (1, \ldots, 1)$.
In particular for s not too large, certain Quasi–Monte Carlo sequences have turned out to be superior to crude probabilistic Monte Carlo techniques in many applications. This is the case for so–called low discrepancy sequences, i.e. sequences for which

$$D_N^*(\mathbf{y}_n) \leq C_s \frac{(\log N)^s}{N} \tag{2.5}$$

holds with an explicitly computable constant C_s. The best known low discrepancy sequences satisfy a discrepancy bound of type (2.5). However, for the L^2–discrepancy bounds of better asymptoptic order in N are known. Recently,

Chen and Skriganov [5] explicitly constructed a sequence $(\mathbf{z}_n)$ such that

$$D_N^{(2)}(\mathbf{z}_n) \leq C'_s \frac{(\log N)^{\frac{1}{2}(s-1)}}{N},$$

which is the best possible asymptotic order by the following general lower bound due to Roth [25]:

$$D_N^{(2)} \geq c_s \frac{(\log N)^{\frac{s-1}{2}}}{N}.$$

From Roth's bound and (2.4) it can immediately be deduced that the average case complexity of numerical integration (on the space of continuous functions equipped with Wiener measure) satisfies a bound of the order $\mathcal{O}(\frac{1}{\epsilon}(\log \frac{1}{\epsilon})^{\frac{s-1}{2}})$, where ϵ is the error in the approximation of $I(f)$ by $I_N(f)$ (see [28]).

For the star–discrepancy only slight improvements of Roth's bound are known. The most recent one is due to Baker [3]:

$$D_N^* \geq c'_s \frac{(\log N)^{\frac{s-1}{2}}}{N} \left(\frac{\log \log N}{\log \log \log N}\right)^{\frac{1}{2s-2}}$$

for $N \geq N_0$.

Bounds for the constants C_s are usually pessimistic and often the actual error made by Quasi–Monte Carlo integration is much lower than the bound implied by C_s (see e.g. [4]).

However, for applications in mathematical finance the dimensions s may be quite big (several hundreds). Thus discrepancy bounds in s and N are of interest. Recently this problem was successfully attacked by Heinrich, Novak, Wasilkowski and Woźniakowski [14]. They considered the quantity

$$disc(N, s) = \inf_{(y_1,\ldots,y_N)\in[0,1]^{Ns}} D_N^*(\mathbf{y}_n)$$

and obtained the estimate

$$disc(N, s) \leq C s^{\frac{1}{2}} N^{-\frac{1}{2}}$$

with an absolute constant C. The proof depends on probabilistic methods and can be generalized to L^p–discrepancies. As a consequence the minimum number $K(s, \epsilon)$ of points $\mathbf{y}_n$ $(1 \leq n \leq N)$ satisfying $disc(N, s) \leq \epsilon$ fulfills a bound of the type $K(s, \epsilon) = \mathcal{O}(s\epsilon^{-2})$. Thus $K(s, \epsilon)$ is of polynomial growth in s and ϵ^{-1}.

The following low discrepancy sequences will be used in the sequel:

- The Halton sequence [13] is defined as a sequence of vectors in U^s based on the digit representation of n in base p_i

$$\xi_n = (b_{p_1}(n), b_{p_2}(n), \ldots, b_{p_s}(n)), \tag{2.6}$$

  where p_i is the ith prime number and $b_p(n)$ is the digit reversal function for base p given by

$$b_p(n) = \sum_{k=0}^{\infty} n_k p^{-k-1}, \qquad n = \sum_{k=0}^{\infty} n_k p^k,$$

  where the n_k are integers. One could also use pairwise coprime base numbers, but the error estimate turns out to be best possible for prime bases p_n. For Halton sequences C_s tends to infinity super–exponentially for $s \to \infty$.

Better error bounds can be obtained for low–discrepancy sequences based on (t, m, s)–nets. These nets are based on the b–adic representation of vectors in U^s. Instead of optimizing the discrepancy itself, only the discrepancy with respect to elementary intervals J in base b is considered, i.e. $J = \prod_{i=1}^{s} [a_i b^{-d_i}, (a_i + 1) b^{-d_i})$ with integers $d_i \geq 0$ and integers $0 \leq a_i < b^{d_i}$ for $1 \leq i \leq s$, and point sequences in U^s are constructed such that the discrepancy with respect to these intervals J is optimal for subsequences of length $N = b^m$. Let $\#(J, N)$ denote the number of points of a sequence $\{\mathbf{y}_n\}_{1 \leq n \leq N}$ that lie in J. A point set $\mathcal{P}$ with $\mathrm{card}(\mathcal{P}) = b^m$ is now called a (t, m, s)–net, if $\#(J, b^m) = b^t$ for every elementary interval J with $\lambda_s(J) = b^{t-m}$. The parameter t is a quality parameter. For $t = 0$ we have the minimal discrepancy of the point set $\mathcal{P}$ with respect to the family of elementary intervals.

For an integer $t \geq 0$, a sequence $\xi_1, \xi_2, \ldots$ of points in U^s is called a (t, s)–sequence in base b, if for all integers $k \geq 0$ and $m > t$, the point set consisting of the ξ_n with $kb^m < n \leq (k+1)b^m$ is a (t, m, s)–net in base b.

- The Sobol sequence is a (t, s)–sequence in base 2 with values t that depend on s. For a construction of this sequence we refer to [26]. The corresponding constant C_s is much lower than for the Halton sequence.

- The Faure sequence [12] is a $(0, s)$–sequence with an even lower constant C_s.

Since Quasi-Monte Carlo methods are especially competitive for low dimensions ($s < 30$), the best numerical performance can be achieved by implementing hybrid Monte Carlo techniques, i.e. Quasi-Monte Carlo sequences for the initial dimensions and the remaining dimensions are then simulated by crude Monte Carlo.

3 A Model for Pricing Catastrophe–Linked Bonds

We now apply the QMC technique to a particular problem of mathematical finance, namely the pricing of a catastrophe–linked bond (CAT bond for short), which is a liability hedge instrument for insurance companies. As an alternative to traditional reinsurance, an insurer can issue (i.e. sell) these bonds under the condition of debt–forgiveness triggers in case the aggregate claims to be paid by the insurer (or a certain composite index of losses among various companies) exceeds a specified threshold K. In that way the issuer of these bonds avoids credit risk that might arise with traditional reinsurance concepts. Since during the last decades the insurance losses due to natural catastrophes have increased dramatically, CAT bonds have become a popular hedging tool and one of the key issues is to correctly price these financial instruments (see e.g. [7,8,18]).
A crucial step in determining a fair price of the CAT bond is to model the dynamics of the assets and the losses of the insurer as well as the dynamics of the (stochastic) interest rate. In this paper we adopt the approach of Lee and Yu [17], which also includes the consideration of default risk (cf. Section 3.2) and basis risk (cf. Section 3.3). For convenience, we briefly repeat the relevant specifications of the model from [17]:
The asset dynamics of the insurer are modeled by

$$\frac{dA_t}{A_t} = \mu_A\, dt + \phi\, dr_t + \sigma_A\, dW_{A,t}, \tag{3.1}$$

where A_t is the value of the insurer's total assets at time t, r_t is the instantaneous interest rate at time t (with ϕ denoting the interest rate elasiticity), $W_{A,t}$ is the Wiener process representing the risk orthogonal to interest rate risk (frequently referred to as credit risk), and μ_A and σ_A are the drift and volatility due to this credit risk of the insurer. The stochastic interest rate r_t is assumed to follow the Cox–Ingersoll–Ross process [6]

$$dr_t = \kappa(m - r_t)\, dt + \nu\sqrt{r_t}\, dZ_t,$$

where κ, m and ν are given parameters and Z_t is a Wiener process independent of $W_{A,t}$. Under the risk–neutral pricing measure Q, the process can thus be written as

$$\frac{dA_t}{A_t} = r_t\,dt + \phi\nu\sqrt{r_t}\,dZ_t^* + \sigma_A\,dW_{A,t}^*, \tag{3.2}$$

where $W_{A,t}^*$ and Z_t^* are independent Wiener processes under Q and

$$dr_t = \kappa^*(m^* - r_t)\,dt + \nu\sqrt{r_t}\,dZ_t^* \tag{3.3}$$

represent the dynamics of the interest rate process under Q with

$$\kappa^* = \kappa^* + \lambda_r, \qquad m^* = \frac{\kappa m}{\kappa + \lambda_r} \quad \text{and} \quad dZ_t^* = dZ_t + \frac{\lambda_r\sqrt{r_t}}{\nu}\,dt,$$

where λ_r denotes the market price of interest rate risk (cf. [17]).
The loss process is assumed to follow a compound Poisson process

$$C_t = \sum_{j=1}^{N(t)} X_j, \tag{3.4}$$

where X_j denotes the amount of losses caused by the jth catastrophe for the issuing insurance company during the specific period and $N(t)$ is a homogeneous Poisson process with intensity λ. Thus C_t represents the aggregate loss at time t for the issuing firm. It is shown in [17] that the above aggregate loss process retains the original distribution characteristics when changing from the physical to the risk–neutral probability measure Q.
Under the risk–neutral pricing measure, the value of the CAT bond price is given by the discounted expectation of its various payoffs in the risk–neutral world. In the following subsections we will specify the payoffs of the CAT bonds under three different scenarios:

3.1 Default–Free CAT Bonds

First we consider the simple case where there is no default risk. The payoff $P_{CAT}(T)$ at maturity time T is then given by

$$P_{CAT}(T) = \begin{cases} 1 & \text{if } C_T \le K \\ p & \text{if } C_T > K, \end{cases} \tag{3.5}$$

where K is the trigger level in the CAT bond provisions and p is the proportion of the principal that has to be paid to bondholders when the amount of losses exceeds K. The face amount of any CAT bond considered in this paper is assumed to be 1. Under the assumptions of the model, the CAT bond price at time 0 is then given by

$$P_{CAT}(0) = A(0,T)e^{-B(0,T)r_0}\left[\sum_{j=0}^{\infty} e^{-\lambda T}\frac{(\lambda T)^j}{j!}F^j(K)+p\left(1-\sum_{j=0}^{\infty} e^{-\lambda T}\frac{(\lambda T)^j}{j!}F^j(K)\right)\right], \tag{3.6}$$

where $F^j(K) = Pr(X_1 + \ldots + X_j \leq K)$ denotes the jth convolution of the claim size distribution F and

$$A(0,T) = \left(\frac{2\gamma e^{(\kappa^*+\gamma)T/2}}{(\kappa^*+\gamma)(e^{\gamma T}-1)+2\gamma}\right)^{2\kappa^* m^*/\nu^2}, \quad B(0,T) = \frac{2(e^{\gamma T}-1)}{(\kappa^*+\gamma)(e^{\gamma T}-1)+2\gamma}$$

with $\gamma = \sqrt{\kappa^{*2}+2\nu^2}$.

3.2 Default–Risky CAT Bonds Without Basis Risk

We now include the possibility of insolvency of the insurer in our model. Typically, CAT bondholders have priority for salvage over other debtholders. However, in this case bondholders still might only be repaid part of the principal and thus the default risk will lower the bond price. The payoff of the CAT bond in a model including default risk is thus given by

$$P_{def}(T) = \begin{cases} 1 & \text{if } C_T \leq K \text{ and } C_T \leq A_T - a\,L \\ p & \text{if } K < C_T < A_T - p\,a\,L, \\ \frac{\max\{A_T - C_T, 0\}}{a\,L} & \text{otherwise,} \end{cases} \tag{3.7}$$

where K and p are defined as in Section 3.1, L is the face amount of the issuing firm's total debts including the face amount of the CAT bonds and a is the ratio of the total CAT bonds' face amount over total outstanding debts.
The CAT bond price at issuing time 0 is then given by

$$P_{def}(0) = E_Q[e^{-\int_0^T r(t)dt}\,P_{def}(T)],$$

where E_Q denotes the expectation with respect to the risk–neutral pricing measure Q.

3.3 Default–Risky CAT Bonds With Basis Risk

In the preceding model the issuing firm obviously has the incentive to settle claims more generously when the incurred loss is close to the trigger level (an effect known as moral hazard). To circumvent this problem, an alternative design of a CAT bond is to define the forgiveness triggers as a function of a composite index of losses (among various firms exposed to the same catastrophic events)

$$C_{index,t} = \sum_{j=1}^{N(t)} X_{index,j}, \tag{3.8}$$

rather than on the individual aggregate loss process C_t of the issuing firm. Here $X_{index,j}$ denotes the amount of losses relevant for the index caused by the jth catastrophe. X_j and $X_{index,j}$, respectively, for $j = 1, \ldots, N(t)$ are assumed to be iid random variables with (marginal) distribution function F, whereas X_j and $X_{index,j}$ are dependent according to some given copula structure (in [17] the dependence structure is modeled by correlation coefficients for the logarithm of lognormal marginals; our copula approach allows to generalize this approach to arbitrary marginal distributions; moreover the well–known deficiency of the correlation coefficient being a measure of linear dependence only is avoided (see e.g. [11])). In this paper we choose Frank's copula and Spearman's rank correlation coefficient ρ as the measure of positive dependence ($0 \le \rho \le 1$), see e.g. [21]. The gap between the insurer's actual loss and the composite index of losses (called the basis risk) may cause the insurer to default on his debt. Thus basis risk affects the bond price. The amount of basis risk is represented by ρ (with larger basis risk for smaller values of ρ).
The payoff of a default–risky CAT bond including basis risk is given by

$$P_{index}(T) = \begin{cases} 1 & \text{if } C_{index,T} \le K \text{ and } C_T \le A_T - a\,L \\ p & \text{if } K < C_{index,T} \text{ and } C_T < A_T - p\,a\,L, \\ \frac{\max\{A_T - C_T, 0\}}{a\,L} & \text{otherwise.} \end{cases} \tag{3.9}$$

Accordingly, the corresponding CAT bond price at issuing time 0 is given by

$$P_{index}(0) = E_Q[e^{-\int_0^T r(t)dt} P_{index}(T)].$$

4 Numerical Analysis

In this section we first briefly discuss some implementation techniques needed to exploit and optimize the numerical advantages of Quasi–Monte Carlo integration for our pricing model of CAT bonds. We then present numerical results and compare the various solution methods in terms of efficiency.

The following refined version of the Koksma–Hlawka inequality (2.3) shows that one can improve the performance of QMC algorithms, if the first dimensions of the QMC sequences (which in most constructions have a lower discrepancy in the lower–dimensional faces, see e.g. [22]) are assigned to those variables which are responsible for most of the integrand's variation:

$$\left|\frac{1}{N}\sum_{n=1}^{N} f(\mathbf{y}_n) - \int_{[0,1)^s} f(u)du\right| \leq \sum_{l=0}^{s-1}\sum_{F_l} D_N^*\left(\mathbf{y}_n^{(F_l)}\right) V^{(s-l)}(f^{(F_l)}), \tag{4.1}$$

where the second sum is extended over all $(s-l)$–dimensional faces F_l of the form $y_{i_1} = \cdots = y_{i_l} = 1$, the discrepancy $D_N^*(\mathbf{y}_n^{(F_l)})$ is computed in the face of $[0,1]^s$ containing $\left(\mathbf{y}_n^{(F_l)}\right)$, and $V^{(s-l)}(f^{(F_l)})$ is the variation (in the sense of Vitali) of the restriction $f^{(F_l)}$ of f to F_l (see e.g. [10]).
For this reason, in the algorithms designed in this paper the first QMC dimension is always used to determine the number $N(T)$ of claims occurring until maturity. Our required distribution is calculated from the uniform distribution by standard inversion techniques (cf. [9]).

Let us first consider the case of default–free CAT bonds. Equations (3.5) and (3.6) show that determining the corresponding price reduces to simulating a compound Poisson process (i.e. the number and size of claims in the given time interval). Since catastrophic events are rare, this problem therefore has rather low dimension from the perspective of numerical integration; for the parameter values chosen in this paper, the dimension typically is between 3 to 5 (note that calculating the expected value of an s–dimensional random variable is equivalent to evaluating an $s-$dimensional integral).

The dimension of the problem increases considerably, when default risk of the bonds is included in the model. Due to the dependence between the asset process A_t and the stochastic interest rate process r_t one has to simulate the interest rates explicitly (opposed to the case of default–free bonds, where a suitable choice of numeraire led to (3.6)). In accordance with [17], we discretize the stochastic differential equation (3.3) to

$$r_{i+1} = r_i + \kappa^*(m^* - r_i)(t_{i+1} - t_i) + \nu\sqrt{r_i(t_{i+1} - t_i)}Z_{i+1}, \tag{4.2}$$

where the time span $t_{i+1} - t_i$ is a week, which for $T = 1$ year introduces 52 dimensions. Here Z_i (i=1..52) are iid random variables with standard normal distribution. A well–known problem in this context is that due to (4.2) negative interest rates are possible. However, this only happens for a negligible fraction of sample paths and thus it seems safe to ignore these paths (i.e. not to use them in the calculations). For the asset process, we solve the stochastic differential equation (3.2) to obtain

$$A_t = A_s \exp\left[(1 - \frac{\phi^2\nu^2}{2}) \int_s^t r_u\, du - \frac{\sigma^2}{2}(t-s) + \phi\nu \int_s^t \sqrt{r_u} dZ_u^* + \sigma W_{t-s}^*\right],$$

where $t > s$, and discretize the solution in the following way

$$A_{i+1} = A_i \exp\left[(1 - \tfrac{\phi^2\nu^2}{2})(t_{i+1} - t_i)\tfrac{r_{i+1}+r_i}{2} - \tfrac{\sigma^2}{2}(t_{i+1} - t_i) + \right.$$
$$\left. + \phi\nu \tfrac{\sqrt{r_{i+1}}+\sqrt{r_i}}{2} \sqrt{(t_{i+1} - t_i)} Z_{i+1} + \sigma\sqrt{(t_{i+1} - t_i)} W_{i+1}\right],$$

where W_i are iid standard normal random variables ($i = 1, \ldots, 52$), mutually independent of Z_i ($i = 1, \ldots, 52$). Therefore we face a problem with dimension larger than 100.

In view of (2.3) and (2.5), theoretically QMC integration should lead to good results for small dimensions only. However, empirically it turns out that QMC techniques can be very effective also for high dimensional numerical integration. One explanation is the low effective dimension of many integrands. This concept (introduced in [4]) formalizes the idea that only a few variables are responsible for most of the variation of the integrand. One well–known technique for reducing the effective dimension of the integrand when simulating a Wiener process is to use Brownian bridges (cf. [20]), which is based on a reordering of the discretization points and using conditional distributions to calculate the corresponding sample paths (see also [16]). We will implement and compare this technique in the sequel.

As indicated in Section 2, in many practical situations QMC methods turn out to be particularly effective, if QMC sequences are used for the "effective" dimensions only and the remaining dimensions are then simulated using pseudo-random numbers. This hybrid Monte Carlo approach increases the simulation speed of pure QMC, while the convergence speed of the relative error in terms of N is nearly unafffected. Throughout this paper, all QMC algorithms are

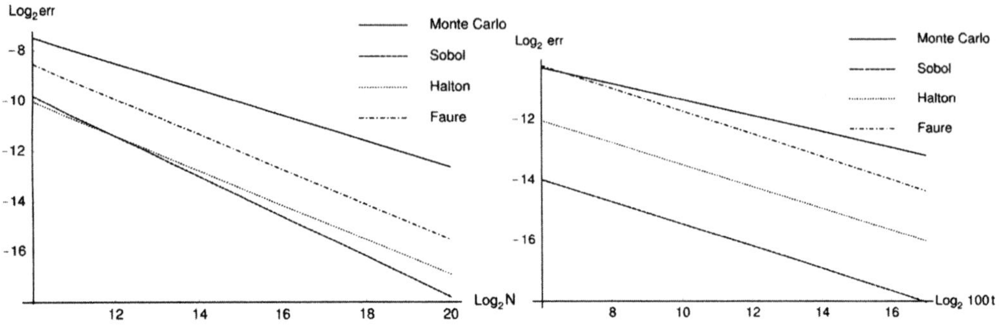

Figure 1: Approximation of default–free CAT bond prices (Lognormal claims): Relative error as a function of the sample size N (left) and calculation time t (right)

hybrid algortihms with 30 QMC dimensions, and we use the *Mersenne Twister* as our pseudorandom number generator, which basically is an optimized version of a multiplicative congruential algorithm (see [19]). In our calculations, the first QMC dimension is always used to generate the number of losses $N(T)$, and the following initial dimensions are used to generate the catastrophic loss amounts X_j (resp. $X_{index,j}$). For consistency of the convergence analysis it is important to fix the number of QMC dimensions reserved for the loss amounts. The remaining QMC dimensions are then assigned to the most important variates of the asset process and the interest rate process, respectively.

The introduction of basis risk makes it necessary not only to generate claim amounts (of the index) $X_{index,j}$, but also random variables X_j dependent on $X_{index,j}$ according to the specified copula structure (see Section 3.3). This can be achieved by implementing a modified inversion technique involving the partial derivatives of the copula function (see e.g. [21], Chapter 2). The dimension of this model is slightly bigger than without basis risk.

Table 1 now gives the parameter values used in the calculations below. For the catastrophic claim size distribution, the heavy–tailed Lognormal and Weibull distributions are chosen with typical parameters in practical situations.

Figure 1 depicts least–squares linear regression fits of the relative error of QMC and Monte Carlo techniques as a function of the sample size N and the calculation time t (measured in seconds), respectively. The "exact" value, in lack of an analytic solution, is obtained by a Monte Carlo simulation over 50 million

T	term of the contract (in years)	1
σ_A	volatility of credit risk	0.05
ϕ	interest elasticity of asset	-3
A_0/L	asset/liability ratio	1.1
r_0	initial instantaneous interest rate	0.05
k	magnitude of mean reverting force	0.2
m	long run mean of interest rates	0.05
ν	volatility of interest rates	0.1
λ_r	market price of interest risk	-0.01
K	trigger level	100
p	ratio of principal paid back in case the trigger has been pulled	0.5
a	ratio of amount of CAT bonds to total debts	0.1
L	insurer's total amount of debts	100

Claim size distribution	density
Lognormal	$f(x) = \frac{1}{\sqrt{2\pi}\cdot 2x}\, e^{-\frac{(\log x - 2)^2}{8}}$
Weibull	$f(x) = 0.32x^{0.96}e^{-0.16\,x^{1.96}}$

Table 1: Parameter values and catastrophic claim size distributions

sample paths. For every N, the estimate of the Monte Carlo error was obtained by averaging the error of 25 runs with N sample paths each. The plot shows that the QMC algorithms outperform crude Monte Carlo significantly, both for a given number of sample paths N and given calculation time t. The latter plot also contains the following information: for a given calculation time t, what is the relative error of the estimate that can be achieved by each of these methods? In the case of default–free CAT bonds, the Sobol sequence seems to be best suited for the corresponding integrand.

Figure 2 shows a comparison of MC and QMC algorithms for the estimation of the price of default–risky CAT bonds without basis risk. The subscript $_{BB}$ refers to the Brownian bridge algorithm of the respective QMC technique. Again the QMC algorithms turn out to be much more effective, both in terms of N and t. The Brownian bridge approach discussed above leads to a slight improvement over the respective plain QMC algorithm. Note that the dimension of the problem in Figure 2 is much higher than that of Figure 1. Especially for large values of N and t, the Sobol and Halton sequences seem to be particularly competitive.

Default–free CAT bonds					
CD	N	MC	Sobol	Halton	Faure
Lognormal	2^{10}	4.7E-03	1.2E-03	7.4E-04	4.6E-03
$\lambda = 2$	2^{15}	9.8E-04	5.3E-04	2.2E-04	1.1E-04
	2^{20}	1.5E-04	4.9E-06	1.6E-05	2.8E-06

Default–risky CAT bonds without basis risk								
CD	N	MC	Sobol	Halton	Faure	S_{BB}	H_{BB}	F_{BB}
LN	2^{10}	1.0E-02	4.9E-03	1.1E-03	1.0E-02	3.8E-03	7.5E-05	9.9E-03
$\lambda = 2$	2^{15}	2.3E-03	5.5E-05	2.2E-04	2.2E-04	1.0E-04	1.8E-04	5.1E-04
	2^{20}	3.9E-04	1.6E-04	1.2E-04	1.0E-04	8.8E-05	8.4E-05	5.5E-06
Weibull	2^{10}	8.6E-03	1.1E-03	7.1E-03	1.8E-02	2.1E-03	5.7E-03	2.0E-02
$\lambda = 4$	2^{15}	1.8E-03	1.5E-03	1.3E-03	8.2E-04	1.8E-03	7.1E-04	6.8E-04
	2^{20}	1.9E-04	2.6E-05	8.9E-05	1-2E-04	8.4E-05	8.4E-05	8.0E-05

Default–risky CAT bonds with basis risk								
CD	N	MC	Sobol	Halton	Faure	S_{BB}	H_{BB}	F_{BB}
LN	2^{11}	1.1E-02	4.4E-03	6.5E-03	9.5E-04	1.3E-03	2.2E-03	2.4E-04
ρ=0.5	2^{15}	2.5E-03	8.8E-05	4.7E-04	1.4E-03	2.9E-04	4.6E-04	7.8E-04
$\lambda = 2$	2^{20}	4.3E-04	3.6E-05	6.3E-05	3.3E-05	5.8E-05	7.0E-05	1.4E-04
Weibull	2^{11}	6.2E-03	4.6E-03	1.9E-03	1.8E-02	3.0E-03	1.4E-04	1.7E-02
ρ=0.8	2^{15}	1.4E-03	4.4E-04	9.0E-04	1.5E-05	1.1E-04	1.5E-03	3.9E-04
$\lambda = 4$	2^{20}	3.2E-04	1.5E-04	1.0E-04	2.7E-05	7.4E-05	2.7E-05	7.9E-05
Weibull	2^{11}	5.5E-03	4.1E-04	2.0E-03	1.6E-02	6.9E-04	6.7E-04	1.5E-03
$\rho = 0.3$	2^{15}	1.6E-03	8.7E-04	1.3E-03	3.2E-03	1.1E-03	1.8E-03	1.4E-04
$\lambda = 4$	2^{20}	3.3E-04	2.7E-05	9.6E-05	6.6E-05	9.9E-05	3.3E-05	1.1E-04

Table 2: Relative errors of MC and QMC algorithms for given claim distributions (CD) and values of ρ and λ.

Figure 3 captures the situation in the case of default–risky CAT bonds including basis risk for various dependence levels ρ. Again, Monte Carlo methods are always beaten by their deterministic alternatives. Some of the QMC sequences outperform the Monte Carlo approach by several orders of magnitude. Note that the size of the relative error in all these plots is very small, which further emphasizes the competitiveness of QMC techniques (e.g. for Weibull distributed claims and $\rho = 0.8$, using the Sobol Brownian bridge technique, a relative error of $5 \cdot 10^{-4}$ can be achieved in about 0.1 seconds!).
As a numerical illustration, Table 2 gives relative errors of our MC and QMC algorithms in all three models for various parameter choices and increasing N.

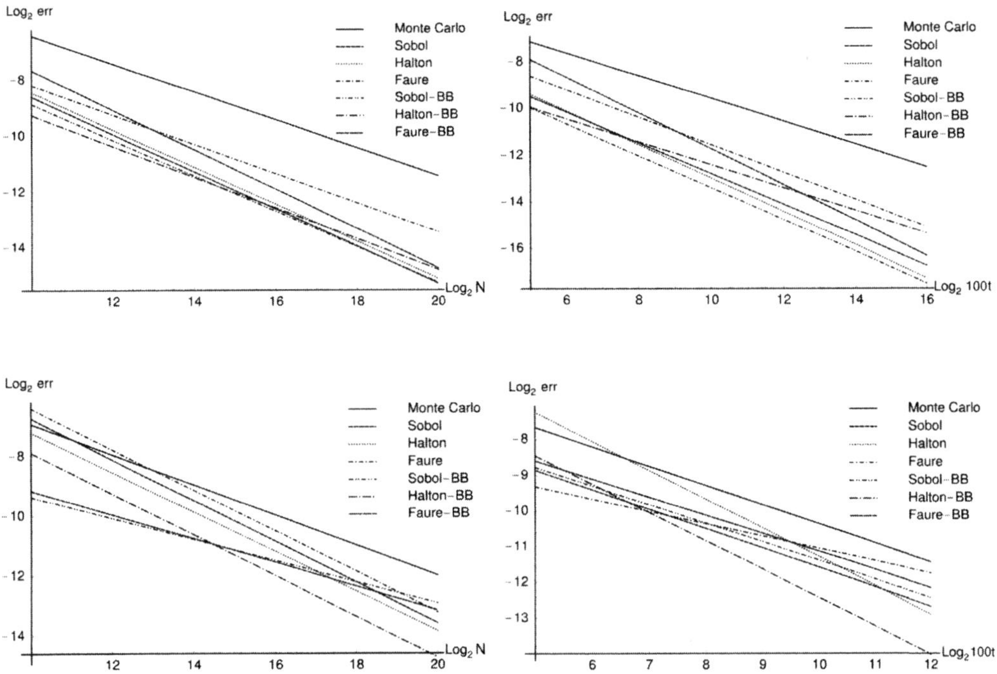

Figure 2: Approximation of default–risky CAT bond prices without basis risk (Lognormal claims (above) and Weibull claims (below)): Relative error as a function of N (left) and calculation time t (right)

Note that it can happen that QMC estimates are closer to the exact value for lower values of N. This may happen since the actual estimate for a given N may be above or below the average behavior represented by the regression fit. This effect does not occur for the Monte Carlo approach, since we already average over 25 simulation runs.

Finally, Figure 4 depicts the CAT bond prices in the presence of basis risk as function of the dependence parameter ρ (representing the amount of basis risk) and the ratio a of the total CAT bonds' face value over total outstanding debts. Moreover the CAT bond prices as a function of ρ and trigger level K are given. Numerical illustrations are given in Table 3. An increase in basis risk leads to considerably lower CAT bond prices, which shows that neglecting basis risk can be very dangerous for holders of these bonds.

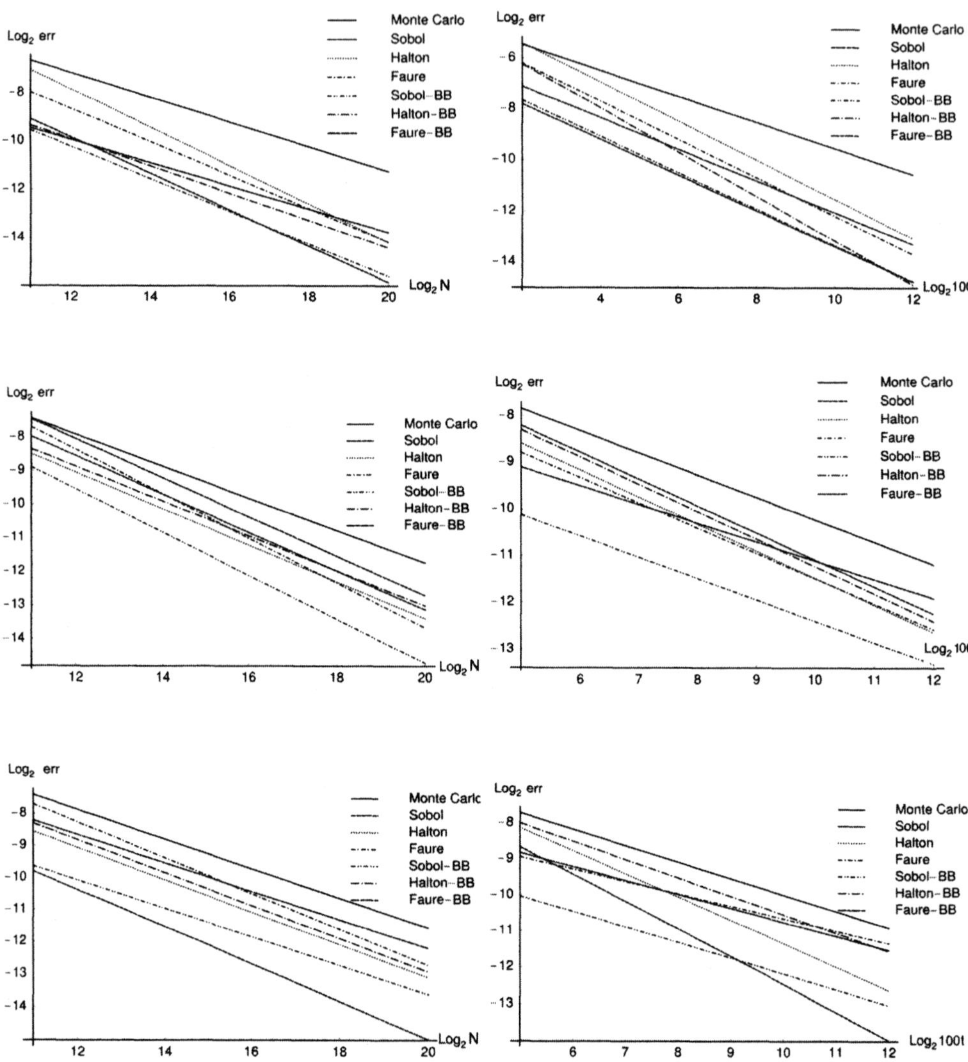

Figure 3: Approximation of default–risky CAT bond prices with different levels of basis risk (Lognormal claims, $\rho = 0.5$ (top); Weibull claims, $\rho = 0.8$ (middle) and Weibull claims, $\rho = 0.3$ (bottom)): Relative error as a function of N (left) and calculation time t (right).

a	Spearman ρ					K	Spearman ρ				
	0	0.2	0.5	0.8	1		0	0.2	0.5	0.8	1
0.1	0.695	0.700	0.709	0.722	0.758	80	0.682	0.687	0.697	0.711	0.740
0.3	0.683	0.688	0.698	0.712	0.747	90	0.689	0.694	0.703	0.717	0.750
0.5	0.668	0.674	0.684	0.699	0.729	100	0.695	0.699	0.708	0.722	0.757
0.7	0.65	0.656	0.666	0.681	0.705	110	0.700	0.704	0.713	0.726	0.762
0.9	0.624	0.631	0.641	0.655	0.672	120	0.704	0.709	0.717	0.730	0.764

Table 3: CAT bond prices for various levels of basis risk (ρ), ratio a and trigger levels K.

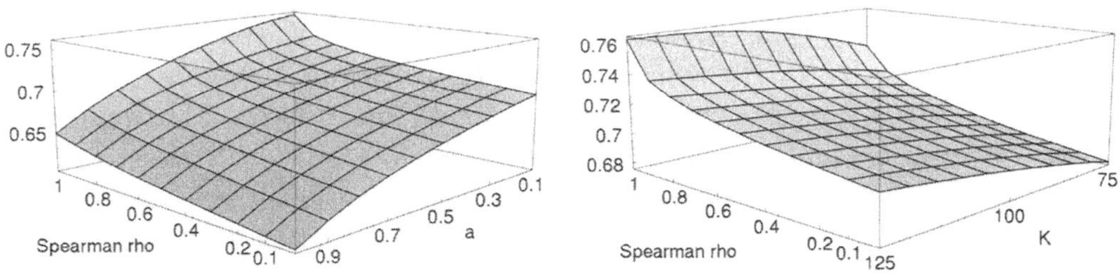

Figure 4: Default–risky CAT bond prices as a function of basis risk (ρ) and ratio a of CAT bonds in total debt (left), respectively trigger level K (right).

References

[1] H. Albrecher, R. Kainhofer: *Risk theory with a non–linear dividend barrier,* Computing **68** (2002), 289–311.

[2] H. Albrecher, R. Kainhofer, R. F. Tichy: *Simulation methods in ruin models with non–linear dividend barriers,* Math. Comput. Simulation (2002), to appear.

[3] R. C. Baker: *On irregularities of distribution II* J. London Math. Soc. (2) **59** (1999), 50–64.

[4] R. E. Caflisch: *Monte Carlo and Quasi-Monte Carlo methods,* Acta Numerica (1998), 1–46.

[5] W. W. Chen, M. M. Skriganov: *Explicit constructions in the classical mean squares problem in irregularities of point distribution,* J. Reine Angew. Math. **545** (2002), 67–95.

[6] J. Cox, J. Ingersoll, S. Ross: *The term structure of interest rates,* Econometrica **53** (1985), 385–407.

[7] S. H. Cox, H. W. Pederson: *Catastrophe risk bonds,* North American Actuarial Journal **4** (2000), 56–82.

[8] N. A. Doherty: *Financial innovation in the management of catastrophe risk,* Journal of Applied Corporate Finance **10** (1997), 134–170.

[9] L. Devroye: *Non Uniform Random Variate Generation,* Springer, 1986.

[10] M. Drmota, R. F. Tichy: *Sequences, Discrepancies and Applications,* Lecture Notes in Mathematics **1651**, Springer, New York, 1997.

[11] P. Embrechts, A. McNeil, D. Straumann: *Correlation and dependence in risk management: Properties and Pitfalls,* In: *Risk Management: Value at Risk and Beyond* (2002), M. Dempster (ed.), Cambridge Univ. Press, 176–223.

[12] H. Faure: *Discrépance de suites associées à un système de numération (en dimension s),* Acta Arithmetica **41** (1982), 337–351.

[13] J. H. Halton: *On the efficiency of certain quasi–random sequences of points in evaluating multidimensional integrals,* Numer. Math. (1960), 84–90.

[14] S. Heinrich, E. Novak, G. Wasilskowski, H. Wozniakowski: *The inverse of the star–discrepancy depends linearly on the dimension,* Acta Arithm. **XCVI** (2001), 279–302.

[15] M. H. Kalos, P. A. Whitlock: *Monte Carlo Methods,* John Wiley, New York, 1986.

[16] G. Larcher, G. Leobacher: *Tractability of the Brownian bridge algorithm,* preprint, 2002.

[17] J. P. Lee, M. T. Yu: *Pricing default–risky CAT bonds with moral hazard and basis risk,* Journal of Risk and Insurance **69** (2002), 25–44.

[18] H. Loubergé, E. Kellezi, M. Gilli: *Using catastrophe–linked securities to diversify insurance risk,* Journal of Insurance Issues **22** (1999), 125–146.

[19] M. Matsumoto and T. Nishimura: *Mersenne twister: A 623-dimensionally equidistributed uniform pseudorandom number generator,* ACM Trans. on Modeling and Computer Simulation **8** (1998), 3–30.

[20] W. Morokoff: *Generating quasi-random paths for stochastic processes,* SIAM Review **40** (1998), 765–788.

[21] R. Nelsen: *An Introduction to Copulas,* Lecture Notes in Statistics **139**, Springer, Berlin, 1999.

[22] H. Niederreiter: *Random Number Generation and Quasi-Monte Carlo Methods,* Society for Industrial and Applied Mathematics, Philadelphia, 1992.

[23] H. Niederreiter, R. F. Tichy, G. Turnwald: *An inequality for differences of distribution functions,* Arch. Math. (Basel), **54** (1990), 166–172.

[24] S. H. Paskov, J. F. Traub: *Faster valuation of financial derivatives,* Journal of Portfolio Management (1995), 113–120.

[25] K. F. Roth: *On irregularities of distribution,* Mathematika **1** (1954), 73–79.

[26] I. M. Sobol': *On the distribution of points in a cube and the approximate evaluation of integrals,* USSR Comput. Math. Math. Phys. (1967), 86–112.

[27] I. M. Sobol': *Computation of improper integrals by means of equidistributed sequences,* Dokl. Akad. Nauk SSSR **210** (1973), 278–281.

[28] H. Woźniakowski: *Average case complexity of multivariate integration,* Bull. Amer. Math. Soc. (N. S.) **24** (1991), 185–194.

Address:

Hansjörg Albrecher, Jürgen Hartinger, and Robert F. Tichy
Graz University of Technology
Department of Mathematics
A–8010 Graz
Austria

International Series of Numerical Mathematics
Vol. 145, ©2003 Birkhäuser Verlag Basel/Switzerland

Note on a Generalized Invariance Principle and its Relevance for Cap Discrepancy and Energy

Johann S. Brauchart

Dedicated to Prof. Dr. Manfred Reimer
on the occasion of his seventieth birthday

Abstract

We give an extension of Stolarsky's invariance principle. For a certain class of energy kernel functions we show relations between the spherical cap discrepancy and the energy of N–point sets on the unit sphere. We see that cap discrepancy and energy are comparable in a sense. In addition an integral representation for the energy which involves discrepancy functions is derived.

Introduction

Given $d \geq 2$, the d–dimensional unit sphere is denoted by

$$S^d \equiv \left\{ \mathbf{x} \in \mathbb{R}^{d+1} : \langle \mathbf{x}, \mathbf{x} \rangle = 1 \right\}$$

with the usual inner product $\langle \cdot, \cdot \rangle$ in $\mathbb{R}^{d+1}$. Throughout σ_d will denote the Lebesgue measure on S^d. We set $\omega_d \equiv \int_{S^d} d\sigma_d$. $|\cdot|$ is the Euclidean distance in $\mathbb{R}^{d+1}$.

In [6] K. B. Stolarsky states his invariance principle. In this note we give an extended version of the principle which is our first main result.

To understand the following we introduce two basic concepts.

A very natural measure for the quality of the distribution of an N–point set $X_N \equiv \{\mathbf{x}_1, \ldots, \mathbf{x}_N\}$ in S^d is the discrepancy with respect to a given system of subsets of S^d. The *spherical cap discrepancy* is defined as

$$D_N(X_N) \equiv \sup_C \left| \frac{\sharp \{\mathbf{x} : \mathbf{x} \in C \cap X_N\}}{N} - \frac{\sigma_d(C)}{\omega_d} \right|.$$

The supremum is extended over all spherical caps $C \subseteq S^d$. Now one can define a sequence of N–point sets $\{X_N\}$ to be *asymptotically equidistributed* if the corresponding sequence of cap discrepancies $\{D_N(X_N)\}$ converges to zero. This means, asymptotically each spherical cap gets its fair share of points. So N–point sets with small cap discrepancy can be seen as to be 'good' point sets.

On the contrary one can define for a suitable function g the *energy*

$$E(g; X_N) \equiv \frac{1}{N^2} \sum_{j \neq k} g\left(\langle \mathbf{x}_j, \mathbf{x}_k \rangle\right)$$

of the N–point set X_N with respect to the *energy kernel* g. This sum can be seen as the discrete analogue of the energy integral

$$\mathcal{I}(g; \beta) \equiv \int_{S^d} \int_{S^d} g\left(\langle \mathbf{x}, \mathbf{y} \rangle\right) \, d\beta(\mathbf{x}) d\beta(\mathbf{y})$$

(β is a Borel probability measure on S^d.)

We will introduce a class of admissible kernel functions. Our aim is to use the extended invariance principle to obtain a lower bound for the cap discrepancy in terms of the energy with respect to this kernel function. Furthermore an upper bound can also be shown. In this sense cap discrepancy and energy are comparable. In addition we will obtain an integral representation of the energy for such kernels which involves discrepancy functions.

1 The Basic Lemma

We adopt the notation in [6]. Let $\int_{a,b}$ denote the integral $\int_a^b$ if $a < b$ and $\int_b^a$ if $a > b$ respectively. Let τ be the normalised Haar measure on the special orthogonal group $SO = SO(d+1)$ acting on S^d. We have $\tau(A\mathcal{M}) = \tau(\mathcal{M})$ for all Borel sets $\mathcal{M} \subseteq S^d$ and for all $A \in SO$.

Definition 1.1. *Let g be integrable and $n \geq 1$. Then we define for all* $\mathbf{x}, \mathbf{y} \in S^d$

$$\rho(\mathbf{x}, \mathbf{y}) \equiv \rho(g; \mathbf{x}, \mathbf{y}) \equiv \int_{SO} \underbrace{\int_{a,b} \cdots \int_{a,b}}_{n} g(\mathbf{t}) \, d\mathbf{t} \, d\tau(A) \qquad a \equiv \langle \mathbf{x}, A\mathbf{e} \rangle, b \equiv \langle \mathbf{y}, A\mathbf{e} \rangle$$

Remarks. **(1)** For all $A \in SO$ we have $\rho(A\mathbf{x}, A\mathbf{y}) = \rho(\mathbf{x}, \mathbf{y})$. **(2)** $\rho(\mathbf{x}, \mathbf{y}) = \rho(\mathbf{y}, \mathbf{x})$. **(3)** ρ is a metric for $g > 0$ although we do not need this property.

Let $p_1 \le \ldots \le p_M$ and $q_1 \le \ldots \le q_N$ be real numbers. Set

$$G(t) \equiv G(p_j, q_k; t) \equiv \sharp\{j : p_j \le t\} - \sharp\{k : q_k \le t\}.$$

Stolarsky used the following useful identity (with $\int_{a,b} \equiv \int_{a,b} g(t)\, dt$)

$$\sum_{j,k} \int_{p_j,q_k} - \sum_{j<k} \int_{p_j,p_k} - \sum_{j<k} \int_{q_j,q_k} = \int_{-\infty}^{\infty} g(t)\, G(t)\, (G(t) - (M-N))\, dt.$$

We will prove an analogous result for double (and multiple) integrals.

Let $\int_{a,b} = \int_{a,b} \cdots \int_{a,b} g(\mathbf{t})\, d\mathbf{t}$, $n \ge 2$. We want to show an identity like

$$\sum_{j,k} \int_{p_j,q_k} - \sum_{j<k} \int_{p_j,p_k} - \sum_{j<k} \int_{q_j,q_k} = \int_{\mathbb{R}^n} g(\mathbf{t})\, G(p_i, q_k; \mathbf{t})\, d\mathbf{t}$$

where $G(p_i, q_k; \mathbf{t})$ 'counts' all integrals on the left side for which the given n–tuple $\mathbf{t} = (t_1, \ldots, t_n)$ appear as an argument for g.

It is sufficient to consider only those points in $\mathbb{R}^n$ with pairwise different components $t_1, \ldots, t_n$ which are distinct from the p_j's and q_k's. Thus there always exist unique indices $1 \le \mu, \nu \le n$ such that $t_\mu = \min\{t_1, \ldots, t_n\}$, $t_\nu = \max\{t_1, \ldots, t_n\}$ and

$$\begin{aligned}
&G(p_i, q_k; \mathbf{t}) \\
&= \sharp\{j : p_j \le t_\mu\}(N - \sharp\{k : q_k \le t_\nu\}) + \sharp\{k : q_k \le t_\mu\}(M - \sharp\{j : p_j \le t_\nu\}) \\
&- \sharp\{j : p_j \le t_\mu\}(M - \sharp\{j : p_j \le t_\nu\}) - \sharp\{k : q_k \le t_\mu\}(N - \sharp\{k : q_k \le t_\nu\}) \\
&= G(t_\mu)\, (G(t_\nu) - (M-N))
\end{aligned}$$

To get an expression independent of the μ's and ν's we sum up the last formula for all μ, ν and use characteristic functions χ_M to single out the appropriate term. We obtain

Lemma 1.2. *Let $p_1 \le \ldots \le p_M$ and $q_1 \le \ldots \le q_N$ be real numbers, $n \ge 1$. Then*

$$\sum_{j,k} \int_{p_j,q_k} - \sum_{j<k} \int_{p_j,p_k} - \sum_{j<k} \int_{q_j,q_k} = \int_{\mathbb{R}^n} g(\mathbf{t})\, G(p_j, q_k; \mathbf{t})\, d\mathbf{t}$$

where

$$\begin{aligned}
\int_{a,b} &\equiv \int_{a,b} \cdots \int_{a,b} g(t_1, \ldots, t_n)\, dt_1 \ldots dt_n, \\
G(p_j, q_k; \mathbf{t}) &\equiv \sum_{\mu,\nu} \chi_{\mu\nu}(\mathbf{t})\, G(t_\mu)\, (G(t_\nu) - (M-N)), \\
\chi_{\mu\nu}(\mathbf{t}) &\equiv \prod_{\mu \ne \ell = 1}^{n} \chi_{(t_\mu,\infty)}(t_\ell) \prod_{\nu \ne \ell = 1}^{n} \chi_{(-\infty,t_\nu)}(t_\ell).
\end{aligned}$$

Note, in the case $n = 1$ we set the empty product to one.

2 The Extension of the Principle

Now we are able to prove the first theorem. We use the ideas developed in [6]. Set $M = N$. Let $X_N \equiv \{\mathbf{x}_1, \ldots, \mathbf{x}_N\}$ and $Y_N \equiv \{\mathbf{y}_1, \ldots, \mathbf{y}_N\}$ be N–point sets in S^d. Averaging over $SO \times SO$ gives

$$\int_{SO}\int_{SO}\sum_{j,k}\rho(A_1\mathbf{x}_j, A_2\mathbf{y}_k)\, d\tau(A_1)d\tau(A_2) = \frac{N^2}{\omega_d^2}\int_{S^d}\int_{S^d}\rho(\mathbf{x},\mathbf{y})\, d\sigma_d(\mathbf{x})d\sigma_d(\mathbf{y}). \tag{2.1}$$

Otherwise we can calculate

$$\sum_{j,k}\rho(A_1\mathbf{x}_j, A_2\mathbf{y}_k)\,,$$

where we use the definition of ρ to obtain

$$\int_{SO}\sum_{j,k}\int_{p_j,q_k}\cdots\int_{p_j,q_k} g(\mathbf{t})\, d\mathbf{t}\; d\tau(B)$$

with $p_j \equiv \langle A_1\mathbf{x}_j, B\mathbf{e}\rangle$ and $q_k \equiv \langle A_2\mathbf{y}_k, B\mathbf{e}\rangle$. We apply Lemma 1.2 and get

$$\sum_{j<k}\rho(A_1\mathbf{x}_j, A_1\mathbf{x}_k) + \sum_{j<k}\rho(A_2\mathbf{y}_j, A_2\mathbf{y}_k) + \int_{SO}\int_{\mathbb{R}^n} g(\mathbf{t})\; G(p_j, q_k; \mathbf{t})\; d\mathbf{t}\, d\tau(B).$$

An integration over $SO \times SO$ yields

$$\sum_{j<k}\rho(\mathbf{x}_j,\mathbf{x}_k) + \sum_{j<k}\rho(\mathbf{y}_j,\mathbf{y}_k) + \int_{\mathbb{R}^n} g(\mathbf{t})\int_{SO}\int_{SO}\int_{SO} G(p_j, q_k; \mathbf{t})\; d\tau(A_1)d\tau(A_2)d\tau(B)\, d\mathbf{t}. \tag{2.2}$$

By the left–invariance of the Haar measure we are allowed to substitute $\tilde{A}_1 = B^{-1}A_1$ and $\tilde{A}_2 = B^{-1}A_2$. This eliminates the dependency on B.

We have to consider the Haar–integral

$$\int_{SO}\int_{SO} G(p_j, q_k; \mathbf{t})\; d\tau(\tilde{A}_1)d\tau(\tilde{A}_2),$$

where $p_j \equiv \langle \tilde{A}_1 \mathbf{x}_j, \mathbf{e}\rangle$ and $q_k \equiv \langle \tilde{A}_2 \mathbf{y}_k, \mathbf{e}\rangle$. We make the following observation

$$\int_{SO} \sharp\{j : \langle \mathbf{x}_j, A\mathbf{e}\rangle \le t\}\ d\tau(A) = N\frac{\sigma_d(C(\mathbf{e};t))}{\omega_d} \tag{2.3}$$

since the surface measure is invariant under rotation and

$$\sharp\{j : \langle \mathbf{x}_j, \mathbf{z}\rangle \le t\} = \sum_{j=1}^{N} \chi_{C(\mathbf{x}_j;t)}(\mathbf{z}), \qquad C(\mathbf{z};t) \equiv \left\{\mathbf{x} \in S^d : \langle \mathbf{z}, \mathbf{x}\rangle \le t\right\}.$$

This motivates the introduction of the discrepancy function

$$D_N^S(C;X) \equiv \frac{\sharp\{\mathbf{x} : \mathbf{x} \in C \cap X\}}{N} - \frac{\sigma_d(C)}{\omega_d}, \qquad C \subseteq S^d.$$

It is easy to see that

$$\frac{1}{N} G(t) = D_N^S(C(\tilde{A}_1^{-1}\mathbf{e};t); X_N) - D_N^S(C(\tilde{A}_2^{-1}\mathbf{e};t); Y_N)$$

and by (2.3) it follows

$$\frac{1}{N^2}\int_{SO}\int_{SO} G(t)G(t')\ d\tau(\tilde{A}_1)d\tau(\tilde{A}_2) = f(X_N;t,t') + f(Y_N;t,t'),$$

where we set

$$f(X_N;t,t') \equiv \frac{1}{\omega_d}\int_{S^d} D_N^S(C(\mathbf{z};t);X_N) D_N^S(C(\mathbf{z};t');X_N)\, d\sigma_d(\mathbf{z}).$$

Hence by the definition of $G(p_j, q_k; \mathbf{t})$ we have

$$\frac{1}{N^2}\int_{SO}\int_{SO} G(p_j,q_k;\mathbf{t})\, d\tau(\tilde{A}_1)d\tau(\tilde{A}_2) = \sum_{\mu,\nu} \chi_{\mu\nu}(\mathbf{t})\left(f(X_N;t_\mu,t_\nu) + f(Y_N;t_\mu,t_\nu)\right). \tag{2.4}$$

Combining (2.1), (2.2) and (2.4) with the observation that the left–hand–side of (2.1) is equal to the expression in (2.2) this yields

Theorem 2.1. **(extended invariance principle)** *Let $n \geq 1$ and g be integrable. For any N–point–set $X_N \equiv \{\mathbf{x}_1, \dots, \mathbf{x}_N\}$ in S^d holds*

$$\frac{1}{N^2} \sum_{j \neq k} \rho(\mathbf{x}_j, \mathbf{x}_k) + 2 \int_{\mathbb{R}^n} \sum_{\mu,\nu} \chi_{\mu\nu}(\mathbf{t})\, g(\mathbf{t})\, f(X_N; t_\mu, t_\nu)\, d\mathbf{t}$$
$$= \frac{1}{\omega_d^2} \int_{S^d} \int_{S^d} \rho(\mathbf{x}, \mathbf{y})\, d\sigma_d(\mathbf{x}) d\sigma_d(\mathbf{y}),$$

where

$$\rho(\mathbf{x}, \mathbf{y}) \equiv \int_{SO} \underbrace{\int_{a,b} \cdots \int_{a,b}}_{n} g(\mathbf{t})\, d\mathbf{t}\, d\tau(A), \qquad a \equiv \langle \mathbf{x}, A\mathbf{e} \rangle, b \equiv \langle \mathbf{y}, A\mathbf{e} \rangle$$

$$\chi_{\mu\nu}(\mathbf{t}) \equiv \prod_{\mu \neq j = 1}^{n} \chi_{(t_\mu, \infty)}(t_j) \prod_{\nu \neq j = 1}^{n} \chi_{(-\infty, t_\nu)}(t_j)$$

$$f(X_N; t, t') \equiv \frac{1}{\omega_d} \int_{S^d} D_N^S(C(\mathbf{z}; t); X_N) D_N^S(C(\mathbf{z}; t'); X_N)\, d\sigma_d(\mathbf{z})$$

and

$$D_N^S(C(\mathbf{z}; t); X_N) \equiv \frac{\sharp\{j : \mathbf{x}_j \in C(\mathbf{z}; t)\}}{N} - \frac{\sigma_d(C(\mathbf{z}; t))}{\omega_d}$$

is the discrepancy function with respect to the spherical cap

$$C(\mathbf{z}; t) \equiv \left\{ \mathbf{x} \in S^d : \langle \mathbf{z}, \mathbf{x} \rangle \leq t \right\}.$$

Remarks. **(1)** Notice that we obtain Stolarsky's original result for $n = 1$. **(2)** The integral over $\mathbb{R}^n$ is in fact an integral over the bounded domain $[-1, 1]^n$. **(3)** The principle says, the sum over mutual distances determined by the N–point set plus the discrepancy is always constant. The constant is the corresponding energy integral.

Theorem 2.1 allows us to obtain a lower bound for the cap discrepancy. The simple upper estimate

$$\left| D_N^S(C(\mathbf{z}; t); X_N) \right| \leq D_N(X_N)$$

implies

$$|f(X_N; t, t')| \leq D_N(X_N)^2.$$

With the notation in Theorem 2.1 we obtain

Corollary 2.2.

$$\sqrt{\left|\frac{1}{N^2}\sum_{j\neq k}\rho(\mathbf{x}_j,\mathbf{x}_k)-\mathcal{I}\left(\rho;\frac{1}{\omega_d}\sigma_d\right)\right|}\leq C_g\,D_N(X_N)$$

with

$$C_g\equiv\sqrt{2\int_{\mathrm{I\!R}^n}\sum_{\mu,\nu}\chi_{\mu\nu}(\mathbf{t})\,|g(\mathbf{t})|\;dt}\;.$$

The case $n=2$ is of special interest. Theorem 2.1 gives

$$\frac{1}{N^2}\sum_{j\neq k}\rho(\mathbf{x}_j,\mathbf{x}_k)+2\int_{-1}^{1}\int_{-1}^{1}g(t,t')\;f(X_N;t,t')\;dtdt' \\ =\frac{1}{\omega_d^2}\int_{S^d}\int_{S^d}\rho(\mathbf{x},\mathbf{y})\;d\sigma_d(\mathbf{x})d\sigma_d(\mathbf{y}). \tag{2.5}$$

For our purpose we may assume that $g(t,t')$ depends on $|t-t'|$. Hence the integrand on the left–hand side is symmetric and we substitute $u\equiv t'-t$, $v\equiv t$:

$$\int_{-1}^{1}\int_{-1}^{1}g(|t-t'|)\;f(X_N;t,t')\;dtdt'=2\int_0^2 g(|u|)\int_{-1}^{1-u}f(X_N;v,u+v)dv\,du. \tag{2.6}$$

Recall the definition of $f(X_N;t,t')$. We rewrite the product of the discrepancy functions as follows

$$D_N^S(C(\mathbf{z};v);X_N)D_N^S(C(\mathbf{z};u+v);X_N)=\left|D_N^S(C(\mathbf{z};v);X_N)\right|^2 \\ +D_N^S(C(\mathbf{z};v);X_N)\left(D_N^S(C(\mathbf{z};u+v);X_N)-D_N^S(C(\mathbf{z};v);X_N)\right)$$

and observe that

$$D_N^S(C(\mathbf{z};u{+}v);X_N)-D_N^S(C(\mathbf{z};v);X_N)=D_N^S(C(\mathbf{z};u{+}v)\backslash C(\mathbf{z};v);X_N). \tag{2.7}$$

The right–hand side is a discrepancy function with respect to the family of differences of two centred spherical caps.

Now feeding (2.7) into (2.6) and combining the result with (2.5) gives

Corollary 2.3. *Let g be integrable and $g(t,t') = g(|t-t'|)$. For any N–point–set $X_N \equiv \{\mathbf{x}_1, \dots, \mathbf{x}_N\}$ in S^d holds*

$$\begin{aligned}
&\frac{1}{N^2}\sum_{j\neq k}\rho(\mathbf{x}_j,\mathbf{x}_k)\\
&+4\int_{-1}^{1}\left(\int_{-1}^{1-v} g(|u|)\,du\right)\frac{1}{\omega_d}\int_{S^d}\left|D_N^S(C(\mathbf{z};v);X_N)\right|^2 d\sigma_d(\mathbf{z})\,dv\\
&+4\int_{0}^{2} g(|u|)\int_{-1}^{1-u} COV(X_N;t,t')\,dvdu\\
&=\frac{1}{\omega_d^2}\int_{S^d}\int_{S^d}\rho(\mathbf{x},\mathbf{y})\,d\sigma_d(\mathbf{x})d\sigma_d(\mathbf{y})
\end{aligned}$$

with

$$\rho(\mathbf{x},\mathbf{y}) \equiv \int_{SO}\int_{a,b}\int_{a,b} g(|t-t'|)\,dtdt'\,d\tau(A), \quad a \equiv \langle \mathbf{x}, A\mathbf{e}\rangle, b \equiv \langle \mathbf{y}, A\mathbf{e}\rangle,$$

$$COV(X_N;v,u{+}v) \equiv \frac{1}{\omega_d}\int_{S^d} D_N^S(C(\mathbf{z};v);X_N)D_N^S(C(\mathbf{z};u{+}v)\backslash C(\mathbf{z};v);X_N)\,d\sigma_d(\mathbf{z}$$

3 Cap Discrepancy and Energy

We introduce the following class of admissible energy kernel functions.

Definition 3.1. *Let $g : [-1,1] \to \mathbb{R}$ be a continuous function which possesses an ultraspherical expansion*

$$g(t) = \sum_{m=0}^{\infty} a_m(g)\,P_m^{(d)}(t), \qquad -1 \le t \le 1$$

with $a_m(g) > 0$ for all $m \ge 1$.

Note that $g(t) - a_0(g)$ is positive definite in the sense of Schoenberg [5]. Here and throughout the rest of this note $P_m^{(d)}(t)$ denotes the ultraspherical polynomial corresponding to S^d, normalised by $P_m^{(d)}(1) = 1$.

Let g be admissible and $X_N \equiv \{\mathbf{x}_1, \dots, \mathbf{x}_N\}$ a N–point set in S^d. Set

$$\psi_m^{(d)}(X_N) \equiv \frac{1}{N^2}\sum_{j,k} P_m^{(d)}(\langle \mathbf{x}_j, \mathbf{x}_k\rangle).$$

Then we get for the g–energy of X_N the expansion

$$E(g; X_N) + \frac{1}{N}\, g(1) = \sum_{m=0}^{\infty} a_m(g)\, \psi_m^{(d)}(X_N). \tag{3.1}$$

Note that by the definition of the ultraspherical polynomials we have

$$\psi_0^{(d)}(X_N) = 1, \qquad \psi_1^{(d)}(X_N) = \Big\| \frac{1}{N} \sum_{j=1}^{N} \mathbf{x}_j \Big\|_2^2 .$$

By applying the Funk–Hecke formula [2, Section 9.7] we see that $\psi_m^{(d)}(X_N)$ is non–negative.

An immediate consequence is the lower bound

$$0 \leq a_1(g) \Big\| \frac{1}{N} \sum_{j=1}^{N} \mathbf{x}_j \Big\|_2^2 \leq E(g; X_N) + \frac{1}{N}\, g(1) - a_0(g) \tag{3.2}$$

which holds for any N–point set X_N in S^d. From that we see that the most interesting N–point sets are *g–minimal energy N–point sets* X_N^* which attain the g–minimal energy

$$\min_{X_N} E(g; X_N) = E(g; X_N^*).$$

Integrating the inequality

$$E(g; X_N^*) + \frac{1}{N}\, g(1) \leq \sum_{m=0}^{\infty} a_m(g)\, \psi_m^{(d)}(X_N)$$

over all points $\mathbf{x}_1, \dots, \mathbf{x}_N \in S^d$ with respect to the normalised surface measure gives the upper bound

$$E(g; X_N^*) + \frac{1}{N}\, g(1) - a_0(g) \leq \frac{1}{N} \left(g(1) - a_0(g) \right). \tag{3.3}$$

Combining (3.2) and (3.3) we get

$$\lim_{N \to \infty} E(g; X_N^*) = a_0(g) \tag{3.4}$$

for all sequences $\{X_N^*\}$ of g–minimal–energy N–point sets. We also have

$$\Big\| \frac{1}{N} \sum_{j=1}^{N} \mathbf{x}_j^* \Big\|_2 \leq \sqrt{\frac{g(1) - a_0(g)}{a_1(g)}}\, \frac{1}{\sqrt{N}} \longrightarrow 0 \quad (N \to \infty)$$

which means that g–minimal–energy point sets are centred at least asymptotically.

The expansion (3.1) together with the non–negativity of $\psi_m^{(d)}(X_N)$ yields the simple estimate for each $m \geq 1$

$$\psi_m^{(d)}(X_N) \leq \frac{1}{a_m(g)} \left(E(g, X_N) + \frac{1}{N} g(1) - a_0(g) \right).$$

Now in [4] an upper bound for the cap discrepancy in terms of $\psi_m^{(d)}(X_N)$ is shown

$$D_N(X_N) \leq \frac{c_1(d)}{M+1} + c_2(d) \sum_{m=1}^{M} m^{\frac{d-3}{2}} \sqrt{\psi_m^{(d)}(X_N)}$$

where $c_1(d)$, $c_2(d)$ are positive constants depending only on d and $M \geq 1$. Hence

Theorem 3.2. *Let g be admissible, $M \geq 1$ and $c_1(d)$, $c_2(d)$ be positive constants depending only on d. Then for any N–point set $X_N \equiv \{\mathbf{x}_1, \dots, \mathbf{x}_N\}$ in S^d holds*

$$D_N(X_N) \leq \frac{c_1(d)}{M+1} + c_2(d)\, M^{\frac{d-1}{2}} \max_{1 \leq m \leq M} \frac{1}{\sqrt{a_m(g)}} \sqrt{E(g, X_N) + \frac{1}{N} g(1) - a_0(g)}.$$

This upper bound holds for any N–point set. But for g–minimal–energy N–point sets the rightmost square root can be made arbitrarily small. So we obtain a result about the distribution property of such point sets

Corollary 3.3. *Let g be admissible. Then each sequence $\{X_N^*\}$ of g–minimal–energy N–point sets in S^d is asymptotically equidistributed.*

Note, this distribution property also follows naturally from standard approximation arguments because the energy is finite.

4 The Lower Bound

The lower bound for the cap discrepancy follows from an invariance principle for the energy which is also an integral representation involving discrepancy functions. The desired representation is obtained in three steps.

STEP 1: SUMS OVER EVEN POWERS OF THE EUCLIDEAN DISTANCE. Recall the definition of $\rho(\mathbf{x}, \mathbf{y})$. For $n = 1$, $g(t) \equiv 1$ the distance function $\rho(\mathbf{x}, \mathbf{y})$ is up to a multiplicative constant the Euclidean distance. Consider for $m \geq 1$ the integral

$$\int_{SO} \left(\int_{\langle \mathbf{x}, A\mathbf{e}\rangle, \langle \mathbf{y}, A\mathbf{e}\rangle} dt \right)^m d\tau(A) = \int_{SO} |\langle \mathbf{y} - \mathbf{x}, A\mathbf{e}\rangle|^m \, d\tau(A)$$

The right–hand side follows from the left invariance of the Haar measure. Changing to the surface measure σ_d and polar coordinates

$$\mathbf{z} = \frac{t}{|\mathbf{y} - \mathbf{x}|}(\mathbf{y} - \mathbf{x}) + \sqrt{1 - t^2}\, \mathbf{z}', \qquad |\mathbf{z}'| = 1, \quad \mathbf{z}' \perp \mathbf{y} - \mathbf{x}$$

we get $\langle \mathbf{y} - \mathbf{x}, A\mathbf{e}\rangle = |\mathbf{y} - \mathbf{x}| t$ and therefore by the rule

$$\frac{1}{\omega_d} \int_{S^d} f(\langle \mathbf{x}, \mathbf{z}\rangle) \, d\sigma_d(\mathbf{z}) = \frac{\omega_{d-1}}{\omega_d} \int_{-1}^{1} f(t)\, (1 - t^2)^{d/2-1} \, dt$$

the result (with $\omega_{d-1}/\omega_d = \Gamma(\frac{d+1}{2})/(\Gamma(\frac{1}{2})\, \Gamma(\frac{d}{2}))$)

$$\int_{SO} \left(\int_{\langle \mathbf{x}, A\mathbf{e}\rangle, \langle \mathbf{y}, A\mathbf{e}\rangle} dt \right)^m d\tau(A) = \frac{\omega_{d-1}}{\omega_d} \frac{\Gamma(\frac{m+1}{2})\, \Gamma(\frac{d}{2})}{\Gamma(\frac{m+d+1}{2})} |\mathbf{y} - \mathbf{x}|^m.$$

Now combining

$$(b - a)^{2m} = -\frac{1}{2} \sum_{\mu=1}^{2m-1} \binom{2m}{\mu} (-1)^\mu \, (b^\mu - a^\mu) \left(b^{2m-\mu} - a^{2m-\mu}\right),$$

and

$$b^\mu - a^\mu = \begin{cases} + \int_{a,b} \mu \, t^{\mu-1} \, dt & (a < b) \\ - \int_{a,b} \mu \, t^{\mu-1} \, dt & (a > b) \end{cases}$$

we obtain

$$\left(\int_{a,b} dt \right)^{2m} = \int_{a,b} \int_{a,b} m(2m - 1)(t - t')^{2(m-1)} dt \, dt'.$$

For $a \in \mathbb{R}$ and $n \geq 1$,

$$(a)_0 = 1, \qquad (a)_n = \prod_{\nu=0}^{n-1} (a + \nu)$$

will denote Pochhammer's shifted factorial. Putting everything together gives

Lemma 4.1. *For all* $\mathbf{x}, \mathbf{y} \in S^d$ *and* $m \in \mathbb{N}$ *we have*

$$\int_{SO} \int_{a,b} \int_{a,b} g(|t-t'|)\, dt dt'\, d\tau(A) = \frac{(\frac{1}{2})_m}{(\frac{d+1}{2})_m} |\mathbf{y} - \mathbf{x}|^{2m}$$

with $a \equiv \langle \mathbf{x}, A\mathbf{e} \rangle$, $b \equiv \langle \mathbf{y}, A\mathbf{e} \rangle$ *and*

$$g(|t-t'|) \equiv m(2m-1)|t-t'|^{2(m-1)}.$$

Theorem 2.1 and Lemma 4.1 yields an invariance principle for sums over even powers of Euclidean distances.

Theorem 4.2. *Let be* $m \geq 1$. *For any* N*–point set* $X_N \equiv \{\mathbf{x}_1, \dots, \mathbf{x}_N\}$ *in* S^d *holds*

$$\frac{1}{N^2} \sum_{j \neq k} |\mathbf{x}_j - \mathbf{x}_k|^{2m} + 2 \frac{(\frac{d+1}{2})_m}{(\frac{1}{2})_m} \int_{-1}^{1} \int_{-1}^{1} g(|t-t'|)\, f(X_N; t, t')\, dt dt' = 2^{2m} \frac{(\frac{d}{2})_m}{(d)_m}$$

where

$$g(u) = m(2m-1) u^{2(m-1)}$$

and $f(X_N; t, t')$ *from Theorem* 2.1.

STEP 2. An ultraspherical polynomial can be expressed as a hypergeometric function,

$$P_m^{(d)}(t) = {}_2F_1\left(\begin{matrix} -m, m+d-1 \\ d/2 \end{matrix} ; \frac{1-t}{2} \right).$$

From this it follows

$$\psi_m^{(d)}(X_N) = \sum_{\mu=0}^{m} \frac{(-m)_\mu\, (m+d-1)_\mu}{(d/2)_\mu} \frac{1}{2^{2\mu}\, \mu!} \left(\frac{1}{N^2} \sum_{j,k} |\mathbf{x}_j - \mathbf{x}_k|^{2\mu} \right),$$

whereon we apply Theorem 4.2 to obtain

$$\begin{aligned} &\psi_m^{(d)}(X_N) \\ &-\frac{1}{2} \int_{-1}^{1} \int_{-1}^{1} \frac{d}{du} {}_3F_2 \left(\begin{matrix} -m, m+d-1, \frac{d+1}{2} \\ d/2, -1/2 \end{matrix} ; u \right) \Bigg|_{u=\left(\frac{t-t'}{2}\right)^2} f(X_N; t, t') dt dt' \\ &= {}_2F_1 \left(\begin{matrix} -m, m+d-1 \\ d \end{matrix} ; 1 \right) \qquad (4.1) \end{aligned}$$

By the linear transformation formula [1, 15.3.3]

$$ {}_2F_1\left(\begin{matrix} a, b \\ c \end{matrix} ; z\right) = (1-z)^{c-a-b} {}_2F_1\left(\begin{matrix} c-a, c-b \\ c \end{matrix} ; z\right) $$

the right–hand side of (4.1) vanishes for $m \geq 1$.

We observe

Lemma 4.3.

$$ {}_3F_2\left(\begin{matrix} -m, m+d-1, \alpha \\ d/2, \beta \end{matrix} ; z\right) = \sum_{\mu=0}^{m} \frac{(\beta-\alpha)_\mu}{(\beta)_\mu} \frac{(-z)^\mu}{\mu!} \left(\frac{d}{dz}\right)^\mu P_m^{(d)}(1-2z)\,. $$

Proof. For suitably chosen α_μ we may write

$$ {}_3F_2\left(\begin{matrix} -m, m+d-1, \alpha \\ d/2, \beta \end{matrix} ; z\right) = \sum_{\mu=0}^{m} \alpha_\mu \frac{(-z)^\mu}{\mu!} \left(\frac{d}{dz}\right)^\mu {}_2F_1\left(\begin{matrix} -m, m+d-1 \\ d/2 \end{matrix} ; z\right) $$

We apply the differentiation formula for hypergeometric functions and use the (finite) series expansion. Then we rearrange the sum in terms of rising powers of z and get for the right hand side

$$ \sum_{\mu=0}^{m} \frac{(-m)_\mu \, (m+d-1)_\mu}{(d/2)_\mu} \left(\sum_{\nu=0}^{\mu} \alpha_\nu \frac{(-\mu)_\nu}{\nu!}\right) \frac{z^\mu}{\mu!}. $$

By comparing this sum with the expansion of ${}_3F_2\left(\begin{matrix} -m, m+d-1, \alpha \\ d/2, \beta \end{matrix} ; z\right)$ we get the relation

$$ \frac{(\alpha)_\mu}{(\beta)_\mu} = \sum_{\nu=0}^{\mu} \alpha_\nu \frac{(-\mu)_\nu}{\nu!}. $$

The application of the Chu–Vandermonde formula

$$ \frac{(c-b)_m}{(c)_m} = {}_2F_1\left(\begin{matrix} -m, b \\ c \end{matrix} ; 1\right) $$

allows us to determine the coefficients α_μ. □

The above lemma allows us to rewrite the weight function in (4.1). We obtain by using the symmetry of the integrand an invariance principle for $\psi_m^{(d)}$

Theorem 4.4. *Let be $m \geq 1$. For any N-point set $X_N \equiv \{\mathbf{x}_1, \dots, \mathbf{x}_N\}$ in S^d holds*

$$\psi_m^{(d)}(X_N) = 2(d+1) \int_0^2 h_m(d;u) \int_{-1}^{1-u} f(X_N; v, u+v) dv\, du$$

with

$$h_m(d;u) \equiv \sum_{\mu=0}^{m-1} \frac{(-1-d/2)_\mu}{(1/2)_\mu} \frac{(1-t)^\mu}{\mu!} \left(\frac{d}{dt}\right)^{\mu+1} P_m^{(d)}(t)\Big|_{t=1-u^2/2}$$

and $f(X_N; t, t')$ from Theorem 2.1.

Note, if d is even ($d = 2D$) then the sum $h_m(d;u)$ has at most $D+1$ terms.

STEP 3. The last theorem gives an integral representation for $\psi_m^{(d)}$. Feeding this into the expansion of the energy we obtain the desired integral representation provided we are allowed to interchange integration with summation and summation with differentiation. The sufficient condition of uniform convergence assures this procedure.

Theorem 4.5. *Let $d = 2D$ be even. Let g be admissible and $g \in C^{D+2}([-1,1])$. Furthermore assume for all $n = 1, 2, \dots, D+2$*

$$\sum_{m=n}^{\infty} a_m(g_\rho) \left(\frac{d}{dt}\right)^n P_m^{(d)}(t) = g^{(n)}(t) \quad \text{uniformly on } [-1,1].$$

Then for any N-point set $X_N \equiv \{\mathbf{x}_1, \dots, \mathbf{x}_N\}$ in S^d holds

$$E(g; X_N) + \frac{1}{N} g(1) - a_0(g) = 2(d+1) \int_0^2 h(d,g;u) \int_{-1}^{1-u} f(X_N; v, u+v) dv du$$

where

$$h(d,g;u) = \sum_{\mu=0}^{D+1} \frac{(-1-D)_\mu}{(1/2)_\mu} \frac{(1-t)^\mu}{\mu!} g^{(\mu+1)}(t)\Big|_{t=1-u^2/2}$$

and $f(X_N; t, t')$ from Theorem 2.1.

The ideas leading to Corollary 2.2 can be used here, too. We obtain the lower bound

Theorem 4.6. *Under the assumptions of Theorem 4.5*

$$\sqrt{E(g; X_N) + \frac{1}{N} g(1) - a_0(g)} \leq C_g\, D_N(X_N)$$

holds for any N-point set $X_N \equiv \{\mathbf{x}_1, \ldots, \mathbf{x}_N\}$ in S^d and

$$C_g \equiv \sqrt{2(d+1) \int_0^2 |h(d,g;u)|(2-u)du} \ .$$

References

[1] M. Abramowitz, I. A. Stegun (eds.): *Handbook of Mathematical Functions with Formulas, Graphs, and Mathematical Tables*, Dover, 1970.

[2] G. Andrews, R. Askey, R. Roy: *Special Functions*, Encyclopedia of Mathematics and Its Applications, Vol. **71**, Cambridge University Press, 1999.

[3] J. S. Brauchart: *Punktverteilungen extremaler diskreter Energie auf Sphären.* Diploma thesis, Technical University Graz, 2001.

[4] P. J. Grabner, R. F. Tichy: *Spherical designs, discrepancy and numerical integration*, Math. Comp. **60** (1993), no. 201, 327–336.

[5] I. J. Schoenberg: *Positive definite functions on spheres*, Duke Math. J. **9** (1942), 96–108.

[6] K. B. Stolarsky: *Sums of distances between points on a sphere II*, Proc. Amer. Math. Soc. **41** (1973), 575–582.

Address:

Johann S. Brauchart
Institut für Mathematik A
Technische Universität Graz
Steyrergasse 30
A–8010 Graz
Austria

International Series of Numerical Mathematics
Vol. 145, ©2003 Birkhäuser Verlag Basel/Switzerland

On Properties of Nonstationary Divided Difference Vector Subdivision Schemes

Maria Charina and Costanza Conti

Dedicated to Professor Manfred Reimer on the occasion of his 70th birthday

Abstract

This paper studies the properties of a nonstationary divided difference vector subdivision scheme S_{B^k} derived from a stationary vector subdivision scheme S_A. We show that a necessary condition for S_A to be C^1 (*i.e.*, its linear polynomial reproduction property) implies the existence of a difference scheme derived from S_{B^k}. This condition turns out to be a necessary condition for the subdivision scheme S_{B^k} to be convergent.

Introduction

Vector subdivision schemes arise, for example, when dealing with refinable vector functions and their corresponding shift invariant spaces [12], [10], when studying multivariate subdivision schemes [7], [1] or in the context of multiwavelets [14], [15], [16]. Many authors have investigated the regularity of vector subdivision schemes, or equivalently, the existence and the regularity of refinable functions associated with matrix refinement masks (see [17] for a comprehensive overview of the subject). The techniques to check the regularity of refinable function vectors can be grouped into those using the transition operator [15], [19], the joint spectral radius [4], [13] and difference operators [7], [8], [16], for example. In [2], [3] the regularity of multivariate vector subdivision schemes has been studied using the difference operator approach without assuming that the so–called Eigenvalue Condition is satisfied by the matrix mask. In this paper, we further investigate the properties of a nonstationary divided difference scheme first introduced in [2]. Note that the existence of common eigenvectors associated with the eigenvalue 1 of nonstationary divided difference scheme provides an alternative approach for checking the necessary condition for the original subdivision scheme to be C^1 (this can be done using a stationary divided difference as well).

The paper is organized as follows: in Section 1 we present some basic facts about vector subdivision schemes. In Section 2, we define nonstationary divided difference subdivision schemes and discuss their convergence. Our main result is presented in Section 3, where we show that a necessary condition for S_A to be C^1 (*i.e.*, its linear polynomial reproduction property) implies the existence of a difference scheme derived from the nonstationary divided difference scheme S_{B^k}. This condition turns out to be a necessary condition for the subdivision scheme S_{B^k} to be convergent.

1 Background

Throughout the paper we make use of the following basic facts (sequences of vectors and sequences of matrices will be written in bold).

A general *nonstationary subdivision scheme* can be represented by an infinite sequence of refinement matrix masks $\{\mathbf{A}^k\}_{k\geq 0}$. More precisely, the *k–level refinement matrix mask* $\mathbf{A}^k = (A^k_\alpha)_{\alpha\in\mathbb{Z}^s}$ is a matrix sequence, whose terms A^k_α are real $(n\times n)$–matrices. We assume that the support of the $\mathbf{A}^k$ are bounded uniformly with respect to k, *i.e.*, $A^k_\alpha = 0$ except for a finite number of multi–indices $\alpha\in\mathbb{Z}^s$.

The *k–level subdivision operator associated with the k–level matrix mask* $\mathbf{A}^k$ is

$$\begin{aligned} &S_{A^k} \;:\; (\ell(\mathbb{Z}^s))^n \to (\ell(\mathbb{Z}^s))^n \;, \\ &(S_{A^k}\,\mathbf{\Lambda})_\alpha := \sum_{\beta\in\mathbb{Z}^s} (A^k_{\alpha-2\beta})^T\,\Lambda_\beta, \quad \alpha\in\mathbb{Z}^s \;, \end{aligned} \tag{1.1}$$

where $(\ell(\mathbb{Z}^s))^n$ denotes the linear space of n–dimensional vector sequences indexed by $\mathbb{Z}^s$. The *nonstationary subdivision scheme* consists of the application of $S_{A^0},\ldots,S_{A^k}$ generating the vector sequences

$$\mathbf{\Lambda}^0 := \mathbf{\Lambda}\;, \quad \mathbf{\Lambda}^{k+1} := S_{A^k}\,\mathbf{\Lambda}^k \;\text{ for }\; k=0,1,2,\ldots\;. \tag{1.2}$$

If the refinement matrix masks $\{\mathbf{A}^k\}_{k\geq 0}$ do not change from step to step, *i.e.*, $\mathbf{A}^k = \mathbf{A}$ for all k, then the scheme is called *stationary.*

The notion of *C–convergence of the subdivision scheme* refers to the fact that for any $\mathbf{\Lambda}$ *in* $(\ell^\infty(\mathbb{Z}^s))^n$, the linear space of bounded n–dimensional vector sequences indexed by $\mathbb{Z}^s$, there exists a continuous vector–valued function $f_{\mathbf{\Lambda}} : \mathbb{R}^d \to \mathbb{R}^n$ (depending on the starting vector sequence $\mathbf{\Lambda}$) satisfying

$$\lim_{k\to\infty} \left\| f_{\mathbf{\Lambda}}(\tfrac{\cdot}{2^k}) - \mathbf{\Lambda}^k \right\|_\infty = 0 \tag{1.3}$$

and $f_{\mathbf{\Lambda}} \neq 0$ for at least some initial data $\mathbf{\Lambda}$. Here, the symbol $f_{\mathbf{\Lambda}}(\frac{\cdot}{2^k})$ abbreviates the vector sequence $(f_{\mathbf{\Lambda}}(\frac{\alpha}{2^k}))_{\alpha\in\mathbb{Z}^s}$ and $\|\cdot\|_\infty$ denotes the sup–norm of the vector sequence $\mathbf{\Lambda} = (\lambda^1, \ldots, \lambda^n)^T$ given by $\left\|\mathbf{\Lambda}\right\|_\infty := \sup_{\alpha\in\mathbb{Z}^s} \|\Lambda_\alpha\|_\infty$.

When dealing with vector subdivision schemes, stationary or nonstationary, it is convenient to consider $n \times n$ matrix–valued functions generated by applying the subdivision scheme to matrix sequences, that is, "simultaneously" to all n vector sequences given by the columns of the matrices of the matrix sequence. In particular, let $\mathbf{\Delta}^0$ be the "delta–matrix sequence" such that $\mathbf{\Delta}^0 = (\delta_{\beta,0} I_n)_{\beta\in\mathbb{Z}^s}$, I_n being the identity matrix of order n. For a convergent subdivision scheme we define the *basic limit function matrix*

$$M_A := \lim_{k\to\infty} S_{A^k} S_{A^{k-1}} \cdots S_{A^0} \mathbf{\Delta}^0 \, . \tag{1.4}$$

By the linearity of S_{A^k} and the compact support of the masks $\{\mathbf{A}^k\}_{k\geq 0}$ any limit of a convergent subdivision satisfies

$$f_{\mathbf{\Lambda}} = \lim_{k\to\infty} S_{A^k} S_{A^{k-1}} \cdots S_{A^0} \mathbf{\Lambda} = \sum_{\alpha\in\mathbb{Z}^s} M_A(\cdot - \alpha)\Lambda_\alpha. \tag{1.5}$$

It is known [6] that the convergence of a stationary vector subdivision scheme implies that the matrices $\bar{A}_\alpha = \sum_{\beta\in\mathbb{Z}^s} A_{\alpha-2\beta}$, $\alpha \in E^s$ (where E^s denotes the set of representatives of the equivalence classes $\mathbb{Z}^s/2\mathbb{Z}^s$) have at least one common left eigenvector v associated with the eigenvalue 1, *i.e.*,

$$v \sum_{\beta\in\mathbb{Z}^s} A_{\alpha-2\beta} = \lambda v \quad \text{for all} \quad \alpha \in E^s \tag{1.6}$$

holds with $\lambda = 1$.

2 Difference and Divided Difference Schemes

Let $\mathbf{A}$ be a refinement matrix mask satisfying (1.6) for $\lambda = 1$, and $1 \le m \le n$. Without loss of generality we may assume that the eigenvectors associated with the eigenvalue $\lambda = 1$ are $e_l^T = (0, \dots, \underbrace{1}_{l}, \dots, 0)^T$ for $l = 1, \dots, m$ after applying some similarity transformation to each term of the matrix sequence $\mathbf{A}$.

Consider a vector sequence $\Lambda \in (\ell(\mathbb{Z}^s))^n$ (with $(\Lambda)_\alpha = (\lambda_\alpha^1, \dots, \lambda_\alpha^n)^T$). In [2] we introduce the difference operator $\delta : (\ell(\mathbb{Z}^s))^n \to (\ell(\mathbb{Z}^s))^{ns}$ at each $\alpha \in \mathbb{Z}^s$, *i.e.*,

$$(\delta\mathbf{\Lambda})_\alpha := \begin{pmatrix} (\delta_{e_1}\mathbf{\Lambda})_\alpha \\ \vdots \\ (\delta_{e_s}\mathbf{\Lambda})_\alpha \end{pmatrix}, \quad \text{where} \quad (\delta_{e_l}\mathbf{\Lambda})_\alpha := \begin{pmatrix} \lambda^1_{\alpha+e_l} - \lambda^1_\alpha \\ \vdots \\ \lambda^m_{\alpha+e_l} - \lambda^m_\alpha \\ \lambda^{m+1}_\alpha \\ \vdots \\ \lambda^n_\alpha \end{pmatrix}, \tag{2.1}$$

i.e., δ_{e_l} for $1 \le l \le s$ is the difference operator in the direction of e_l acting on the first m components of each Λ_α. The same type of a difference operator and associated difference subdivision scheme were introduced in [18] for studying polynomial reproduction properties of stationary vector subdivision schemes. In [3] we also give a constructive proof of the following theorem.

Theorem 2.1 *Let S_A be a subdivision operator with a mask satisfying* (1.6). *Then, there exists a matrix mask $\mathbf{B} = \{B_\alpha\, , \alpha \in \mathbb{Z}^s\}$ with B_α being $(n\cdot s)\times(n\cdot s)$ matrices ($s \times s$ block matrices with the blocks being $n \times n$ matrices) with the property*

$$(\delta S_A \mathbf{\Lambda})_\alpha = \sum_{\beta\in\mathbb{Z}^s} (B_{\alpha-2\beta})^T (\delta\mathbf{\Lambda})_\beta \, . \tag{2.2}$$

It follows that there exists an associated subdivision operator S_B with the matrix mask $\mathbf{B}$ such that

$$\delta S_A = S_B \delta \, . \tag{2.3}$$

To define an associated divided difference scheme we define the diagonal $(n \times n)$–matrix $I_{2^k} := diag\{\underbrace{2^k, \dots, 2^k}_{m}, 1, \dots, 1\}$ and D^k the block diagonal matrix of dimension $(n \cdot s) \times (n \cdot s)$ of the form $D^k := (\underbrace{I_{2^k}, \ldots, I_{2^k}}_{s})^T$.

Definition 2.1 *The k–level divided difference operator* $d^k : (\ell(\mathbb{Z}^s))^n \to (\ell(\mathbb{Z}^s))^{ns}$ *is*

$$(d^k \mathbf{\Lambda}^k)_\alpha := D^k (\delta \mathbf{\Lambda}^k)_\alpha \,. \tag{2.4}$$

The nonstationary divided difference subdivision scheme associated with matrix masks $\{\mathbf{B}^k := D^{-k}\mathbf{B}D^{k+1}\}_{k \geq 0}$ *is given by*

$$\mathbf{\Lambda}^0 := \mathbf{\Lambda}\,, \quad \mathbf{\Lambda}^{k+1} := S_{B^k}\, \mathbf{\Lambda}^k \quad \textit{for} \quad k = 0, 1, 2, \ldots\,,$$

and it has the property

$$(d^{k+1} S_A \mathbf{\Lambda})_\alpha = \sum_{\beta \in \mathbb{Z}^s} ({B^k}_{\alpha - 2\beta})^T (d^k \mathbf{\Lambda})_\beta, \quad i.e., \quad d^{k+1} S_A = S_{B^k} d^k \,. \tag{2.5}$$

The following theorem, proven in [2], links the convergence of the divided difference subdivision scheme to the regularity of the original scheme S_A.

Theorem 2.2 *Let* S_A *be uniformly convergent and* S_{B^k} *be defined as in* (2.5). *If the subdivision scheme generated by* $\{\mathbf{B}^k\}_{k \geq 0}$ *converges uniformly, then* $(S_A)^\infty \mathbf{\Lambda} \in C^1$ *for all initial data* $\mathbf{\Lambda} \in (\ell^\infty(\mathbb{Z}^s))^n$.

To check the convergence of the nonstationary subdivison scheme associated with matrix masks $\{\mathbf{B}^k\}_{k \geq 0}$ we use the following theorem. Here,the symbol $\otimes$ stands for the Kronecker product of vectors.

Theorem 2.3 *Let the sequence* $\{\mathbf{B}^k\}_{k \geq 0}$ *of* $(n \cdot s) \times (n \cdot s)$ *matrix masks satisfy*

$$(e_i \otimes e_j)^T \sum_{\beta \in \mathbb{Z}^s} B^k_{\alpha - 2\beta} = (e_i \otimes e_j)^T, \quad 1 \leq i \leq s, \ 1 \leq j \leq m\,, \quad \alpha \in E^s. \tag{2.6}$$

Then, there exists another sequence of $(n \cdot s^2) \times (n \cdot s^2)$ *matrix masks* $\{\mathbf{C}^k\}_{k \geq 0}$ *with the property*

$$\delta S_{B^k} = S_{C^k} \delta.$$

Furthermore, the subdivision scheme associated with nonstationary masks $\{\mathbf{B}^k\}_{k \geq 0}$ *converges uniformly if and only if the subdivision scheme associated with matrix masks* $\{\mathbf{C}^k\}_{k \geq 0}$ *applied to* $\delta \mathbf{\Lambda}$ *converges uniformly to zero for any* $\mathbf{\Lambda}$ *in* $(\ell^\infty(\mathbb{Z}^s))^{ns}$.

This theorem is a special case of a result proven in [3] for general nonstationary subdivision schemes. In the next section we show that (2.6) is indeed a necessary condition for S_A to be C^1 and is a necessary condition for the subdivision scheme associated with nonstationary masks $\{\mathbf{B}^k\}_{k\geq 0}$ to be convergent.

3 A Necessary Condition

We start by studying the properties of the original subdivision scheme S_A that yield the existence of linearly independent vectors v_i such that

$$v_i^T \sum_{\beta\in\mathbb{Z}^s} B_{\alpha-2\beta} = \frac{1}{2} v_i^T, \quad 1 \leq i \leq m\cdot s\ , \ \alpha \in E^s. \tag{3.1}$$

Condition (3.1) for some specific vector v_i is equivalent to $S_B \mathbf{V}_i = \frac{1}{2}\mathbf{V}_i$ for some constant vector sequence $\mathbf{V}_i = (v_i)_{\alpha\in\mathbb{Z}^s}$. On the other hand, if $S_A \mathbf{W}_i = \frac{1}{2}\mathbf{W}_i$ and if $\mathbf{V}_i := \delta\mathbf{W_i}$ is constant, then (3.1) holds true, since $\delta S_A = S_B \delta$.

Proposition 3.1 *The matrix mask* $\mathbf{B}$ *satisfies* (3.1) *if there exist* $m\cdot s$ *vector sequences* $\mathbf{W}_{j,e_l} \in (\ell(\mathbb{Z}^s))^n$ *such that* $S_A \mathbf{W}_{j,e_l} = \frac{1}{2}\mathbf{W}_{j,e_l}, 1 \leq j \leq m, 1 \leq l \leq s$.

Proof. Condition (3.1) is equivalent to the existence of $m\cdot s$ constant sequences $\mathbf{V}_i \in (\ell(\mathbb{Z}^s))^{ns}$ preserved by the subdivision scheme S_B, *i.e.*, $S_B\mathbf{V}_i = \frac{1}{2}\mathbf{V}_i$ for $1 \leq i \leq m\cdot s$. Note that $\delta S_A = S_B\delta$ implies that the existence of vector sequences $\mathbf{W}_{j,e_l} \in (\ell(\mathbb{Z}^s))^n$ for $1 \leq j \leq m$ and $1 \leq l \leq s$ such that $S_A\mathbf{W}_{j,e_l} = \frac{1}{2}\mathbf{W}_{j,e_l}$ guarantees that $S_B\mathbf{V}_i = \frac{1}{2}\mathbf{V}_i$ for $1 \leq i \leq m\cdot s$ with $\mathbf{V}_i = \delta\mathbf{W}_{j,e_l}$, for $1 \leq j \leq m$ and $1 \leq l \leq s$ $(i = j\cdot l)$. □

Remark 3.1 *In practice, we construct the vector sequences* $\mathbf{W}_{j,e_l} \in (\ell(\mathbb{Z}^s))^n$ *from the mask* $\mathbf{A}$ *by applying the algorithm given in* [11], [4], *for example. This algorithm yields the vectors* w_{e_l} *generating the sequences* $\mathbf{W}_{j,e_l} = \{e_j\alpha^{e_l} + w_{e_l}\}_{\alpha\in\mathbb{Z}^s}$ *for* $1 \leq j \leq m$ *and* $1 \leq l \leq s$.

Theorem 3.1 is a multidimensional analogue and a straightforward extension of a part of [8, Theorem 4.6]. Itgives us a sufficient condition for vector sequences $\mathbf{W}_{j,e_l} \in (\ell(\mathbb{Z}^s))^n$ to exist.

Theorem 3.1 *Let S_A be a C^1 uniformly convergent vector subdivision scheme with $m \geq 1$. Then $\lambda = 1/2$ is an eigenvalue of S_A with the geometric multiplicity at least $m \cdot s$ and for any $0 \neq \mathbf{W} \in (\ell(\mathbb{Z}^s))^n$ satisfying $S_A \mathbf{W} = \lambda \mathbf{W}$ it follows that $(S_A^\infty \mathbf{W})(x) = x^{e_l} \rho$, $\rho \in \mathbb{R}^n$ with $\rho_k = 0$, $k = n-m+1, \ldots, n$. Also, there exist $m \cdot s$ linearly independent eigenvectors $\mathbf{W}_{1,e_l}, \ldots, \mathbf{W}_{m,e_l}$ corresponding to $\lambda = 1/2$ and satisfying $(S_A^\infty \mathbf{W}_{j,e_l})(x) = x^{e_l} \mu_{e_l}^j$, $l = 1, \ldots, s$ such that $\mu_{e_l}^1, \ldots, \mu_{e_l}^m$ are linearly independent.*

Theorem 2.2 and the definition of $\{\mathbf{B}^k\}_{k \geq 0}$ yield a necessary condition for convergence of the nonstationary divided difference subdivision scheme defined in (2.5).

Theorem 3.2 *Let the subdivision scheme generated by $\{\mathbf{B}^k\}_{k \geq 0}$ converge uniformly. Then there exist $m \cdot s$ linearly independent vectors v_i such that for all $k \geq 0$*

$$v_i^T D^k \sum_{\beta \in \mathbb{Z}^s} B^k_{\alpha - 2\beta} = \frac{1}{2} v_i^T D^{k+1}, \quad \textit{for all } \alpha \in \mathbb{Z}^s \textit{ and } 1 \leq i \leq m \cdot s.$$

Proof. By Theorem 2.2. the subdivision scheme S_A is C^1, thus by Theorem 3.1 there exist $m \cdot s$ non–trivial vector sequences $\mathbf{W}_{j,e_l}$ such that $S_A \mathbf{W}_{j,e_l} = \frac{1}{2} \mathbf{W}_{j,e_l}$. It follows that there exist $m \cdot s$ linearly independent $v_i \in \mathbb{R}^{n \cdot s}$ such that

$$\delta \mathbf{W}_{j,e_l} = (\delta_{e_1} \mathbf{W}_{j,e_l}), \ldots, (\delta_{e_s} \mathbf{W}_{j,e_l})^T = (v_i)_{\alpha \in \mathbb{Z}^s} \ .$$

In fact, by Remark 3.1 and definition (2.1) we get

$$(\delta_{e_i} \mathbf{W}_{j,e_l})_\alpha = (0, \ldots, 0, \underbrace{\delta_{i,l}}_{j-th}, 0, \ldots, 0, \underbrace{*, \ldots, *}_{n-m})^T, \ 1 \leq j \leq m, \ 1 \leq l \leq s, \quad (3.2)$$

i.e., , v_i are linearly independent. The property of the δ operator, *i.e.*, , $\delta S_A = S_B \delta$, implies that for all $k \geq 0$ the vectors v_i satisfy

$$v_i^T \sum_{\beta \in \mathbb{Z}^s} B_{\alpha - 2\beta} = \frac{1}{2} v_i^T, \ 1 \leq i \leq m \cdot s \ \text{ for all } \alpha \in \mathbb{Z}^s.$$

Multiplying both sides of the above equation from the right by D^{k+1} we obtain the desired result. □

According to Theorem 2.3. the second difference scheme can be computed when (2.6) is satisfied which is not always the case according to (3.2). Therefore, we have to apply some similarity transformation to $\mathbf{B}$ first. Using the structure of the vectors v_i, $1 \leq i \leq m \cdot s$ given in (3.2) we define, as in [9] for $i = 1$, an $n \cdot s \times n \cdot s$ block matrix

$$U := \begin{pmatrix} T_1 & O_1 & \dots & O_1 \\ O_2 & T_2 & \dots & O_2 \\ & & \dots & \\ O_s & O_s & \dots & T_s \end{pmatrix}$$

where the $n \times n$ matrices

$$T_l := \begin{pmatrix} I_{m\times m} & -*_{m\times(n-m)} \\ 0_{n-m\times m} & I_{n-m\times n-m} \end{pmatrix}, O_l := \begin{pmatrix} 0_{m\times m} & -*_{m\times(n-m)} \\ 0_{n-m\times m} & 0_{(n-m)\times(n-m)} \end{pmatrix}$$

are such that the $-*_{m\times(n-m)}$ blocks are exactly the same for the same index l. Furthermore, the rows of the $-*_{m\times(n-m)}$ blocks correspond to the last $n-m$ entries of the vectors in (3.2) for $1 \leq l \leq s$. It is easy to check that U is invertible and the inverse is given by

$$U^{-1} := \begin{pmatrix} \tilde{T}_1 & \tilde{O}_1 & \dots & \tilde{O}_1 \\ \tilde{O}_2 & \tilde{T}_2 & \dots & \tilde{O}_2 \\ & & \dots & \\ \tilde{O}_s & \tilde{O}_s & \dots & \tilde{T}_s \end{pmatrix}$$

with the $n \times n$ matrices

$$\tilde{T}_l := \begin{pmatrix} I_{m\times m} & *_{m\times(n-m)} \\ 0_{n-m\times m} & I_{n-m\times n-m} \end{pmatrix}, \; \tilde{O}_l := \begin{pmatrix} 0_{m\times m} & *_{m\times(n-m)} \\ 0_{n-m\times m} & 0 \end{pmatrix}$$

for $1 \leq l \leq s$ and the $*_{m\times(n-m)}$ blocks with the corresponding entries being the entries of $-*_{m\times(n-m)}$ in T_l and O_l multiplied by -1. Note that the first m rows of each row block

$$(\tilde{O}_l \; \dots \; \tilde{O}_l \; \underbrace{\tilde{T}_l}_{l-th} \; \tilde{O}_l \; \dots \; \tilde{O}_l)$$

of U^{-1} are the vectors $v^T_{m\cdot(l-1)+1}, \dots, v^T_{m\cdot l}$. Note also that $v_i^T U = (e_l \otimes e_j)^T$ for $1 \leq l \leq s$ and $1 \leq j \leq m$. In the next proposition we show that the transformation U preserves the structure of the limit of the nonstationary divided difference schemes introduced in Definition 2.1. By definition of U, the structure of v_i's and (3.1) we get

$$(e_i \otimes e_j)^T U^{-1} \sum_\beta B_{\alpha-2\beta} U = \frac{1}{2}(e_i \otimes e_j)^T, \; 1 \le i \le s, \; 1 \le j \le m. \tag{3.3}$$

Define $\tilde{\mathbf{B}}^k := D^{-k} U^{-1} \mathbf{B} U D^{k+1}$. It follows from (3.3) that the submasks of $\tilde{\mathbf{B}}^k$ satisfy

$$(e_i \otimes e_j)^T \sum_\beta \tilde{B}^k_{\alpha-2\beta} = (e_i \otimes e_j)^T, \; \alpha \in \mathbb{Z}^d, \; 1 \le i \le s, \; 1 \le j \le m, \tag{3.4}$$

for all $k \ge 0$. Next, we would like to study the effect of the transformation U on the subdivision scheme associated with the masks $\{\mathbf{B}^k\}_{k\ge0}$.

Proposition 3.2 *The subdivision scheme generated by $\{\mathbf{B}^k\}_{k\ge0}$ is convergent if and only if the subdivision scheme generated by $\{\tilde{\mathbf{B}}^k\}_{k\ge0}$ is convergent.*

Proof. Note that the α component of the sequences we get at the level k applying the subdivision associated with masks $\{\mathbf{B}^k\}_{k\ge0}$ is of the form

$$D^{k+1} \sum_\beta (B_{\alpha-2\beta})^T (\delta \Lambda^k)_\beta ,$$

where $\mathbf{\Lambda}^k$ is the vector sequence we get applying S^k_A to some starting sequence $\mathbf{\Lambda}^0$. The structure of $\tilde{\mathbf{B}}^k$ then implies that the corresponding component of the sequences we get iterating $S_{\tilde{\mathbf{B}}^k}$, $k \ge 0$ is

$$D^{k+1} U^T \sum_\beta (B_{\alpha-2\beta})^T (\delta \Lambda^k)_\beta .$$

It is easy to see that due to the structure of U and Remark 3.2 the multiplication by U^T leaves the resulting sequence components that contain differences unchanged. The other elements of the sequences are replaced by linear combinations of the terms that converge to zero as $k \to \infty$. This tells us that the convergence of the difference scheme generated by $\tilde{\mathbf{B}}^k$ is equivalent to the convergence of the scheme generated by $\mathbf{B}^k$. □

Then, Proposition 3.2 and Theorem 3.2 yield a necessary condition for the subdivision scheme associated with $\mathbf{B}^k$ to be convergent.

Corollary 3.3 *Let the subdivision scheme generated by the masks $\{\tilde{\mathbf{B}}^k\}_{k\ge0}$ be convergent, then for all $k \ge 0$*

$$(e_i \otimes e_j)^T \sum_{\beta \in \mathbb{Z}^s} \tilde{B}^k_{\alpha-2\beta} = (e_i \otimes e_j)^T, \quad \alpha \in E^s, \; 1 \le i \le s, \; 1 \le j \le m .$$

4 Example

In this section we consider an example of a C^1 subdivision scheme based on the C^1 subdivision scheme presented in [5]. It shows how the theory presented in Section 3 can be used. Consider a matrix mask $\mathbf{A}$

$$\begin{array}{cccccc}
 & \vdots & \vdots & \vdots & \vdots & \\
\cdots & \begin{pmatrix} 0 & 0 \\ 0 & 0 \end{pmatrix} & \begin{pmatrix} \frac{1}{8} & 0 \\ 0 & \frac{1}{8} \end{pmatrix} & \begin{pmatrix} \frac{3}{8} & \frac{-1}{8} \\ \frac{1}{4} & 0 \end{pmatrix} & \begin{pmatrix} \frac{1}{4} & \frac{-1}{8} \\ \frac{1}{4} & \frac{-1}{8} \end{pmatrix} & \cdots \\
\cdots & \begin{pmatrix} \frac{1}{8} & 0 \\ 0 & 0 \end{pmatrix} & \begin{pmatrix} \frac{1}{2} & 0 \\ 0 & \frac{1}{4} \end{pmatrix} & \begin{pmatrix} \frac{3}{4} & \frac{-1}{8} \\ \frac{1}{4} & \frac{1}{4} \end{pmatrix} & \begin{pmatrix} \frac{3}{8} & \frac{-1}{8} \\ \frac{1}{4} & 0 \end{pmatrix} & \cdots \\
\cdots & \begin{pmatrix} \frac{1}{8} & \frac{1}{8} \\ 0 & 0 \end{pmatrix} & \begin{pmatrix} \mathbf{\frac{1}{2}} & \mathbf{\frac{1}{8}} \\ \mathbf{0} & \mathbf{\frac{1}{8}} \end{pmatrix} & \begin{pmatrix} \frac{1}{2} & 0 \\ 0 & \frac{1}{4} \end{pmatrix} & \begin{pmatrix} \frac{1}{8} & 0 \\ 0 & \frac{1}{8} \end{pmatrix} & \cdots \\
\cdots & \begin{pmatrix} 0 & \frac{1}{8} \\ 0 & 0 \end{pmatrix} & \begin{pmatrix} \frac{1}{8} & \frac{1}{8} \\ 0 & 0 \end{pmatrix} & \begin{pmatrix} \frac{1}{8} & 0 \\ 0 & 0 \end{pmatrix} & \begin{pmatrix} 0 & 0 \\ 0 & 0 \end{pmatrix} & \cdots \\
 & \vdots & \vdots & \vdots & \vdots &
\end{array}$$

with the submasks $\sum_\beta A_{\alpha-2\beta}$ satisfying

$$(1,0)\sum_\beta A_{\alpha-2\beta} = (1,0)\frac{1}{8}\begin{pmatrix} 8 & 0 \\ 2 & 2 \end{pmatrix} = (1,0), \quad \alpha \in E^2 .$$

Using the algorithm given in [11] we get $\mathbf{W}_{1,e_l} := \{(1,0)^T\alpha^{e_l} + w_{e_l}\}_{\alpha\in\mathbb{Z}^2}$ with $w_{(1,0)} = w_{(0,1)} = \begin{pmatrix} \frac{1}{2} \\ \frac{1}{2} \end{pmatrix}$ satisfying

$$w_{e_l}^T(\mathbf{H}(\pi e) - \frac{1}{2}\delta_{e,0}I) = -(1,0)(2i)^{-1}(D^{e_l}\mathbf{H})(\pi e) \text{ for all } e \in E^2 ,\ l = 1,2,$$

where $\mathbf{H}(\xi) = \frac{1}{4}\sum_{\alpha\in\mathbb{Z}^2} A_\alpha e^{-i\alpha\cdot\xi}$ is the symbol of $\mathbf{A}$.

It follows that the two sequences $\mathbf{W}_{1,(1,0)}$ and $\mathbf{W}_{1,(0,1)}$ are

$$(\mathbf{W}_{1,(1,0)})_\alpha = \begin{pmatrix} \alpha_1 + \frac{1}{2} \\ \frac{1}{2} \end{pmatrix} \quad (\mathbf{W}_{1,(1,0)})_\alpha = \begin{pmatrix} \alpha_2 + \frac{1}{2} \\ \frac{1}{2} \end{pmatrix},$$

i.e., the eigenvectors of the submasks of $\mathbf{B}$ are given by

$$v_1 = (\delta \mathbf{W}_{1,(1,0)})_\alpha = \begin{pmatrix} 1 \\ \frac{1}{2} \\ 0 \\ \frac{1}{2} \end{pmatrix} \qquad v_2 = (\delta \mathbf{W}_{1,(0,1)})_\alpha = \begin{pmatrix} 0 \\ \frac{1}{2} \\ 1 \\ \frac{1}{2} \end{pmatrix}.$$

It is also possible to get v_1, v_2 directly from the submasks of B (*i.e.*, $\bar{B}_\alpha = \sum_{\alpha \in \mathbb{Z}^2} B_{\alpha - 2\beta}$ for α being extreme points of $[0,1]^2$) given below

$$\bar{B}_{(0,0)} = \bar{B}_{(1,0)} = \frac{1}{8} \begin{pmatrix} 4 & 1 & 0 & 1 \\ 0 & 0 & 0 & 0 \\ 0 & 1 & 4 & 1 \\ 0 & 2 & 0 & 2 \end{pmatrix}, \quad \bar{B}_{(0,1)} = \bar{B}_{(1,1)} = \frac{1}{8} \begin{pmatrix} 4 & 1 & 0 & 1 \\ 0 & 4 & 0 & 4 \\ 0 & 1 & 4 & 1 \\ 0 & -2 & 0 & -2 \end{pmatrix}.$$

Thus, the existence of v_1 and v_2 tells us that the necessary condition for the convergence of the subdivision scheme associated with $\{\mathbf{B}^k\}_{k \geq 0}$ is satisfied.

References

[1] A. S. Cavaretta, W. Dahmen, C. A. Micchelli: *Stationary Subdivision*, Memoirs Amer. Math. Soc. **93**, No. 453 (1991).

[2] M. Charina, C. Conti: *Regularity of multivariate vector subdivision schemes*, Manuscript (December 2001).

[3] M. Charina, C. Conti: *Convergence of non–stationary vector subdivision schemes*, to appear in Math. Computers in Simulation.

[4] D. R. Cheng, R. Q. Jia, S. D. Riemenschneider: *Convergence of vector subdivision schemes in Sobolev spaces*, Applied Comp. Harm. Anal. **12** (2002), 128–149.

[5] C. Conti, K. Jetter: *A new subdivision method for bivariate splines on the four–directional mesh*, J. Comp. Appl. Math. **119** (2000), 81–96.

[6] W. Dahmen, C. A. Micchelli: *Biorthogonal wavelet expansions*, Constr. Approx. **13** (1997), 293–328.

[7] N. Dyn: *Subdivision schemes in* CAGD, in: *Advances in Numerical Analysis, vol.* II: *Wavelets, Subdivision Algorithms and Radial Basis Functions*, W. A. Light (ed.), pp. 36–104, Oxford University Press, 1992.

[8] N. Dyn, D. Levin: *Matrix subdivision–analysis by factorization*, in: *Approximation Theory*, B. D. Bojanov (ed.), pp. 187–211, DARBA, Sofia, 2002.

[9] B. Han: *Vector cascade algorithms and refinable function vectors in Sobolev spaces*, preprint, 40 p. (July 2002).

[10] K. Jetter, G. Plonka: *A survey on L_2–approximation order from shift–invariant spaces*, in: *Multivariate Approximation and Applications* N. Dyn, D. Leviatan, D. Levin, A. Pinkus (eds.), pp. 73–111, Cambridge University Press, 2001.

[11] K. Jetter, G. Zimmermann: *Constructing polynomial surfaces from vector subdivision schemes*, in: *Constructive Functions Theory*, Varna 2002, B. D. Bojanov (ed.), pp. 327–6332, DARBA, Sofia, 2003.

[12] R. Q. Jia, S. D. Riemenschneider, D. X. Zhou: *Approximation by multiple refinable functions*, Can. J. Math. **49** (1997), 944–962.

[13] R. Q. Jia, S. D. Riemenschneider, D. X. Zhou: *Vector subdivision schemes and multiple wavelets*, Math. Comp. **67** (1998), 1533–1563.

[14] R. Q. Jia, S. D. Riemenschneider, D. X. Zhou: *Smoothness of multiple refinable functions and multiple wavelets*, SIAM J. Matrix Anal. Appl. **21** (1999), 1–28.

[15] Q. T. Jiang: *Multivariate matrix refinable functions with arbitrary matrix*, Trans. Amer. Math. Soc. **351** (1999), 2407–2438.

[16] C. A. Micchelli, T. Sauer: *Regularity of multiwavelets*, Adv. Comp. Math. **7** (1997), 455–545.

[17] A. Ron, Z. Shen: *The Sobolev regularity of refinable functions*, J. Approx. Theory **106** (2000), 185–225.

[18] T. Sauer: *Stationary vector subdivision – quotient ideals, differences and approximation power*, Rev. R. Acad. Cienc. Serie A. Mat. **229** (2002), 621–674.

[19] Z. Shen: *Refinable functions vectors*, *SIAM J. Math. Anal.* **29** (1998), 235–250.

Addresses:

Maria Charina
Institut für Angewandte Mathematik
Universität Dortmund
D–44221 Dortmund
Germany

Costanza Conti
Universitá degli Studi di Firenze
Dipartimento di Energetica "Sergio Stecco"
I–50133 Firenze
Italy

International Series of Numerical Mathematics
Vol. 145,

Multivariate Balanced Vector-Valued Refinable Functions

Charles K. Chui and Qingtang Jiang

Dedicated to Manfred Reimer on the occasion of his 70th birthday

Abstract

Vanishing moments of sufficiently high order and compact supports of reasonable size contribute to the great success of wavelets in various areas of applications, particularly in signal and image processing. However, for multi–wavelets, polynomial preservation of the refinable function vectors does not necessarily imply annihilation of discrete polynomials by the high–pass filters of the corresponding orthogonal or bi–orthogonal multi–wavelets. This led to the introduction of the notion of "balanced" multi–wavelets by Lebrun and Vetterli, and later, generalization to higher–order balancing by Selesnick. Selesnick's work is concerned only with orthonormal refinable function vectors and orthonormal multi–wavelets. In this paper after giving a brief overview of the state–of–the–art of vector–valued refinable functions in the preliminary section, we will discuss our most recent contribution to this research area. Our goal is to derive a set of necessary and sufficient conditions that characterize the balancing property of any order for the general multivariate matrix–dilation setting. We will end the second section by demonstrating our theory with examples of univariate splines and bivariate splines on the four–directional mesh.

Introduction

Orthonormal multi–wavelets associated with certain refinable function vectors (also called scaling function vectors) have several advantages in comparison with scalar wavelets. For example, an orthonormal multi–wavelet can possess all the desirable properties of orthogonality, short support, high order of smoothness and vanishing moments, and symmetry/anti–symmetry [8, 10, 11, 12, 33]. Thus, orthonormal multi–wavelets offer the possibility of superior performance for signal/image processing applications, particularly for feature extraction, pattern recognition, and noise reduction (removal) [52, 53].

However, various mathematical difficulties arise in applying multi-wavelets to process scalar–valued data. Although pre–filtering methods have been suggested and work well for some applications [1, 16, 37, 52, 53, 57, 58], the extra effort requires additional computational cost and often leads to other complications. More recently, the notion of "balanced" orthonormal multi–wavelets was introduced in [40] in an attempt to eliminate the need of pre–filtering. After a follow–up work [48] was completed, the terminology "balancing" in [40] is now called "1–balancing," since this notion is generalized to "K–balancing" for any $K \geq 1$ for better performance. However, with the exception of a few examples constructed in [48, 49], all the balanced orthonormal multi–wavelets in the open literature are only 1–balanced [19, 43, 54, 55, 56].

Nonetheless, as compared with scalar–valued wavelets, even 1–balanced multi–wavelets already achieve at least comparable results when applied to image compression. For example, it was demonstrated in [19, 43, 56], that 1–balanced orthonormal multi–wavelets already perform better than scalar–valued orthogonal wavelets in several comparison tests among a wide variety of natural images, and even outperform the bi–orthogonal 9/7 wavelet for certain images such as "Barbara". It is therefore convincing that K–balanced wavelets, for $K \geq 2$, should provide a much more powerful tool. In this regard, it should be mentioned that the bi–orthogonal 9/7 wavelet was adopted by the JPEG–2000 standard for image compression and by the MPEG–4 standard as an option for compression of I–frames for videos.

In this paper we generalize the notion of K–balanced multi–wavelets from the orthonormal setting in [48] to the bi–orthogonal and multivariate settings in order to give more flexibility for construction (particularly for the univariate setting) and to allow for a broader range of applications (when multivariate multi–wavelets are preferred). More precisely, suppose that $\mathbf{a}_1, \cdots, \mathbf{a}_r \in \mathbb{R}^s$. A compactly supported vector–valued function $F = [f_1, \cdots, f_r]^T \in (L^2)^r := (L^2(\mathbb{R}^s))^r$ is said to be *K-balanced* relative to $(\mathbf{a}_1, \cdots, \mathbf{a}_r)$, if

$$\int_{\mathbf{R}^s} f_l(\mathbf{x})(\mathbf{x} - \mathbf{a}_l)^\alpha d\mathbf{x} = \int_{\mathbf{R}^s} f_i(\mathbf{x})(\mathbf{x} - \mathbf{a}_i)^\alpha d\mathbf{x},$$

for $1 \leq l, i \leq r, |\alpha| < K$. Let $\Phi, \widetilde{\Phi}$ be refinable function vectors that are bi–orthogonal duals to each other. We will give several characterizations for K–balancing of $\widetilde{\Phi}$. In particular, a characterization formulation in terms of the vectors $\widetilde{\mathbf{y}}_\alpha$ in the definition of the sum rule order of the mask of Φ (see §1.3 about sum rule order) will be given. More precisely, we will see that $\widetilde{\Phi}$ is K–balanced relative to $(\mathbf{a}_1, \cdots, \mathbf{a}_r) \subset \mathbb{R}^s$ if and only if $\widetilde{\mathbf{y}}_\alpha = (y_{\alpha,1}, \cdots, y_{\alpha,r})$

satisfy

$$y_{\alpha,l} = \sum_{\beta \leq \alpha} \binom{\alpha}{\beta} (\mathbf{a}_l - \mathbf{a}_1)^\beta y_{\alpha-\beta,1}, \quad 2 \leq l \leq r, |\alpha| < K.$$

This condition enables us to decide easily the balancing order K of the bi–orthogonal dual of a refinable function vector, and helps us to construct orthonormal and biothogonal refinable function vectors.

This paper is organized as follows. We will first discuss preliminary results on refinable function vectors in Section 1. K–balancing of the refinable function vectors and its characterizations are discussed in Section 2, where some examples of balanced refinable function vectors are also given.

1 Preliminary Results

The study of polynomial preservation (or polynomial reproduction) in the linear algebraic span of integer–shifts of a finite number of compactly supported functions has been a popular area of investigation in the Approximation Theory community. We first give a very brief summary in §1.1 of several equivalent statements of this problem in the distribution setting in $\mathbb{R}^s$, $s \geq 1$. A general approach for constructing such compactly supported distributions that are refinable relative to a given expansive dilation matrix and some matrix–valued Laurent polynomial symbol $P(\mathbf{z})$ is discussed in §1.2. The notion of sum rules to be satisfied by $P(\mathbf{z})$ is introduced in §1.3, where sum rules and polynomial preservation are shown to be intimately related for compactly supported refinable distributions. In §1.4, we show that these distributions are indeed functions in L^2, provided that the transition operator associated with $P(\mathbf{z})$ satisfies Condition E and that $P(\mathbf{z})$ itself satisfies the sum rule of at least the first order. Smoothness of these L^2 functions and methods, along with available software, to determine the order of Sobolev and Hölder smoothness are discussed in §1.5. Finally, the characterization of bi–orthogonal duals by the bi–orthogonal condition of their corresponding two–scale Laurent polynomial symbols is formulated in terms of the sum rule and Condition E in §1.6.

1.1 Polynomial Preservation

Let $\phi_\ell, \ell = 1, \cdots, r$, be compactly supported distributions in $\mathbb{R}^s$, $s \geq 1$, and $\Phi := [\phi_1, \cdots, \phi_r]^T$. We say that Φ has the property of *polynomial preservation* of order m (or $\Phi \in \mathrm{PP}_m$ for short) if there exists a (finite) linear combination φ of integer shifts of $\phi_1, \cdots, \phi_r$, such that

$$\sum_{\mathbf{k}\in\mathbf{Z}^s} q(\mathbf{k})\varphi(\cdot - \mathbf{k}) = q, \qquad q \in \pi^s_{m-1},$$

holds in the distribution sense (i.e. equality holds upon taking inner product with any test function), where π^s_{m-1} denotes the space of all polynomials of total degree $< m$ in $\mathbb{R}^s$. It follows from the Poisson summation formula that the above formulation is equivalent to the (modified) Strang–Fix conditions:

$$D^\alpha \hat{\varphi}(2\pi\mathbf{k}) = \delta_{0,\alpha}\delta_{0,\mathbf{k}} \qquad |\alpha| < m, \quad \mathbf{k} \in \mathbb{Z}^s. \tag{1.1}$$

It is also easily seen that Φ satisfies (1.1) if and only if there exists a $1 \times r$ vector $\mathbf{t}(\omega) = \sum_{|\alpha|<m} \mathbf{t}_\alpha e^{-i\alpha\omega}$ of trigonometric polynomials such that

$$D^\alpha(\mathbf{t}\hat{\Phi})(2\pi\mathbf{k}) = \delta_{0,\alpha}\delta_{0,\mathbf{k}}, \qquad |\alpha| < m, \mathbf{k} \in \mathbb{Z}^s, \tag{1.2}$$

which, in turn, is equivalent to

$$\mathbf{x}^\alpha = \sum_{\mathbf{k}\in\mathbf{Z}^s} \{\sum_{\beta\le\alpha} \binom{\alpha}{\beta} \mathbf{k}^{\alpha-\beta}\mathbf{y}_\beta\}\Phi(\mathbf{x}-\mathbf{k}), \ |\alpha| < m, \tag{1.3}$$

in the distribution sense, with

$$\mathbf{y}_\alpha := (-iD)^\alpha \mathbf{t}(0), \qquad |\alpha| < m. \tag{1.4}$$

For $r = 1$, this formula was discussed in [6], and for arbitrary $r \ge 1$, the reader is referred to [25].

1.2 Construction of Compactly Supported Φ

Let A be an expansive matrix (i.e. all eigenvalues λ of A satisfy $|\lambda| > 1$) with integer entries. For any finite sequence $\{P_\mathbf{k}\}$ of square matrices of dimension r, we consider its corresponding matrix–valued Laurent polynomial symbol

$$P(\mathbf{z}) := \frac{1}{a} \sum_{\mathbf{k}\in\mathbf{Z}^s} P_\mathbf{k}\mathbf{z}^\mathbf{k}, \quad a := |\det(A)|, \tag{1.5}$$

and say that $P(\mathbf{1})$, $\mathbf{1} := (1, \cdots, 1)$, satisfies *Condition E* (or $P(\mathbf{1}) \in E$ for short [51]), if the value 1 is a simple eigenvalue of the matrix $P(\mathbf{1})$ and all other eigenvalues of $P(\mathbf{1})$ lie in $|z| < 1$. Under the assumption that $P(\mathbf{1}) \in E$, it follows that the infinite product

$$\prod_{j=1}^{\infty} P(e^{-i(A^T)^{-j}\omega})$$

converges uniformly on every compact subset of $\mathbb{R}^s$. This was first proved in [17] for $s = 1$ and $A = [2]$, and extended to $A = 2I_s$ in [42]. The proof for our more general matrix A follows the same argument given in [42]. Let $\mathbf{v}_0$ be the right eigen—vector of $P(\mathbf{1})$ corresponding to the eigenvalue 1, i.e.

$$P(\mathbf{1})\mathbf{v}_0 = \mathbf{v}_0,$$

such that $|\mathbf{v}_0| = 1$, and define

$$\widehat{\Phi}(\omega) = \prod_{j=1}^{\infty} P(e^{-i(A^T)^{-j}\omega})\mathbf{v}_0. \tag{1.6}$$

Then its inverse Fourier transform Φ is a compactly supported distribution that satisfies the so–called refinement equation

$$\Phi(\mathbf{x}) = \sum_{\mathbf{k}\in\mathbf{Z}^s} P_{\mathbf{k}}\Phi(A\mathbf{x} - \mathbf{k}). \tag{1.7}$$

We call the sequence $\{P_{\mathbf{k}}\}$ and its corresponding symbol $P(\mathbf{z})$ the mask and two–scale symbol of Φ, respectively. We also call Φ a normalized solution of the refinement equation (1.7).

On the other hand, if Φ is a compactly supported distribution vector that satisfies (1.7) for some finite sequence $\{P_{\mathbf{k}}\}$, we say that Φ is a refinable distribution vector. For such Φ, if each entry of the matrix $\sum_{\mathbf{k}\in\mathbf{Z}^s}(\widehat{\Phi}\widehat{\Phi}^*)(\omega)$ is a bounded function and that the matrix

$$\sum_{\mathbf{k}\in\mathbf{Z}^s} (\widehat{\Phi}\widehat{\Phi}^*)(2\mathbf{k}\pi)$$

is non–singular, then the symbol $P(\mathbf{z})$ defined by (1.5) satisfies $P(\mathbf{1}) \in E$ (see e.g. [34]).

1.3 Sum Rules

In this section, we will always assume that $P(\mathbf{1}) \in E$. We will study conditions on the finite sequence $\{P_{\mathbf{k}}\}$ under which the compactly supported distribution Φ introduced in (1.6) satisfies $\Phi \in \mathrm{PP}_m$ as discussed in §1.1.

Let A be defined as in §1.2, and let ω_h for $h = 0, 1, ..., a-1$, with $a = |\det A|$ and $\omega_0 = 0$, be the representors of $\mathbb{Z}^s/A^T\mathbb{Z}^s$. We say that P satisfies the *sum rule* of order m (or $P \in \mathrm{SR}_m$ for short) if there exist constant vectors $\widetilde{\mathbf{y}}^\alpha$, with $\widetilde{\mathbf{y}}_0 \neq \mathbf{0}$, such that

$$\sum_{\beta\le\alpha}(-1)^{|\beta|}\binom{\alpha}{\beta}\widetilde{\mathbf{y}}_{\alpha-\beta}J_{\beta,\gamma_h} = \sum_{|\tau|=|\alpha|}\left[\sum_{|\beta|=|\alpha|}(A^{-1}\beta)^\beta t^\alpha_{\beta\tau}\right]\frac{\widetilde{\mathbf{y}}_\tau}{\tau!}, \tag{1.8}$$

for all $|\alpha| < m, \gamma_h \in \mathbb{Z}^s/A^T\mathbb{Z}^s$, where

$$J_{\beta,\gamma_h} := \sum_{\mathbf{k}}(\mathbf{k} + A^{-1}\gamma_h)^\beta P_{A\mathbf{k}+\gamma_h},$$

and $\left[t^\alpha_{\beta\tau}\right]$ is the inverse of the matrix $\left[\frac{\beta^\tau}{\tau!}\right]_{|\tau|,|\beta|=|\alpha|}$. For the case $A = 2I_s$, (1.8) can be simplified to be

$$\sum_{\beta\le\alpha}(-1)^{|\beta|}\binom{\alpha}{\beta}\widetilde{\mathbf{y}}_{\alpha-\beta}J_{\beta,\gamma} = 2^{-\alpha}\widetilde{\mathbf{y}}_\alpha, \tag{1.9}$$

for all $\gamma \in \{0,1\}^s, |\alpha| < m$, with

$$J_{\beta,\gamma} := \sum_{\mathbf{k}}(\mathbf{k} + 2^{-1}\gamma)^\beta P_{2\mathbf{k}+\gamma}.$$

Of course when $r = 1$, we can further simplify (1.9) to

$$\begin{cases} \sum_{\mathbf{k}\in\mathbf{Z}^s} P_{\mathbf{k}} = 2^s, \\ \sum_{\mathbf{k}\in\mathbf{Z}^s}(2\mathbf{k}+\gamma)^\alpha P_{2\mathbf{k}+\gamma} = \sum_{\mathbf{k}\in\mathbf{Z}^s}(2\mathbf{k})^\alpha P_{2\mathbf{k}}, \quad \gamma \in \{0,1\}^s, |\alpha| < m, \end{cases}$$

for some suitable choice of $\{\widetilde{y}_\alpha\}$.

Now, consider an r–vector $\widetilde{\mathbf{t}}(\omega) = \sum_{\mathbf{k}\in\mathbf{Z}^s}\widetilde{\mathbf{t}}_{\mathbf{k}}e^{-i\mathbf{k}\omega}$ of trigonometric polynomials that gives $\widetilde{\mathbf{y}}_\alpha = (-iD)^\alpha\widetilde{\mathbf{t}}(0)$ for all $|\alpha| < m$. It was shown in [25] that (1.8) is equivalent to

$$D^\alpha(\widetilde{\mathbf{t}}(A^T\omega)P(e^{-i\omega}))|_{\omega=2\pi A^{-T}\omega_h} = \delta_{h,0}D^\alpha\widetilde{\mathbf{t}}(0), \quad |\alpha| < m, \tag{1.10}$$

for all $0 \le h < a$.

The following result shows that if $P \in \mathrm{SR}_m$, then the normalized solution Φ of (1.7) has the property of polynomial preservation of order m ($\Phi \in \mathrm{PP}_m$).

Theorem 1.1. *Let $P \in SR_m$ for some integer $m \geq 1$ and Φ be a compactly supported normalized solution of* (1.7). *Then Φ satisfies* (1.3) *with $\mathbf{y}_\alpha = \widetilde{\mathbf{y}}_\alpha$; or equivalently, $\Phi \in PP_m$.*

Conversely, if Φ is some compactly supported refinable distribution vector with finite mask $\{P_\mathbf{k}\}$ that satisfies PP_m such that the matrices $\sum_\mathbf{k}(\widehat{\Phi}\widehat{\Phi}^*)(2\mathbf{k}\pi + 2\pi A^{-T}\omega_h)$ are non-singular for all $0 \leq h < a$, then $P \in SR_m$ with $\widetilde{\mathbf{y}}_\alpha = \mathbf{y}_\alpha$.

Theorem 1.1 was obtained in [21] for $r = 1$, and independently in [2] and [34] for any $r \geq 1$. The equivalence of PP_m and SR_m in the vector setting were derived earlier under certain stronger assumptions on Φ, such as linear independence of $\{\phi_\ell(\cdot - \mathbf{k}) : \mathbf{k} \in \mathbb{Z}^s, \ell = 1, \cdots, r\}$ in [18] for $s = 1$ and [3, 4] for $s \geq 1$, and stability of Φ in [46] for $s = 1$.

For a given finite mask $\{P_\mathbf{k}\}$ and dilation matrix A, (1.8) can be used to determine the maximum sum rule order m of P and to find the vectors $\widetilde{\mathbf{y}}_\alpha$, and hence, the maximum polynomial preservation order and the vectors $\mathbf{y}_\alpha = \widetilde{\mathbf{y}}_\alpha$ in (1.3) for generating π^s_{m-1}. Matlab routines for calculating sum rule orders can be downloaded from the second author's website: `www.math.umsl.edu/~jiang`.

1.4 From Compactly Supported Distributions to Function Vectors

To assure that the compactly supported distribution vector Φ constructed in §1.2 is a function vector in $(L^2)^r$, we consider the transition operator T_P associated with the symbol $P(\mathbf{z})$ in (1.5). Let $C_0(\mathbb{T}^s)^{r\times r}$ denote the space of all $r \times r$ matrices with trigonometric polynomial entries. The transition operator T_P associated with P is defined on $C_0(\mathbb{T}^s)^{r\times r}$ as

$$T_P F(\omega) = \sum_{0\leq h<a} P(e^{-iA^{-T}(\omega+2\pi\omega_h)})F(A^{-T}(\omega + 2\pi\omega_h))P(e^{-iA^{-T}(\omega+2\pi\omega_h)})^*,$$

The representation matrix of T_P, also denoted by T_P for convenience, is

$$T_P := [B_{A\mathbf{k}-\mathbf{j}}]_{\mathbf{k},\mathbf{j}}, \tag{1.11}$$

where

$$B_\mathbf{j} = \frac{1}{a}\sum_\mathbf{k} P_{\mathbf{k}-\mathbf{j}} \otimes \overline{P}_\mathbf{k},$$

and $\otimes$ denotes the Kronecker product of A and B, namely $A \otimes B = [a_{ij}B]$ (see e.g. [34]). Hence, we use the notation $T_P \in E$ if the matrix defined in (1.11) satiefies Condition E, as defined in §1.2.

Next, consider the refinement operator

$$U_P \mathbf{f} := \sum_{\mathbf{k}\in \mathbf{Z}^s} P_{\mathbf{k}} \mathbf{f}(A\cdot - \mathbf{k}), \quad \mathbf{f} \in (L^2)^r, \tag{1.12}$$

that defines the so-called *cascade algorithm*

$$\begin{cases} \Phi_n := U_P \Phi_{n-1}, \ n = 1, 2, \cdots, \\ \Phi_0 \in (L^2)^r. \end{cases} \tag{1.13}$$

It is known (see e.g. [5]) that for the cascade algorithm (1.13) with compactly supported initial function vector $\Phi_0 \in (L^2)^r$ to converge in the $(L^2)^r$–norm, it is necessary that Φ_0 satisfies

$$\sum_{\mathbf{k}\in \mathbf{Z}^s} \mathbf{y}_0 \Phi_0(\cdot - \mathbf{k}) = c_0 \neq 0, \tag{1.14}$$

where c_0 is a nonzero constant and $\mathbf{y}_0$ is a left eigenvector of $P(\mathbf{1})$ corresponding to the eigenvalue 1. The significance of the convergence of the cascade algorithm is that the limit Φ, which is necessarily an $(L^2)^r$ function vector, is a refinable function vector that satisfies (1.7). The following result is established in [5] for the more general Sobolev setting.

Theorem 1.2. *A necessary and sufficient condition for the cascade algorithm to converge in the $(L^2)^r$–norm for any initial compactly supported $\Phi_0 \in (L^2)^r$ that satisfies* (1.14) *is that*

$$P \in SR_1 \text{ and } T_P \in E. \tag{1.15}$$

We remark that special cases of this theorem have been considered in earlier works [14, 28, 39, 50, 51]. In particular, Theorem 1.2 itself was already established in [50] for $A = 2I_s$. The convergence in the L^p–norm has also been studied in [15, 20, 28].

The cascade algorithm (1.13) is closely related to the theory of stationary vector subdivision. The vector subdivision operator S_P

$$(S_P \mathbf{u})_{\mathbf{k}} := \sum_{\mathbf{j}\in \mathbf{Z}^s} \mathbf{u}_{\mathbf{j}} P_{\mathbf{k}-A\mathbf{j}}, \quad \mathbf{k} \in \mathbb{Z}^s, \tag{1.16}$$

defined on $\ell_0(\mathbb{Z}^s)^{1\times r}$, where $\mathbf{u} = (\mathbf{u}_{\mathbf{j}})_{\mathbf{j}\in \mathbf{Z}^s}$, is used to generate $\mathbf{u}^n := S_P \mathbf{u}^{n-1}$, $n = 1, 2, \cdots$, with initial $\mathbf{u}^0 \in \ell_0(\mathbb{Z}^s)^{1\times r}$. The characterization of L^p convergence of $(\mathbf{u}^n)$ (for $s = 1, r \geq 1$) was studied in [45].

To verify (1.15), the difficulty is to study $T_P \in E$. In this regard, it is sufficient to consider certain truncation of T_P. More precisely, consider

$$\Omega := \{\sum_{k=1}^{\infty} A^{-k} x_k : \quad x_k \in [-N, N]^s, k \in \mathbb{N}\},$$

where $N = N_2 - N_1$, with N_1 and N_2 denoting the lower and upper coordinate degrees of P, respectively. Also, consider the invariant subspace

$$H_\Omega := \{F(\omega) \in C_0(\mathbb{T}^s)^{r\times r} : \quad F(\omega) = \sum_{\mathbf{k}\in\Omega\cap\mathbf{Z}^s} F_{\mathbf{k}} e^{-i\mathbf{k}\omega}\},$$

of $C_0(\mathbb{T}^s)$ under T_P. It was shown in [22] and [34] that eigenfunctions of T_P corresponding to the nonzero eigenvalues lie in H_Ω. Thus, to study the eigenvalues and eigenfunctions of T_P, one needs only consider those of the restriction $T_{P,\Omega}$ of T_P to the finite–dimensional space H_Ω, or equivalently those of

$$T_{P,\Omega} = [B_{A\mathbf{k}-\mathbf{j}}]_{\mathbf{k},\mathbf{j}\in(\Omega\cap\mathbf{Z}^s)}.$$

For $\Phi = [\phi_1, \cdots, \phi_r]^T \in (L^2)^r$, we say that Φ is $(L^2)^r$–stable, provided that there exist positive constants $0 < C_1 \le C_2 < \infty$ such that

$$C_1\|\mathbf{c}\|_{\ell^2(\mathbf{Z}^s)^{1\times r}} \le \|\sum_{\mathbf{k}\in\mathbf{Z}^s} \mathbf{c}_{\mathbf{k}}\Phi(\cdot - \mathbf{k})\|_{L^2(\mathbf{R}^s)} \le C_2\|\mathbf{c}\|_{\ell^2(\mathbf{Z}^s)^{1\times r}}, \quad \forall \mathbf{c} \in \ell^2(\mathbb{Z}^s)^{1\times r}.$$

The following theorem is concerned with the stability of the refinable vector–valued functions (see [50] for $A = 2I_s$ and a straightforward generalization to arbitrary A in [34]).

Theorem 1.3. *The normalized solution Φ of* (1.7) *is $(L^2)^r$–stable if and only if* (1.15) *is satisfied and that the matrix–valued eigenfunction of T_P corresponding to the eigenvalue* 1 *is either positive or negative definite.*

From Theorems 1.2 and 1.3, we see that stability implies the convergence of the cascade algorithm.

1.5 Smoothness of Refinable Function Vectors

The Sobolev smoothness of the refinable vector–valued functions was discussed in [9, 27, 29, 34, 44, 47, 50]. The results for $s = 1$ in [9, 44] are based on the factorization of the symbol $P(z)$. In general, it is not possible to generalize these results to the multivariate setting. On the other hand, the smoothness estimates in [34, 50] for $\mathbb{R}^s$ are related to both the eigenvalues and the corresponding eigenvectors of the transition operator T_P, while those in [27, 47] are related to the spectral radius of the operator T_P restricted to an invariant subspace of T_P. Since computation of the eigenvectors of a large matrix is not as stable as that of the eigenvalues, it is difficult to decide whether an eigenvalue of $T_{P,\Omega}$ should be kept or thrown away when the methods in [34, 50] are used for matrices $T_{P,\Omega}$ of high dimensions. On the other hand, due to roundoff error, it is difficult to find the invariant subspaces in [27, 29, 47] when the size of matrix $T_{P,\Omega}$ is large. A characterization of the Sobolev smoothness in terms solely of the eigenvalues of the transition operator is desired. For the scalar case, such a characterization is available in [31] when the dilation matrix A is isotropic, meaning that all eigenvalues of A have the same magnitude. Recently a characterization of the smoothness for vector–valued Φ in terms of the eigenvalues of T_P was obtained in [26].

More precisely, suppose that A is istropic with eigenvalues $\sigma_1, \cdots \sigma_s$, and that the mask $\{P_{\mathbf{k}}\}$ satisfies the sum rule of order m. Let $\sigma = (\sigma_1, \cdots, \sigma_s)$, and $\lambda_j, 1 \le j \le r$ with $\lambda_1 = 1$ and $|\lambda_j| < 1, j = 2, \cdots r$, be the eigenvalues of $P(\mathbf{1})$. Set

$$S_m := \operatorname{spec}(T_{P,\Omega}) \backslash \widetilde{S}_m \,, \tag{1.17}$$

with $\widetilde{S}_m$ being the set of all

$$\sigma^{-\alpha}\overline{\lambda_j}, \overline{\sigma^{-\alpha}}\lambda_j, \sigma^{-\beta}, \quad \alpha, \beta \in \mathbb{Z}_+^s, |\alpha| < m, |\beta| < 2m, 2 \le j \le r,$$

where multiplicities of the above $\sigma^{-\alpha}\overline{\lambda_j}, \overline{\sigma^{-\alpha}}\lambda_j, \sigma^{-\beta}$ are taken into account in the definition of $\widetilde{S}_m$. Also let $\nu_2(\Phi)$ denote its critical Sobolev exponent, defined by

$$\nu_2(\Phi) := \sup\{\nu : \ \phi_j \in W^\nu(\mathbb{R}^s), j = 1, \cdots, r\},$$

where $W^\nu(\mathbb{R}^s)$ is the Sobolev space defined by

$$W^\nu(\mathbb{R}^s) := \{f : \quad \int_{\mathbb{R}^s} (1 + |\omega|^2)^\nu |\hat{f}(\omega)|^2 d\omega < \infty\}.$$

Then we have following theorem [26].

Theorem 1.4. *Let $\Phi \in (L^2)^r$ be the normalized solution of* (1.7) *with mask $\{P_{\mathbf{k}}\}$ and some isotropic dilation matrix A. Also, let m be the largest positive integer for which $P \in SR_m$, and $\rho_0 = \max\{|\lambda| : \lambda \in S_m\}$, where S_m is defined in* (1.17). *Then*

$$\nu_2(\Phi) \ge -\frac{s}{2} \log_a \rho_0 \,. \tag{1.18}$$

If, in addition, Φ *is* $(L^2)^r$*–stable, then*

$$\nu_2(\Phi) = -\frac{s}{2}\log_a \rho_0 \,. \tag{1.19}$$

For the scalar case ($r = 1$), $\widetilde{S}_m$ is given by

$$\widetilde{S}_m = \{\sigma^{-\beta} : |\beta| < 2m\}.$$

Theorem 1.4 for $r = 1$ and $s \geq 1$ is contained in [31]. This theorem leads to some efficient ways for computing highly accurate values for the Sobolev smoothness $\nu_2(\Phi)$ by standard eigenvalue solvers.

The Hölder smoothness of Φ can be characterized in terms of the uniform joint spectral radius of the family $\mathcal{A}_m := \{A_{P,\gamma}|_{V_m}, \gamma \in \Gamma\}$, where Γ is a fixed set of representators for $\mathbb{Z}^s/A\mathbb{Z}^s$, and $A_{P,\gamma}, \gamma \in \Gamma$ are the operators defined by

$$(A_{P,\gamma}\mathbf{v})_{\mathbf{k}} = \sum_{\mathbf{j}\in\mathbf{Z}^s} P_{\gamma+A\mathbf{k}-\mathbf{j}}\mathbf{v}_{\mathbf{j}} \,, \qquad \mathbf{k} \in \mathbb{Z}^s, \quad v \in \ell_0(\mathbb{Z}^s)^r,$$

and V_m is the subspace of $\ell_0(\mathbb{Z}^s)^r$ consisting of $\{\mathbf{v}_{\mathbf{k}}\}$ with support in an appropriate finite set $\Omega_1 \cap \mathbb{Z}^s$, and satisfying

$$D^\alpha(\widetilde{\mathbf{t}}(\omega)\mathbf{v}(\omega))|_{\omega=0} = 0, \quad |\alpha| < m, \tag{1.20}$$

where $\mathbf{v}(\omega) := \sum_{\mathbf{k}\in\Omega_1\cap\mathbf{Z}^s} \mathbf{v}_{\mathbf{k}} e^{-i\mathbf{k}\omega}$ and $\widetilde{\mathbf{t}}(\omega)$ is the vector in (1.10). (See Theorem 5 in [36] for the details.) For $s = 1$, a different characterization of Hölder smoothness of Φ was obtained in [29] using an invariant subspace of $A_{P,\gamma}$ different from V_m. For the subspace V_m defined in (1.20), since we know $\widetilde{\mathbf{t}}(\omega)$, we can easily give its basis. Therefore we can give the representating matrices of the operators $A_{P,\gamma}|_{V_m}$. The reader is referred to [36] for the Matlab routines for calculation of the order of smoothness.

1.6 Bi–orthogonal Duals

Let $\Phi = [\phi_1, \cdots, \phi_r]^T \in (L^2)^r$ and $\widetilde{\Phi} = [\widetilde{\phi}_1, \cdots, \widetilde{\phi}_r]^T \in (L^2)^r$ be compactly supported refinable vector–valued functions with dilation matrix A and with finite masks $\{P_{\mathbf{k}}\}$ and $\{G_{\mathbf{k}}\}$, respectively. We say that Φ and $\widetilde{\Phi}$ are bi–orthogonal duals of each other if $\{\Phi(\mathbf{x} - \mathbf{k})\}$ and $\{\widetilde{\Phi}(\mathbf{x} - \mathbf{k})\}$ are bi-orthogonal, meaning that

$$\int_{\mathbf{R}^s} \Phi(\mathbf{x})\widetilde{\Phi}(\mathbf{x} - \mathbf{k})^* \, d\mathbf{x} = \delta_{0,\mathbf{k}} I_r, \quad \mathbf{k} \in \mathbb{Z}^s. \tag{1.21}$$

Hence, if $\widetilde{\Phi} = \Phi$, we say that Φ is orthonormal.

A necessary condition for Φ and $\widetilde{\Phi}$ to be bi–orthogonal duals is that their two–scale symbols satisfy

$$\sum_{h=0}^{a-1} P(\mathbf{z}e^{-i2\pi A^{-T}\omega_h})G(\mathbf{z}e^{-i2\pi A^{-T}\omega_h})^* = I_r, \quad \mathbf{z} = e^{-i\omega}, \omega \in \mathbb{R}^s. \tag{1.22}$$

Under certain mild conditions, this condition is also sufficient, as follows.

Theorem 1.5. *A necessary and sufficient condition for* Φ *and* $\widetilde{\Phi}$*, with finite maskes* $\{P_{\mathbf{k}}\}$ *and* $\{G_{\mathbf{k}}\}$*, respectively, to be bi–orthogonal duals is that*

(i) $P(\mathbf{z})$ *and* $Q(\mathbf{z})$ *are bi–orthogonal matrix–valued Laurent polynomials as defined in* (1.22),

(ii) $P(\mathbf{1}), G(\mathbf{1}) \in E$,

(iii) $P, G \in SR_1$, *and*

(iv) $T_P, T_G \in E$.

The proof this theorem can be formulated as a generalization of the proof in [23] for $s = 1, A = [2]$, and the proof in [24] for $s \geq 1$. When $\Phi = \widetilde{\Phi}$ and $P = G$, Theorem 1.5 gives a characterization of the orthonormality of Φ in terms of its mask $\{P_{\mathbf{k}}\}$.

Suppose $Q^h(\mathbf{z})$ and $H^h(\mathbf{z}), 1 \leq h < a$, are matrix–valued Laurent polynomials that satisfy

$$\begin{cases} \sum_{0\leq n<a} P(\mathbf{z}e^{-i2\pi A^{-T}\omega_n})H^h(\mathbf{z}e^{-i2\pi A^{-T}\omega_n})^* = 0, & 1 \leq h < a \\ \sum_{0\leq n<a} Q^h(\mathbf{z}e^{-i2\pi A^{-T}\omega_n})G(\mathbf{z}e^{-i2\pi A^{-T}\omega_n})^* = 0, & 1 \leq h < a \\ \sum_{0\leq n<a} Q^h(\mathbf{z}e^{-i2\pi A^{-T}\omega_n})H^\ell(\mathbf{z}e^{-i2\pi A^{-T}\omega_n})^* = \delta_{h,\ell}I_r, & 1 \leq h, \ell < a, \end{cases} \tag{1.23}$$

for $\omega \in \mathbb{R}^s$, where $\mathbf{z} = e^{-i\omega}$—. Let $\Psi^h, \widetilde{\Psi}^h$ be the vector–valued functions defined by

$$\widehat{\Psi}^h(A^T\omega) = Q^h(\mathbf{z})\hat{\Phi}(\omega), \quad \hat{\widetilde{\Psi}}^h(A^T\omega) = H^h(\mathbf{z})\hat{\widetilde{\Phi}}(\omega). \tag{1.24}$$

Then $\{a^{-\frac{n}{2}}\Psi^h(A^{-n}\mathbf{x}-\mathbf{k}), a^{-\frac{n}{2}}\widetilde{\Psi}^h(A^{-n}\mathbf{x}-\mathbf{k}), 1 \leq h < a, n \in \mathbb{Z}, \mathbf{k} \in \mathbb{Z}^s\}$ forms a bi–orthogonal system (cf. [30]).

We use $\{G, H^h, 1 \leq h < a\}$ and $\{P, Q^h, 1 \leq h < a\}$ for wavelet decomposition and reconstruction respectively. More precisely, the decomposition algorithm is given by

$$\mathbf{c}_{\mathbf{n}}^{(j-1)} = \frac{1}{\sqrt{m}} \sum_{\mathbf{k}} G_{\mathbf{k}-A\mathbf{n}} \mathbf{c}_{\mathbf{k}}^{(j)}, \quad \mathbf{d}_{\mathbf{n},h}^{(j-1)} = \frac{1}{\sqrt{m}} \sum_{\mathbf{k}} H_{\mathbf{k}-A\mathbf{n}}^{h} \mathbf{c}_{\mathbf{k}}^{(j)}, \quad \mathbf{n} \in \mathbb{Z}^s, \tag{1.25}$$

and the reconstruction algorithm is given by

$$\mathbf{c}_{\mathbf{n}}^{(j)} = \frac{1}{\sqrt{m}} \sum_{\mathbf{k}} P_{\mathbf{n}-A\mathbf{k}}^{T} \mathbf{c}_{\mathbf{k}}^{(j-1)} + \frac{1}{\sqrt{m}} \sum_{1 \le h < a} \sum_{\mathbf{k}} (Q_{\mathbf{n}-A\mathbf{k}}^{h})^T \mathbf{d}_{\mathbf{k},h}^{(j-1)}, \quad \mathbf{n} \in \mathbb{Z}^s. \tag{1.26}$$

2 Balanced Refinable Vectors

The notion of balancing is generalized to the multivariate bi–orthogonal setting in this section. With balanced bi–orthogonal duals, discrete polynomials are preserved by the lowpass filters and annihilated by the highpass filters of the corresponding multi–wavelets. Several important characterization results for these two properties are discussed in this section. Examples are given in some details, first for univariate quadratic and cubic spline function vectors, then for continuous piecewise linear polynomials on the four–directional mesh $\triangle^2$ in $\mathbb{R}^2$, and finally for C^1 cubic bivariate splines also on $\triangle^2$. Both the dilation matrix $2I_2$ and the quincunx matrix are considered in the bivariate examples.

2.1 The Notion of Balancing

Let $\mathbf{a}_1, \cdots, \mathbf{a}_r \in \mathbb{R}^s$. A compactly supported vector–valued function $F = [f_1, \cdots, f_r]^T \in (L^2)^r$ is said to be *K–balanced*, $K \ge 1$, relative to $(\mathbf{a}_1, \cdots, \mathbf{a}_r)$, if

$$\int_{\mathbf{R}^s} f_l(\mathbf{x})(\mathbf{x} - \mathbf{a}_l)^{\alpha} d\mathbf{x} = \int_{\mathbf{R}^s} f_i(\mathbf{x})(\mathbf{x} - \mathbf{a}_i)^{\alpha} d\mathbf{x}, \tag{2.1}$$

for $1 \le l, i \le r, |\alpha| < K$. This is an extension of the notion of K–balancing introduced for the orthonormal univariate setting by Lebrun and Vetterli [40] for $K = 1$ and Selesnick [48] for $K \ge 1$. (See their other work [41, 49].) The motivation of their consideration is for processing scalar–valued digital data by $r \times r$ matrix–valued lowpass filters $\{P_k\}$ without the need of pre–filtering as studied in [1, 16, 37, 52, 53, 57, 58], so that independent of shifts in the formulation of the r–vector blocks of the scalar–valued input data, the filtering process yields polynomial output upon lowpass filtering of any polynomial input data in $\pi_{K-1} := \pi_{K-1}^1$. This is significant in wavelet signal processing to allow for the discrete wavelet transform (DWT), with vanishing moments of order K or higher, to extract detailed features of the input signal. Our extension from $\mathbb{R}^1$ to $\mathbb{R}^s$ and from orthogonal Φ to the bi-orthogonal setting facilitates the study of arbitrary dilation matrix A and achieving certain desirable filter and multi-wavelet design criteria. For example, the dilation matrix A with determinant of A equal to ± 2 reduces implementation complexity and contributes to memory and cost reduction.

For the univariate and orthonormal setting, Selesnick [48] gave a complete characterization of compactly supported K-balanced orthonormal refinable function vector Φ with two-scale symbol $P(z) = [p_{j,k}(z)]$ in terms of an associated polynomial

$$\sum_{j=0}^{r-1} j(j+r)\cdots(j+(K-1)r)\{\sum_{\ell}^{K-1} \frac{(-1)^\ell}{(j+\ell r)}\binom{K-1}{\ell} z^{r\ell}\} \sum_{k=1}^{r} p_{(j+1),k}(z^r)z^{k-1} \tag{2.2}$$

in that Φ is K-balanced relative to centers

$$a_1 = 0, \quad a_2 = \frac{1}{r}, \quad \cdots, \quad a_r = \frac{r-1}{r}$$

if and only if the polynomial (2.2) is divisible by $(1 + z + \cdots z^{2r-1})^K$. Hence, to construct $P(z)$ and hence Φ, it is necessary to solve for the coefficients of $p_{j,k}(z)$ so that the QMF condition

$$P(z)P(z)^* + P(-z)P(-z)^* = I_r, \quad |z| = 1,$$

is satisfied.

In the next section we will give a necessary and sufficient condition for (2.1) in terms of an intimate relation between these centers and the vectors $\mathbf{y}_\alpha$ in (1.8), or equivalently $\mathbf{y}_\alpha$ in (1.3), where $K \le m$. We will also show that K-balancing of $\widetilde{\Phi}$ is equivalent to discrete polynomial preservation of total degree $K-1$ by its mask $\{G_\mathbf{k}\}$, and that K-balancing of $\widetilde{\Phi}$ implies discrete polynomial annihilation of total degree $K-1$ by the highpass filters $\{H^h_\mathbf{k}\}$ corresponding to their lowpass filter $\{G_\mathbf{k}\}$.

2.2 Characterization Theorems

Suppose that Φ and $\widetilde{\Phi}$ are compactly supported bi-orthogonal dual refinable vector-valued functions in $(L^2)^r$ with dilation matrix A and finite masks $\{P_\mathbf{k}\}$ and $\{G_\mathbf{k}\}$, respectively. Let $Q^h(\mathbf{z})$ and $H^h(\mathbf{z}), 1 \le h < a$, be the matrix-valued

Laurent polynomials satisfying (1.23) and that $\Psi^h, \widetilde{\Psi}^h$ be the vector–valued functions defined by (1.24).

Suppose Φ satisfies the sum rule with order m and $\mathbf{y}_\alpha, |\alpha| < m$ in (1.8). We have the following characterizations of K–balancing for $\widetilde{\Phi}$, $K \le m$.

Theorem 2.1. *$\widetilde{\Phi}$ is K–balanced relative to $(\mathbf{a}_1, \cdots, \mathbf{a}_r) \subset \mathbb{R}^s$ if and only if $\mathbf{y}_\alpha = [y_{\alpha,1}, \cdots, y_{\alpha,r}]$ satisfy*

$$y_{\alpha,l} = \sum_{\beta \le \alpha} \binom{\alpha}{\beta} (\mathbf{a}_l - \mathbf{a}_1)^\beta y_{\alpha-\beta,1}, \quad 2 \le l \le r, |\alpha| < K. \tag{2.3}$$

For the case $s = 1$ with $a_\ell = \frac{\ell-1}{r}$, (2.3) reduces to

$$y_{\alpha,l} = \sum_{\beta \le \alpha} \binom{\alpha}{\beta} (\frac{\ell-1}{r})^\beta y_{\alpha-\beta,1}, \quad 2 \le l \le r, 0 \le \alpha < K. \tag{2.4}$$

Theorem 2.1 tells us that $\widetilde{\Phi}$ is K–balanced if and only if $\mathbf{y}_\alpha$ can be so chosen that both (1.8) and (2.3) are satisfied for $|\alpha| < K$. Theorem 2.1 also enables us to decide the possible balanced order of the bi–orthogonal dual scaling vectors for a given refinable function vector, and tells us how to construct high balancing orthogonal multi–wavelets and the masks of the primal refinable function vectors with high balanced bi–orthogonal duals. We also have have the following:

Theorem 2.2. *$\widetilde{\Phi}$ is K–balanced relative to $(\mathbf{a}_1, \cdots, \mathbf{a}_r) \subset \mathbb{R}^s$ if and only if*

$$\sum_{\mathbf{k}} [(\mathbf{k} + \mathbf{a}_1)^\alpha, \cdots, (\mathbf{k} + \mathbf{a}_r)^\alpha] \Phi(\mathbf{x} - \mathbf{k}) \in \pi^s_{|\alpha|},$$

for all $|\alpha| < K$.

The following result tells us that K–balancing of $\widetilde{\Phi}$ implies the preservation of π^s_K for G and the annihilation of π^s_K for H^h.

Theorem 2.3. *$\widetilde{\Phi}$ is K–balanced relative to $(\mathbf{a}_1, \cdots, \mathbf{a}_r) \subset \mathbb{R}^s$ if and only if*

$$\sum_{\mathbf{k}} G_{\mathbf{k}-A\mathbf{j}} \begin{bmatrix} (\mathbf{k} + \mathbf{a}_1)^\alpha \\ \vdots \\ (\mathbf{k} + \mathbf{a}_r)^\alpha \end{bmatrix} = \begin{bmatrix} p_\alpha(\mathbf{j} + \mathbf{a}_1) \\ \vdots \\ p_\alpha(\mathbf{j} + \mathbf{a}_r) \end{bmatrix}, \quad |\alpha| < K, \mathbf{j} \in \mathbb{Z}^s,$$

where $p_\alpha \in \pi^s_{|\alpha|}$. *If* $\widetilde{\Phi}$ *is* K*–balanced relative to* $(\mathbf{a}_1, \cdots, \mathbf{a}_r) \subset \mathbb{R}^s$, *then*

$$\sum_{\mathbf{k}} H^h_{\mathbf{k}-A\mathbf{j}} \begin{bmatrix} (\mathbf{k}+\mathbf{a}_1)^\alpha \\ \vdots \\ (\mathbf{k}+\mathbf{a}_r)^\alpha \end{bmatrix} = \begin{bmatrix} 0 \\ \vdots \\ 0 \end{bmatrix}, \quad |\alpha| < K, \mathbf{j} \in \mathbb{Z}^s, 1 \le h < a.$$

We also have the following result that gives the intimate relationship of polynomial preservation of total degree $K-1$ by the mask of Φ and K–balancing of its bi–orthogonal dual $\widetilde{\Phi}$.

Theorem 2.4. $\widetilde{\Phi}$ *is* K*–balanced relative to* $(\mathbf{a}_1, \cdots, \mathbf{a}_r) \subset \mathbb{R}^s$ *if and only if*

$$\sum_{\mathbf{k}} P^T_{\mathbf{j}-A\mathbf{k}} \begin{bmatrix} (\mathbf{k}+\mathbf{a}_1)^\alpha \\ \vdots \\ (\mathbf{k}+\mathbf{a}_r)^\alpha \end{bmatrix} = \begin{bmatrix} q_\alpha(\mathbf{j}+\mathbf{a}_1) \\ \vdots \\ q_\alpha(\mathbf{j}+\mathbf{a}_r) \end{bmatrix}, \quad \mathbf{j} \in \mathbb{Z}^s,$$

for all $|\alpha| < K$, *where* $q_\alpha \in \pi^s_{|\alpha|}$.

The proofs of Theorems 2.1–2.4 are given in our paper [7].

2.3 Univariate B-Splines

In this section we discuss univariate balanced bi–orthogonal dual $\widetilde{\Phi}$ (relative to $a_1 = 0, a_2 = \frac{1}{2}$) of Φ with spline components. We will use (1.8) to determine the sum rule order of the masks. For bi–orthogonal filters P, G, we will apply Theorem 1.5 to decide whether they generate bi–orthogonal refinabe function vectors, and Theorem 1.4 to calculate the Sobolev smoothness of the refinable function vectors.

Example 1. Let $N_{3,0}, N_{3,1}$ be the normalized quadratic B–splines with knots $0,0,1,1$ and $0,1,1,2$, respectively, i.e.,

$$N_{3,0}(x) = 2x(1-x)\chi_{[0,1]}, \quad N_{3,1}(x) = x^2\chi_{[0,1]} + (2-x)^2\chi_{[1,2]}.$$

Then $\Phi_1 := [N_{3,0}, N_{3,1}]^T$ is refinable with two–scale symbol given by

$$P_1(z) = \frac{1}{8}\left(\begin{bmatrix} 2 & 2 \\ 0 & 1 \end{bmatrix} + \begin{bmatrix} 2 & 0 \\ 2 & 4 \end{bmatrix} z + \begin{bmatrix} 0 & 0 \\ 2 & 1 \end{bmatrix} z^2 \right),$$

which satisfies the sum rule of order 3 with

$$\mathbf{y}_0 = [1,1], \quad \mathbf{y}_1 = [\frac{1}{2}, 1], \quad \mathbf{y}_2 = [0,1],$$

(cf. [46]). It is clear that these vectors don't satisfy (2.4) with $K = 3$, but $\mathbf{y}_0, \mathbf{y}_1$ satisfy (2.4) with $K = 2$. Thus it is possible to construct a 2–balanced (but not 3–balanced) bi–orthogonal dual of Φ_1 by applying Theorem 2.1.

On the other hand, for

$$\phi_1 = 3N_{3,0} + N_{3,1}, \quad \phi_2 = -N_{3,0} + N_{3,1},$$

$\Phi_2 := [\phi_1, \phi_2]^T$ is refinable with two–scale symbol given by

$$P_2(z) = \frac{1}{32}\left(\begin{bmatrix} 13 & 15 \\ -3 & -1 \end{bmatrix} + \begin{bmatrix} 12 & 4 \\ 4 & 12 \end{bmatrix} z + \begin{bmatrix} 3 & 1 \\ 3 & 1 \end{bmatrix} z^2 \right),$$

which satisfies the sum rule of order 3 with

$$\mathbf{y}_0 = [1, 1], \quad \mathbf{y}_1 = [\frac{3}{4}, \frac{5}{4}], \quad \mathbf{y}_2 = [\frac{1}{2}, \frac{3}{2}].$$

One can easily verify that these vectors satisfy (2.4) with $K = 3$. Therefore it is possible to construct a 3-balanced bi–orthogonal dual $\widetilde{\Phi}$ of Φ. Indeed, we can construct such a $\widetilde{\Phi}_2$ in $W^{1.1442}(\mathbb{R})$ with supp($\widetilde{\Phi}_2$)=$[-1, 3]$. Its mask $\{G_k\}$ is given by

$$G_{-1} = \begin{bmatrix} -0.18960780661003 & -0.18960780661003 \\ 0.24341593393929 & 0.24341593393929 \end{bmatrix},$$

$$G_0 = \begin{bmatrix} 0.81782419542379 & 0.58025231948912 \\ -1.08301023392189 & -0.64562424126293 \end{bmatrix},$$

$$G_1 = \begin{bmatrix} 1.07469506607298 & -0.12915190410120 \\ 1.14314871878723 & 1.68587976841759 \end{bmatrix},$$

$$G_2 = \begin{bmatrix} 0.12120349037547 & -0.07884298044345 \\ 1.38787180492559 & -0.86142245396542 \end{bmatrix},$$

$$G_3 = \begin{bmatrix} -0.04574279673035 & 0.03897822313370 \\ -0.58328696638954 & 0.46961173553079 \end{bmatrix},$$

where $G_k = \mathbf{0}$ for $k < -1$ or $k > 3$. □

Example 2. Let $N_{4,0}, N_{4,1}$ be the normalized cubic B–splines with knots $0, 0, 1, 1, 2$ and $0, 1, 1, 2, 2$, respectively, i.e.,

$$N_{4,0}(x) = \frac{1}{2}x^2(3 - \frac{5}{2}x)\chi_{[0,1]} + \frac{1}{4}(2 - x)^3\chi_{[1,2]},$$
$$N_{4,1}(x) = \frac{1}{4}x^3\chi_{[0,1]} + \frac{1}{2}(2 - x)^2(\frac{5}{2}x - 2)\chi_{[1,2]}.$$

Then $\Phi_3 := [N_{4,0}, N_{4,1}]^T$ is refinable with two–scale symbol given by

$$P_3(z) = \frac{1}{8}(\begin{bmatrix} 2 & 5 \\ 0 & 1 \end{bmatrix} + \begin{bmatrix} 6 & 2 \\ 2 & 6 \end{bmatrix} z + \begin{bmatrix} 1 & 0 \\ 5 & 2 \end{bmatrix} z^2),$$

which satisfies the sum rule of order 4 with

$$\mathbf{y}_0 = [1, 1], \quad \mathbf{y}_1 = [\frac{2}{3}, \frac{4}{3}], \quad \mathbf{y}_2 = [\frac{1}{3}, \frac{5}{3}], \quad \mathbf{y}_3 = [0, 2].$$

However, it is clear that $\mathbf{y}_0, \mathbf{y}_1$ do not satisfy (2.4) with $K = 2$, but $\mathbf{y}_0$ does with $K = 1$. Thus we can only construct a 1–balanced (but not 2–balanced) bi–orthogonal dual.

On the other hand, for

$$\varphi_1 = 2N_{4,0} + 2N_{4,1}, \quad \varphi_2 = \frac{2}{3}(N_{4,1} - N_{4,0}),$$

that constitute the canonical Hermite cubic basis, the refinement mask corresponding to $\Phi_4 := [\varphi_1, \varphi_2]^T$ satisfies the sum rule of order 4 with

$$\mathbf{y}_0 = [1, 0], \quad \mathbf{y}_1 = [1, 1], \quad \mathbf{y}_2 = [1, 2], \quad \mathbf{y}_3 = [1, 3].$$

Hence, it follows from Theorem 2.1 that Φ_4 does not have a balanced bi–orthogonal dual.

Next, consider the splines

$$\phi_1 = \frac{1}{6}(7N_{4,0} - N_{4,1}), \quad \phi_2 = \frac{1}{6}(-N_{4,0} + 7N_{4,1}),$$

and $\Phi_5 := [\phi_1, \phi_2]^T$ with two–scale symbol given by

$$P_5(z) = \frac{1}{32}(\begin{bmatrix} 11 & 21 \\ -1 & 1 \end{bmatrix} + \begin{bmatrix} 24 & 8 \\ 8 & 24 \end{bmatrix} z + \begin{bmatrix} 1 & -1 \\ 21 & 11 \end{bmatrix} z^2),$$

which satisfies the sum rule of order 4 with

$$\mathbf{y}_0 = [1, 1], \quad \mathbf{y}_1 = [\frac{3}{4}, \frac{5}{4}], \quad \mathbf{y}_2 = [\frac{1}{2}, \frac{3}{2}], \quad \mathbf{y}_3 = [\frac{1}{4}, \frac{7}{4}].$$

One can easily verify that these vectors don't satisfy (2.4) with $K = 4$, but $\mathbf{y}_0, \mathbf{y}_1, \mathbf{y}_2$ satisfy (2.4) with $K = 3$. It is therefore possible to construct a 3–balanced bi–orthogonal dual $\widetilde{\Phi}_5$ of Φ_5. Indeed we can construct such a $\widetilde{\Phi}_5$ in

$W^{0.8252}(\mathbb{R})$ with $\text{supp}(\widetilde{\Phi}_5)=[-1,3]$ and mask given by

$$G_{-1} = \frac{1}{512}\begin{bmatrix} 133 & -277 \\ -29 & 61 \end{bmatrix}, \quad G_0 = \frac{1}{512}\begin{bmatrix} -161 & 815 \\ 33 & -175 \end{bmatrix},$$

$$G_1 = \frac{1}{256}\begin{bmatrix} 395 & -83 \\ -83 & 395 \end{bmatrix}, \quad G_2 = \frac{1}{512}\begin{bmatrix} -175 & 33 \\ 815 & -161 \end{bmatrix},$$

$$G_3 = \frac{1}{512}\begin{bmatrix} 61 & -29 \\ -277 & 133 \end{bmatrix}, \quad G_k = \mathbf{0}, k \in \mathbb{Z}\backslash\{-1,0,1,2,3\}. \qquad \square$$

2.4 The Parametric Approach

The following two examples are based on the parametric expression of the bi–orthogonal two–scale symbols $P(z), G(z)$ in [32]. For the parametric expression of the symmetric filters which could generate high balanced orthonormal Φ, the reader is referred to [35].

Example 3. Consider the bi–orthogonal two–scale symbols

$$P(z) = \frac{1}{2}\begin{bmatrix} \widetilde{a} \\ -\widetilde{b} \end{bmatrix}[d,-c] + \frac{1}{2}\begin{bmatrix} \widetilde{e} & 0 \\ 0 & \widetilde{f} \end{bmatrix} z + \frac{1}{2}\begin{bmatrix} \widetilde{a} \\ \widetilde{b} \end{bmatrix}[d,c]z^2,$$

$$G(z) = \frac{1}{2}\begin{bmatrix} a \\ -b \end{bmatrix}[c,-d] + \frac{1}{2}\begin{bmatrix} e & 0 \\ 0 & f \end{bmatrix} z + \frac{1}{2}\begin{bmatrix} a \\ b \end{bmatrix}[c,d]z^2,$$

with $b = 7/(16dc\widetilde{b})$. If we choose

$$e = \widetilde{e} = 1, f = \widetilde{f} = \frac{1}{2}, ac = \widetilde{a}d = \frac{1}{2}, c = d/4/(d\widetilde{b}-1),$$

then P, G satisfy the sum rule of order 2 with the vectors $\mathbf{y}_0, \mathbf{y}_1$ for P given by

$$\mathbf{y}_0 = [1,0], \quad \mathbf{y}_1 = [c_0, \frac{1}{4}],$$

for some $c_0 \in \mathbb{R}$. Let

$$R_0 := \frac{\sqrt{2}}{2}\begin{bmatrix} 1 & -1 \\ 1 & 1 \end{bmatrix}.$$

Then for suitable choices of $\widetilde{b}, d$, the symbols $R_0 G(z) R_0^T$, $R_0 P(z) R_0^T$ generate a 2–balanced bi–orthogonal dual $\widetilde{\Phi}$ of a 1–balanced Φ. For $(\widetilde{b}, d) = -(6/13, 3/2)$,

$\widetilde{\Phi} \in W^{0.736924}(\mathbb{R}), \Phi \in W^{1.356559}(\mathbb{R})$, and for $(\widetilde{b}, d) = (3/10, 13/6)$, $\widetilde{\Phi} \in W^{0.4421}(\mathbb{R}), \Phi \in W^{1.6027}(\mathbb{R})$. □

Example 4. Let $P = \frac{1}{2}\sum_{k=0}^{3} P_k z^k, G = \frac{1}{2}\sum_{k=0}^{3} G_k z^k$ be bi–orthogonal two–scale symbols with

$$P_0 = \frac{1}{2}\begin{bmatrix} 1 - \frac{x_3}{by} & \frac{1}{4y} \\ 1/b - x_1 & -\frac{1}{4} \end{bmatrix}, \quad P_1 = \frac{1}{2}\begin{bmatrix} 1 + \frac{x_3}{by} & \frac{1}{4y} \\ -1/b - x_1 & \frac{1}{4} \end{bmatrix},$$

$$P_2 = \frac{1}{2}\begin{bmatrix} 1 + \frac{x_3}{by} & -\frac{1}{4y} \\ 1/b + x_1 & \frac{1}{4} \end{bmatrix}, \quad P_3 = \frac{1}{2}\begin{bmatrix} 1 - \frac{x_3}{by} & -\frac{1}{4y} \\ -1/b + x_1 & -\frac{1}{4} \end{bmatrix},$$

$$G_0 = \frac{1}{2}\begin{bmatrix} 1 - x_1 b & -x_2 \\ b - x_3/y & x_2/y \end{bmatrix}, G_1 = \frac{1}{2}\begin{bmatrix} 1 + x_1 b & -x_2 \\ -b - x_3/y & -x_2/y \end{bmatrix},$$

$$G_2 = \frac{1}{2}\begin{bmatrix} 1 + x_1 b & x_2 \\ b + x_3/y & -x_2/y \end{bmatrix}, G_3 = \frac{1}{2}\begin{bmatrix} 1 - x_1 b & x_2 \\ -b + x_3/y & x_2/y \end{bmatrix}.$$

By setting

$$x_3 = \frac{b+1}{2}, \quad x_1 = \frac{16b+1}{2b(4b+5)}, \quad x_2 = -\frac{3b-1}{b(4b+5)}, \quad y = x_1 x_3 - \frac{x_2}{4},$$

P, G satisfy the sum rule of order 3 and 2 respectively, and the vectors $\mathbf{y}_0, \mathbf{y}_1, \mathbf{y}_2$ for $R_0 P R_0^T$ have the structure (2.3) for $K = 3$. With $b = 28/13$, $\widetilde{\Phi} \in W^{0.7222}(\mathbb{R}), \Phi \in W^{2.5524}(\mathbb{R})$, and with $b = 2$, $\widetilde{\Phi} \in W^{0.7857}(\mathbb{R}), \Phi \in W^{2.4559}(\mathbb{R})$. In both cases, $R_0\widetilde{\Phi}$ is 3–balanced. □

2.5 Bivariate Linear Splines

Example 5. Let $\triangle^2$ denote the four–directional mesh with grid lines given by $x = j, y = k, x + y = \ell$, and $x - y = m$, where $j, k, \ell, m \in \mathbb{Z}$. Let ϕ_1 denote the bivariate piecewise linear hat function with $\phi_1(0,0) = 1$ and support given by the square with vertices $(1,0), (0,1), (-1,0)$ and $(0,-1)$, and ϕ_2 be the bivariate piecewise linear hat function with $\phi_2(\frac{1}{2}, \frac{1}{2}) = 1$ and support given by the square $[0,1] \times [0,1]$. See Figure 1 for the Bézier–nets of ϕ_1, ϕ_2.

Then $\{\phi_\ell(\cdot - \mathbf{k}) : \ell = 1, 2, \mathbf{k} \in \mathbb{Z}^2\}$ is a basis of the space $S_1^0(\triangle^2)$ of continuous bivariate piecewise linear splines on $\triangle^2$. Then $\Phi = [\phi_1, \phi_2]^T$ is refinable with respect to the quincunx dilation matrix

$$A = \begin{bmatrix} 1 & 1 \\ 1 & -1 \end{bmatrix}. \tag{2.5}$$

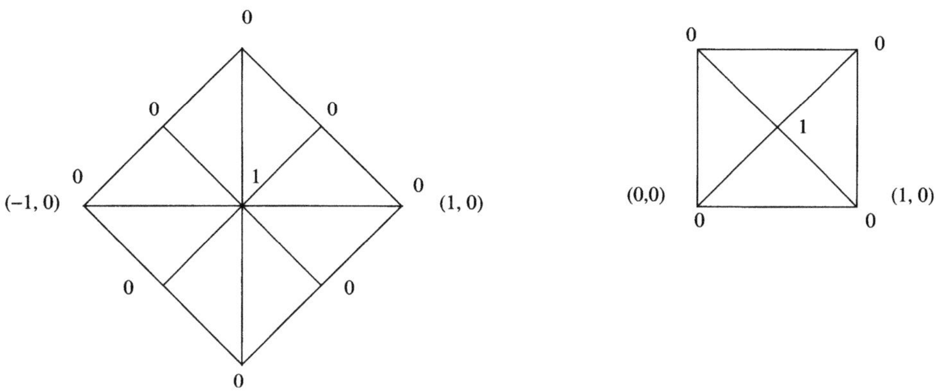

Figure 1: Bézier–nets for ϕ_1 and ϕ_2

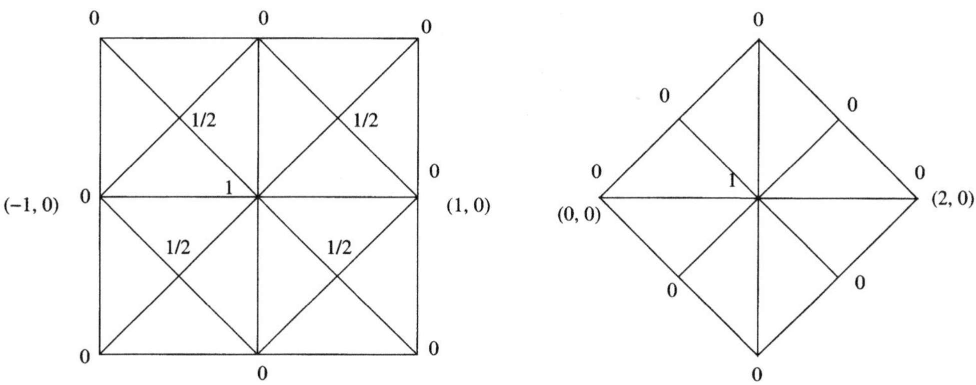

Figure 2: Bézier–nets for $\phi_1(A^{-1}\cdot), \phi_2(A^{-1}\cdot)$

See Figure 2 for the Bézier–nets of $\phi_1(A^{-1}\cdot), \phi_2(A^{-1}\cdot)$. The two–scale symbol of Φ is given by

$$P_A(\mathbf{z}) = \frac{1}{2}\begin{bmatrix} 1 & \frac{1}{2}(1+z_1^{-1})(1+z_2^{-1}) \\ z_1 & 0 \end{bmatrix}.$$

One can easily verify that P satisfies the sum rule of order 2 with

$$\mathbf{y}_{0,0} = [1, 1], \quad \mathbf{y}_{0,1} = [0, \frac{1}{2}], \quad \mathbf{y}_{1,0} = [0, \frac{1}{2}].$$

By Theorem 2.1, it is possible to construct a 2–balanced bi–orthogonal dual $\widetilde{\Phi}$ with balanced centers $\mathbf{a}_1 = \mathbf{0}, \mathbf{a}_2 = (\frac{1}{2}, \frac{1}{2})$. Indeed we can construct such $\widetilde{\Phi}$

in $W^{1.5420}(\mathbb{R}^2)$ with mask $\{G_{\mathbf{k}}\}$ supported in $[-2,2]^2 \cap \mathbb{Z}^2$. The mask is not provided here. □

Example 6. Consider the two–scale symbol $P_A(\mathbf{z})$ in Example 5, and observe that $\Phi = [\phi_1, \phi_2]^T$ is also refinable with respect to the dilation matrix $2I_2$ (cf. [13]), with two–scale symbol $P(\mathbf{z})$ given by

$$P(\mathbf{z}) = P_A(e^{-iA\omega})P_A(\mathbf{z}), \quad \mathbf{z} = e^{-i\omega}.$$

Then P satisfies the sum rule of order 2 with the same $\mathbf{y}_\alpha, |\alpha| < 2$. Φ has a 2–balanced bi–orthogonal dual in $W^{0.2783}(\mathbb{R}^2)$ with mask $\{G_{\mathbf{k}}\}$ supported in $[-2,2]^2 \cap \mathbb{Z}^2$. □

2.6 Bivariate C^1 Cubic Splines

Example 7. Let $S_3^1(\triangle^2)$ denote the bivariate C^1 cubic spline space on the four–directional mesh $\triangle^2$ and let $\phi_1, \cdots, \phi_5$ be its basis functions, as constructed in [38]. See Figure 3 for the Bézier–nets for ϕ_j, $1 \le j \le 5$. In Figure 3A and Figure 3B, only portions of the Bézier–nets of ϕ_1, ϕ_2 are displayed due to the symmetry of ϕ_1, ϕ_2: Here, ϕ_1 is symmetric about lines $x = y, y = -x$ and $x = 0$, $y = 0$, and ϕ_2 is symmetric about the x–axis and anti–symmetric about the y–axis. In Figure 3C, the Bézier coefficients that are not displayed are supposed to be zeros. Then $\Phi = [\phi_1, \cdots, \phi_5]^T$ is refinable with respect to the quincunx dilation matrix A given in (2.5). See Figures 4, 5 and 6 for the Bézier–nets for $\phi_1(A^{-1}\cdot)$, $\phi_2(A^{-1}\cdot)$ and $\phi_4(A^{-1}\cdot)$. Again due to the symmetric properties of ϕ_1, ϕ_2, only portions of the Bézier–nets of $\phi_1(A^{-1}\cdot)$ and $\phi_2(A^{-1}\cdot)$ are displayed in Figures 4 and 5, while in Figure 6, the Bézier coefficients that are not displayed are supposed to be zeros. This mask is given by

$$P_{0,-2} = \frac{1}{16}\begin{bmatrix} 0 & 0 & 0 & 0 & 0 \\ 0 & 0 & 0 & 2 & 0 \\ 0 & 0 & 0 & 2 & 0 \\ 0 & 0 & 0 & 1 & 0 \\ 0 & 0 & 0 & 1 & 0 \end{bmatrix}, \quad P_{-1,0} = \frac{1}{16}\begin{bmatrix} 4 & 12 & 0 & -24 & 0 \\ -1 & -3 & 1 & 2 & 2 \\ -1 & -3 & -1 & 10 & -2 \\ 0 & 0 & 0 & 0 & 0 \\ \frac{1}{6} & \frac{1}{2} & -\frac{1}{2} & 1 & -1 \end{bmatrix},$$

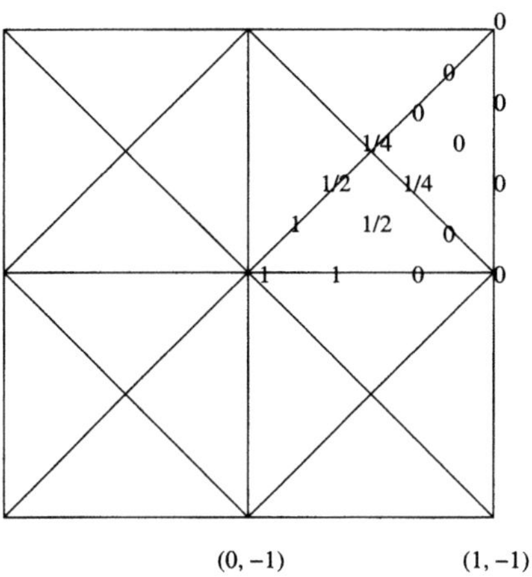

Fig.3A

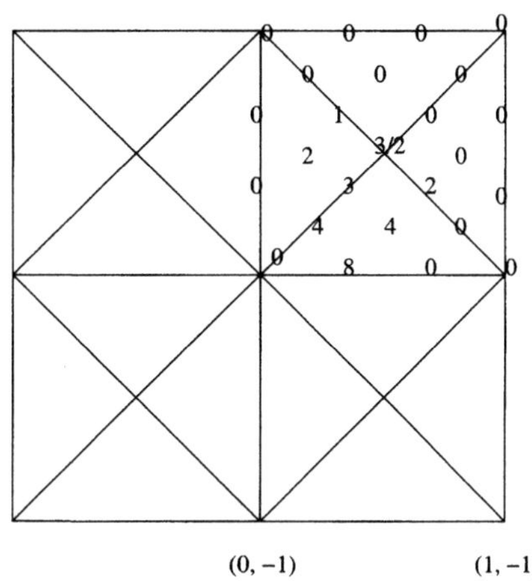

Fig.3B

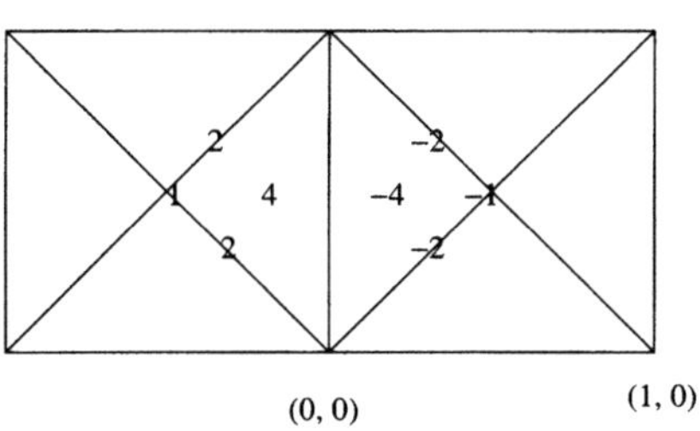

Fig.3C

Figure 3: Bézier–nets for ϕ_1, $24\phi_2$ and $96\phi_4$ with $\phi_3(x,y) = \phi_2(y,x)$, $\phi_5(x,y) = -\phi_4(-y,-x)$

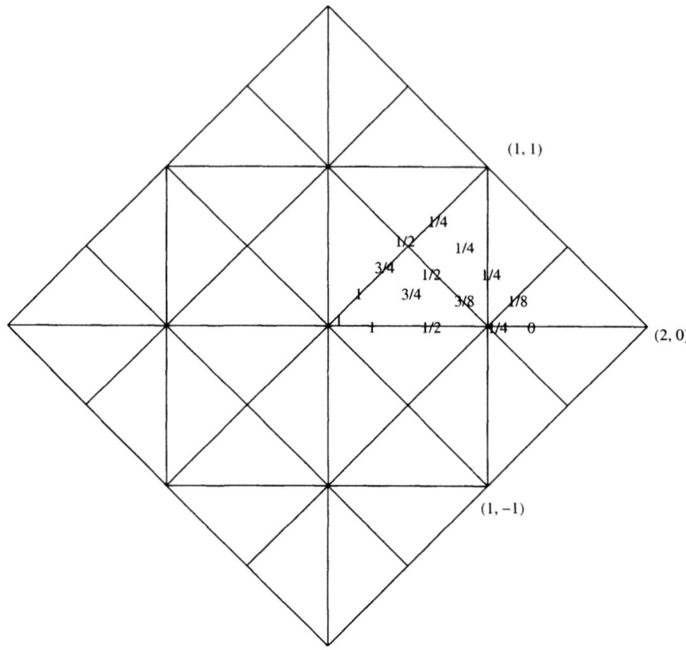

Figure 4: Bézier–nets for $\phi_1(A^{-1}\cdot)$

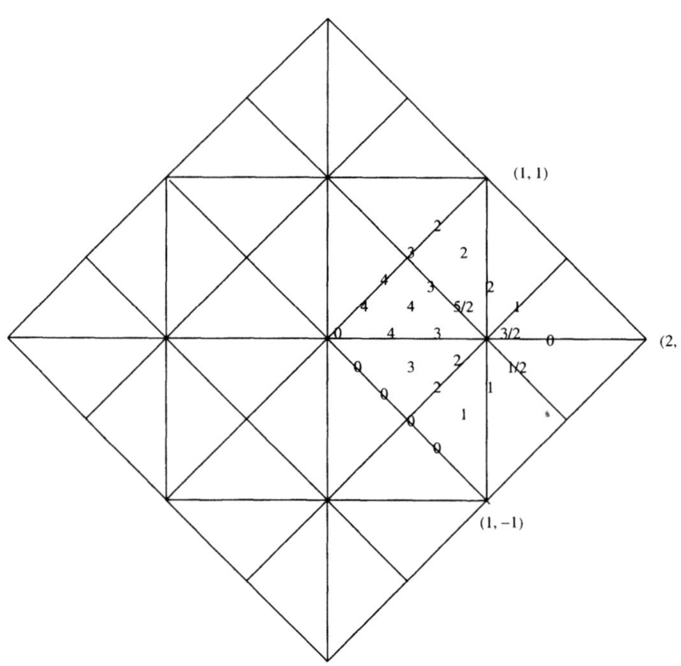

Figure 5: Bézier–nets for $24\phi_2(A^{-1}\cdot)$

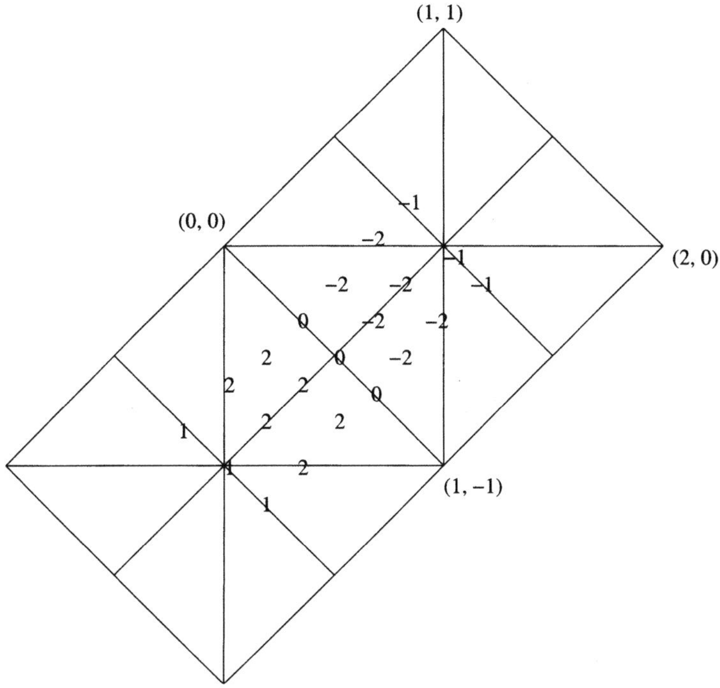

Figure 6: Bézier–nets for $96\phi_4(A^{-1}\cdot)$

$$P_{0,-1} = \frac{1}{16}\begin{bmatrix} 4 & 0 & 12 & 0 & -24 \\ -1 & 1 & -3 & 10 & 10 \\ 1 & 1 & 3 & 10 & -2 \\ \frac{1}{6} & \frac{1}{2} & \frac{1}{2} & -3 & 1 \\ -\frac{1}{6} & \frac{1}{2} & -\frac{1}{2} & -3 & 3 \end{bmatrix}, \quad P_{-1,-1} = \frac{1}{16}\begin{bmatrix} 0 & 0 & 0 & -24 & 0 \\ 0 & 0 & 0 & 10 & 0 \\ 0 & 0 & 0 & 2 & 0 \\ 0 & 0 & 0 & 0 & 0 \\ 0 & 0 & 0 & -3 & 0 \end{bmatrix},$$

$$P_{1,-1} = \frac{1}{16}\begin{bmatrix} 0 & 0 & 0 & 24 & -24 \\ 0 & 0 & 0 & 2 & 2 \\ 0 & 0 & 0 & 10 & -10 \\ 0 & 0 & 0 & -3 & -3 \\ 0 & 0 & 0 & 0 & -1 \end{bmatrix}, \quad P_{0,0} = \frac{1}{16}\begin{bmatrix} 16 & 0 & 0 & 0 & 0 \\ 0 & 8 & 8 & 10 & 10 \\ 0 & 8 & -8 & 10 & -10 \\ 0 & 0 & 0 & 0 & 0 \\ 0 & 0 & 0 & 0 & 3 \end{bmatrix},$$

$$P_{0,1} = \frac{1}{16}\begin{bmatrix} 4 & 0 & -12 & 0 & 24 \\ 1 & 1 & -3 & 2 & 2 \\ -1 & 1 & 3 & 2 & -10 \\ 0 & 0 & 0 & 0 & 0 \\ 0 & 0 & 0 & 0 & 0 \end{bmatrix}, \quad P_{1,1} = \frac{1}{16}\begin{bmatrix} 0 & 0 & 0 & 0 & 24 \\ 0 & 0 & 0 & 0 & 10 \\ 0 & 0 & 0 & 0 & -2 \\ 0 & 0 & 0 & 0 & 0 \\ 0 & 0 & 0 & 0 & 0 \end{bmatrix},$$

$$P_{1,0} = \frac{1}{16}\begin{bmatrix} 4 & -12 & 0 & 24 & 0 \\ 1 & -3 & 1 & 10 & 10 \\ 1 & -3 & -1 & 2 & -10 \\ -\frac{1}{6} & \frac{1}{2} & \frac{1}{2} & 1 & -3 \\ 0 & 0 & 0 & 0 & 0 \end{bmatrix}, \quad P_{2,0} = \frac{1}{16}\begin{bmatrix} 0 & 0 & 0 & 0 & 0 \\ 0 & 0 & 0 & 0 & 2 \\ 0 & 0 & 0 & 0 & -2 \\ 0 & 0 & 0 & 0 & 1 \\ 0 & 0 & 0 & 0 & 0 \end{bmatrix}$$

and $P_{\mathbf{k}} = \mathbf{0}$ for other indices $\mathbf{k}$.

Remark 2.1. *Observe that for each j, supp $\phi_j(\cdot - \mathbf{k})$ for $\mathbf{k} \in \mathbb{Z}^2$ with $|\mathbf{k}| = 1$ is not a subset of $\cup_{j=1}^{5}$supp $\phi_j(A^{-1}\cdot)$. However Φ is still refinable with a finite mask.*

Of course $\Phi = [\phi_1, \cdots, \phi_5]^T$ is also refinable with $2I_2$. In this case the two–scale symbol, denoted by $R(\mathbf{z})$, is given by $R(\mathbf{z}) = P(e^{-iA\omega})P(\mathbf{z})$, where P is the two–scale symbol for the quincunx dilation matrix A given above, and the nonzero matrix coefficients $R_{\mathbf{k}}$ of $R(\mathbf{z})$ are given by

$$R_{-1,-2} = \frac{1}{8}\begin{bmatrix} 0 & 0 & 0 & 6 & 0 \\ 0 & 0 & 0 & -1 & 0 \\ 0 & 0 & 0 & -2 & 0 \\ 0 & 0 & 0 & 0 & 0 \\ 0 & 0 & 0 & 0 & 0 \end{bmatrix}, R_{-1,-1} = \frac{1}{8}\begin{bmatrix} 2 & 3 & 3 & -6 & 6 \\ -\frac{1}{2} & -\frac{1}{2} & -1 & 1 & -2 \\ -\frac{1}{2} & -1 & -\frac{1}{2} & 2 & -1 \\ 0 & 0 & 0 & 0 & 0 \\ \frac{1}{12} & 0 & \frac{1}{4} & 0 & \frac{1}{2} \end{bmatrix}$$

$$R_{-1,0} = \frac{1}{8}\begin{bmatrix} 4 & 6 & 0 & -6 & 0 \\ -1 & -1 & 0 & 1 & 0 \\ 0 & 0 & 2 & -2 & 0 \\ 0 & 0 & 0 & \frac{1}{2} & 0 \\ 0 & 0 & -\frac{1}{2} & 0 & 1 \end{bmatrix}, R_{0,-1} = \frac{1}{8}\begin{bmatrix} 4 & 0 & 6 & 0 & -6 \\ 0 & 2 & 0 & 0 & 2 \\ -1 & 0 & -1 & 0 & 1 \\ 0 & 0 & 0 & 0 & 0 \\ 0 & 0 & 0 & 0 & \frac{1}{2} \end{bmatrix},$$

$$R_{-1,1} = \frac{1}{8}\begin{bmatrix} 2 & 3 & -3 & 6 & -6 \\ -\frac{1}{2} & -\frac{1}{2} & 1 & -1 & 2 \\ \frac{1}{2} & 1 & -\frac{1}{2} & 2 & -1 \\ \frac{1}{12} & \frac{1}{4} & 0 & \frac{1}{2} & 0 \\ -\frac{1}{12} & 0 & \frac{1}{4} & 0 & \frac{1}{2} \end{bmatrix}, R_{0,1} = \frac{1}{8}\begin{bmatrix} 4 & 0 & -6 & 0 & 6 \\ 0 & 2 & 0 & 0 & -2 \\ 1 & 0 & -1 & 0 & 1 \\ 0 & -\frac{1}{2} & 0 & 1 & 0 \\ 0 & 0 & 0 & 0 & \frac{1}{2} \end{bmatrix}$$

$$R_{1,-2} = \frac{1}{8}\begin{bmatrix} 0 & 0 & 0 & -6 & 0 \\ 0 & 0 & 0 & -1 & 0 \\ 0 & 0 & 0 & 2 & 0 \\ 0 & 0 & 0 & 0 & 0 \\ 0 & 0 & 0 & 0 & 0 \end{bmatrix}, R_{1,-1} = \frac{1}{8}\begin{bmatrix} 2 & -3 & 3 & 6 & -6 \\ \frac{1}{2} & -\frac{1}{2} & 1 & 1 & -2 \\ -\frac{1}{2} & 1 & -\frac{1}{2} & -2 & 1 \\ 0 & 0 & 0 & 0 & 0 \\ 0 & 0 & 0 & 0 & 0 \end{bmatrix},$$

$$R_{1,0} = \frac{1}{8}\begin{bmatrix} 4 & -6 & 0 & 6 & 0 \\ 1 & -1 & 0 & 1 & 0 \\ 0 & 0 & 2 & 2 & 0 \\ 0 & 0 & 0 & \frac{1}{2} & 0 \\ 0 & 0 & 0 & 0 & 0 \end{bmatrix}, R_{1,1} = \frac{1}{8}\begin{bmatrix} 2 & -3 & -3 & -6 & 6 \\ \frac{1}{2} & -\frac{1}{2} & -1 & -1 & 2 \\ \frac{1}{2} & -1 & -\frac{1}{2} & -2 & 1 \\ -\frac{1}{12} & \frac{1}{4} & 0 & \frac{1}{2} & 0 \\ 0 & 0 & 0 & 0 & 0 \end{bmatrix},$$

$$R_{2,-1} = \frac{1}{8}\begin{bmatrix} 0 & 0 & 0 & 0 & 6 \\ 0 & 0 & 0 & 0 & 2 \\ 0 & 0 & 0 & 0 & -1 \\ 0 & 0 & 0 & 0 & 0 \\ 0 & 0 & 0 & 0 & 0 \end{bmatrix}, R_{2,1} = \frac{1}{8}\begin{bmatrix} 0 & 0 & 0 & 0 & -6 \\ 0 & 0 & 0 & 0 & -2 \\ 0 & 0 & 0 & 0 & -1 \\ 0 & 0 & 0 & 0 & 0 \\ 0 & 0 & 0 & 0 & 0 \end{bmatrix},$$

$$R_{0,0} = \mathrm{diag}(1, \tfrac{1}{2}, \tfrac{1}{2}, \tfrac{1}{8}, \tfrac{1}{8}).$$

One can also verify that both P and R satisfy the sum rule of order 4 with the same vectors

$$\begin{aligned} &\mathbf{y}_{0,0} = [1,0,0,0,0], \quad \mathbf{y}_{1,0} = [0,1,0,0,0], \quad \mathbf{y}_{0,1} = [0,0,1,0,0], \\ &\mathbf{y}_{2,0} = \mathbf{y}_{1,1} = \mathbf{y}_{2,0} = \mathbf{y}_{3,0} = \mathbf{y}_{0,3} = [0,0,0,0,0], \\ &\mathbf{y}_{2,1} = [0,0,0,0,2], \quad \mathbf{y}_{1,2} = [0,0,0,2,0]. \end{aligned}$$

However, from the structure of $\mathbf{y}_\alpha$, we know that Φ does not have a balanced bi–orthogonal dual. □

Acknowledgement. The research of the first author was supported by NSF Grants #CCR–9988289 and #CCR–0098331 and ARO Grant # DAAD 19–00–1–0512.

References

[1] K. Attakitmongcol, D. P. Hardin, D. M. Wilkes: *Multiwavelet prefilters II: Optimal orthogonal prefilters*, IEEE Tran. Image Proc., to appear.

[2] C. de Boor, R. DeVore, A. Ron: *Approximation orders of FSI spaces in* $L_2(\mathbb{R}^d)$, Constr. Approx. **14** (1998), 631–652.

[3] C. Cabrelli, C. Heil, U. Molter: *Accuracy of lattice translates of several multidimensional refinable functions*, J. Approx. Theory **95** (1998), 5–52.

[4] C. Cabrelli, C. Heil, U. Molter: *Accuracy of several multidimensional refinable distributions*, J. Fourier Anal. Appl. **6** (2000), 483–502.

[5] D. R. Chen, R. Q. Jia, S. D. Riemenschneider: *Convergence of vector subdivision schemes in Sobolev spaces*, Appl. Comput. Harmonic Anal. **12** (2002), 128–149.

[6] C. K. Chui: *Multivariate Splines*, NSF–CBMS Series #54, SIAM, Philadelphia, 1988.

[7] C. K. Chui, Q. T. Jiang: *Characterization and construction of balanced multi–wavelets in* $\mathbb{R}^s$, manuscript.

[8] C. K. Chui, J. Lian: *A study of orthonormal multi–wavelets*, J. Appl. Numer. Math. **20** (1996), 273–298.

[9] A. Cohen, I. Daubechies, G. Plonka: *Regularity of refinable function vectors*, J. Fourier Anal. Appl. **3** (1997), 295–324.

[10] G. Donovan, J. Geronimo, D. P. Hardin, P. Massopust: *Construction of orthogonal wavelets using fractal interpolation functions*, SIAM J. Math. Anal. **27** (1996), 1158–1192.

[11] G. C. Donovan, J. S. Geronimo, D. P. Hardin: *Orthogonal polynomials and the construction of piecewise polynomial smooth wavelets*, SIAM J. Math. Anal. **30** (1999), 1029–1056.

[12] J. Geronimo, D. P. Hardin, P. Massopust: *Fractal functions and wavelet expansions based on several scaling functions*, J. Approx. Theory **78** (1994), 373–401.

[13] T. N. T. Goodman: *Pairs of refinable bivariate splines*, In: *Advanced Topics in Multivariate Approximation*, F. Fontanella, K. Jetter and L.L. Schumaker (eds.), World Sci. Publ. Co., Singapore, 1996.

[14] T. N. T. Goodman, S. L. Lee: *Convergence of cascade algorithms*, In: *Mathematical Methods for Curves and Surfaces II*, 191–212, M. Dæhlen, T. Lyche, and L.L. Schumaker (eds.), Vanderbilt University Press, Nashville, 1998.

[15] B. Han, R. Q. Jia: *Multivariate refinement equations and convergence of subdivision schemes*, SIAM J. Math. Anal. **29** (1998), 1177–1999.

[16] D. P. Hardin, D. W. Roach: *Multiwavelet prefilters I: Orthogonal prefilters preserving approximation order* $p \leq 2$, IEEE Tran. Circuits and System–II **45** (1998), 1119–1125.

[17] C. Heil, D. Colella: *Matrix refinement equations: existence and uniqueness*, J. Fourier Anal. Appl. **2** (1996), 363–377.

[18] C. Heil, G. Strang, V. Strela: *Approximation by translates of refinable functions*, Numer. Math. **73** (1996), 75–94.

[19] L. R. Iyer, A. E. Bell: *Improving image compression performance with balanced multiwavelets*, in *Proceedings of the 35th Asilomar Conference on Signals*, Systems and Computers (Asilomar 2001), 773–777, Pacific Grove, CA, 2001.

[20] R. Q. Jia: *Subdivision schemes in* L_p *spaces*, Adv. Comp. Math. **3** (1995), 309–341.

[21] R. Q. Jia: *Approximation properties of multivariate wavelets*, Math. Comp. **67** (1998), 647–665.

[22] R. Q. Jia: *Characterization of smoothness of multivariate refinable functions in Sobolev spaces*, Trans. Amer. Math. Soc. **351** (1999), 4089–4112.

[23] R. Q. Jia: *Convergence of vector subdivision schemes and construction of biorthogonal multiple wavelets*, In: *Advances in Wavelets*, Ka–Sing Lau (ed.), 199–227, Springer, Singapore, 1999.

[24] R. Q. Jia: *Cascade algorithms in wavelet analysis*, preprint, Univ. of Alberta, 2002.

[25] R. Q. Jia, Q. T. Jiang: *Approximation power of refinable vectors of functions*, In: *Wavelet analysis and applications*, Studies Adv. Math. #25, 155–178, Amer. Math. Soc., Providence, RI, 2002.

[26] R. Q. Jia, Q. T. Jiang: *Spectral analysis of transition operators and its applications to smoothness analysis of wavelets*, SIAM J. Matrix Anal. Appl., to appear.

[27] R. Q. Jia, K. S. Lau, D. X. Zhou: L_p *solutions of refinement equations*, J. Fourier Anal. Appl. **7**(2001), 143–167.

[28] R. Q. Jia, S. D. Riemenschneider, D. X. Zhou: *Vector subdivision schemes and multiple wavelets*, Math. Comp. **67** (1998), 1533–1563.

[29] R. Q. Jia, S. D. Riemenschneider, D. X. Zhou: *Smoothness of multiple refinable functions and multiple wavelets*, SIAM J. Matrix Anal. Appl. **21** (1999), 1–28.

[30] R. Q. Jia, Z. W. Shen: *Multiresolution and wavelets*, Proc. Edinburgh Math. Soc. **37** (1994), 271–300.

[31] R. Q. Jia, S. R. Zhang: *Spectral properties of the transition operator associated to a multivariate refinement equation*, Linear Algebra Appl. **292** (1999), 155–178.

[32] Q. T. Jiang: *Biorthogonal multiwavelets*, Tech. Report, National University of Singapore, 1997.

[33] Q. T. Jiang: *Orthogonal multiwavelets with optimum time–frequency resolution*, IEEE Trans. Signal Proc., **46** (1998), 830–844.

[34] Q. T. Jiang: *Multivariate matrix refinable functions with arbitrary matrix dilation*, Trans. Amer. Math. Soc. **351** (1999), 2407–2438.

[35] Q. T. Jiang: *Parameterization of symmetric orthogonal multifilter banks with different filter length*, Linear Algebra Appl. **311** (2000), 79–96.

[36] Q.T. Jiang, P. Oswald: *On the analysis of* $\sqrt{3}$*–subdivision*, J. Comp. Appl. Math., to appear.

[37] B. R. Johnson: *Multiwavelet moments and projection prefilters*, IEEE Trans. Signal Proc. **48** (2000), 3100–3108.

[38] M. J. Lai: *Approximation order from bivariate* C^1 *cubics on a four–directional mesh is full*, Computer Aided Geometric Design **11** (1994), 215–223.

[39] W. Lawton, S. L. Lee, Z. W. Shen: *Convergence of multidimensional cascade algorithm*, Numer. Math. **78** (1998), 427–438.

[40] J. Lebrun, M. Vetterli: *Balanced multiwavelets: Theory and design*, IEEE Trans. Signal Processing **46** (1998), 1119–1125.

[41] J. Lebrun, M. Vetterli: *High order balanced multiwavelets*, In: *Proc. IEEE Int. Conf. Acoust., Speech, Signal Processing* (ICASSP), Seattle, 1998.

[42] R. L. Long, W. Chen, S. L. Yuan: *Wavelets generated by vector multiresolution analysis*, Appl. Comput. Harmonic Anal. **4** (1997), 317–350.

[43] M.B. Martin, A. E. Bell: *New image compression techniques Using multiwavelets and multiwavelet packets*, IEEE Trans. Image Proc. **10** (2001), 500–511.

[44] C. A. Micchelli, T. Sauer: *Regularity of multiwavelets*, Adv. Comp. Math. **7** (1997), 455–545.

[45] C. A. Micchelli, T. Sauer: *On vector subdivision*, Math. Z. **229** (1998), 621–674.

[46] G. Plonka: *Approximation order provided by refinable function vectors*, Constr. Approx. **13** (1997), 221–244.

[47] A. Ron, Z. W. Shen: *The Sobolev regularity of refinable functions*, J. Approx. Theory **106** (2000), 185–225.

[48] I. W. Selesnick: *Multiwavelets with extra approximation properties*, IEEE Trans. Signal Proc. **46** (1998), 2898–2909.

[49] I. W. Selesnick: *Balanced multiwavelet bases based on symmetric FIR filters*, IEEE Trans. Signal Proc. **48** (2000), 184–191.

[50] Z. W. Shen: *Refinable function vectors* , SIAM J. Math. Anal. **29** (1998), 235–250.

[51] G. Strang: *Eigenvalues of* $(\downarrow 2)H$ *and convergence of cascade algorithm*, IEEE Trans. Signal Proc. **44** (1996), 233–238.

[52] V. Strela, P. Heller, G. Strang, P. Topiwala, C. Heil: *The application of multiwavelet filter banks to image processing*, IEEE Trans. Image Proc. **8** (1999), 548–563.

[53] V. Strela, A. Walden: *Signal and image denoising via wavelet thresholding: orthogonal and biorthogonal, scalar and multiple wavelet transforms*, In: *Nonlinear and nonstationary signal processing*, 395–441, Cambridge Univ. Press, Cambridge, 2000.

[54] J. Y. Tham, L. X. Shen, S. L. Lee, H. H. Tan: *A general approach for analysis and application of discrete multiwavelet transforms*, IEEE Trans. Signal Proc. **48** (2000), 457–464.

[55] C. Weidmann, J. Lebrun, M. Vetterli: *Significance tree image coding using balanced multiwavelets*, In: *Proceedings of IEEE ICIP*, Vol. **1**, 97–101, 1998.

[56] T. Xia, Q. T. Jiang: *Optimal multifilter banks: Design, related symmetric extension transform and application to image compression*, IEEE Trans. Signal Proc. **47** (1999), 1878–1889.

[57] X. G. Xia: *A new prefilter design for discrete multiwavelet transform*, IEEE Trans. Signal Proc. **46** (1998), 1558–1570.

[58] X. G. Xia, D. P. Hardin, J. S. Geronimo, B. Suter: *Design of prefilters for discrete multiwavelet transforms*, IEEE Trans. Signal Proc. **44** (1996), 25–35.

Addresses:

Charles K. Chui
University of Missouri–St. Louis
Department of Mathematics and Computer Science
St. Louis, MO 63121
USA
and
Stanford University
Department of Statistics
Stanford, CA 94305
USA

Qingtang Jiang
University of Missouri–St. Louis
Department of Mathematics and Computer Science
St. Louis, MO 63121
USA

International Series of Numerical Mathematics
Vol. 145, ©2003 Birkhäuser Verlag Basel/Switzerland

Transfinite Interpolation by Blending Functions, Best One–Sided L^1–Approximation, and Cubature Formulae

Dimiter P. Dryanov

Dedicated to Manfred Reimer on the occasion of his 70th birthday

Abstract

The main purpose of this paper is to present results on blending–type interpolation, best multivariate L^1–approximation, and cubature formulae for approximate integration of multiple integrals. Since the classical concept of the Haar–Chebyshev condition is essentially restricted to the univariate case we try to give a new treatment of the notion for a Haar–Chebyshev system in the multivariate case.

1 Blending Functions of Order (m, n)

The classical univariate algebraic polynomials have a natural multivariate extension by the so called *blending functions.* Let $I^2 := [-1, 1]^2$, and

$$C^{m,n}(I^2) := \left\{ f : I^2 \to \mathbb{R} \ : \ D^{i,j} f \in C(I^2), \ \ 0 \le i \le m, \ 0 \le j \le n \right\}$$

denote the class of all real valued functions having continuous derivatives up to order (m, n). The real vector space $B^{m,n}(I^2)$ of all blending functions of order (m, n) on the unit square I^2 is defined as follows:

$$B^{m,n}(I^2) := \left\{ h \in C^{m,n}(I^2) \ : \ D^{m,n} h := \frac{\partial^{m+n}}{\partial x^m \partial y^n} h = 0 \right\} .$$

The vector space $B^{m,n} := B^{m,n}(I^2)$ is *infinite–dimensional* in contrast to the fact that the linear space of all univariate algebraic polynomials of degree less than or equal to $m - 1$

$$\pi_{m-1} := \left\{ p \in C^m(I) \ : \ \frac{d^m}{dx^m} p = 0 \right\}$$

is of finite dimension m.

Any blending function (or we may say *extended polynomial of order* (m,n)) $h \in B^{m,n}$ can be represented in the form

$$h(x,y) = \sum_{k=0}^{m-1} a_k(y)\, x^k + \sum_{l=0}^{n-1} b_l(x)\, y^l,\ a_k \in C^n\,(I)\,,\ b_l \in C^m\,(I)\ ,$$

and conversely, each function of this form is a $B^{m,n}$–blending function, so here is another characterization of the real vector space $B^{m,n}$:

$$B^{m,n} = \left\{ h \in C^{m,n}\left(I^2\right)\ :\ h(x,y) = \sum_{k=0}^{m-1} a_k(y) x^k + \sum_{l=0}^{n-1} b_l(x) y^l \right\},$$

where $a_k \in C^n(I)$, $b_l \in C^n(I)$, $0 \le i \le m$, $0 \le j \le n$. Note, that *the above representation is not unique.* Moreover, to see the difference between *the finite dimensional univariate polynomial space of order* m and *the infinite dimensional vector space* $B^{m,n}$ we consider for a given continuous function $f \in C\,(\,I^2\,)$ the following extremal problem :

$$\inf_{h \in B^{m,n}} \|\, f \,-\, h \,\|_{L^2(I^2)}.$$

The unique blending function $h_2^f \in B^{m,n}$ of minimal L^2–deviation is explicitly given by

$$\begin{aligned} h_2^f(x,y) \;&=\; \sum_{r=0}^{m-1} \left(f(\cdot,y), L_r\right) L_r(x) + \sum_{s=0}^{n-1} \left(f(x,\cdot), L_s\right) L_s(y) \\ &\quad - \sum_{r=0}^{m-1} \sum_{s=0}^{n-1} \left(f(\cdot,\cdot), L_r\, L_s\right) L_r(x)\, L_s(y)\,, \end{aligned}$$

where L_j denotes the Legendre polynomial of degree j normalized such that the scalar product $(L_j, L_j) = 1$. The proof in the L^2 case follows from the fact that the best approximant to f must be an orthogonal projection of f on the approximating subspace, see [25], [27]. However, even the existence of a blending function of minimal L^p–deviation $(1 \le p \le \infty\,, p \ne 2)$ is *an open problem and not an easy task*, because the approximating space is infinite dimensional. We like to mention also that in contrast to the univariate polynomial theory, the following result holds [18], [29]: *For a function* $f \in C^{1,1}\,(I^2)$ *satisfying* $D^{1,1}f(x,y) \ge 0$ *for* $(x,y) \in I^2$, *the best uniform approximant to* f *from* $B^{1,1}\,(I^2)$ *is never unique if* $D^{1,1}f > 0$ *at some point in* I^2.

2 Transfinite Interpolation by Blending Functions

Transfinite Interpolation

Interpolation schemes that yield interpolants which *match (interpolate) a given function on a nondenumerable set of points* are said to be transfinite [4], [5], [21]. Such schemes require the ability to construct interpolants that pass through a specified network of curves, where more classical interpolation schemes are restricted to a finite or denumerable number of points of interpolation. The classical Dirichlet problem is an example of a transfinite interpolation scheme. Transfinite interpolation schemes based on blending functions have their application to various engineering problems related to finite difference and finite element analysis and computer aided geometric design.

Lagrange Transfinite Interpolation Scheme

Approximate recovery of a univariate function can be obtained by the univariate Lagrange interpolation polynomial: Let $f \in C[-1,1]$ and let $\{x_1, \ldots, x_m\}$ be distinct points in $[-1,1]$. Then there exists a uniquely determined polynomial p^* of degree not exceeding $m-1$ such that

$$p^*(x_k) = f(x_k) \quad (1 \le k \le m).$$

Being defined as the kernel of the differential operator $D^{m,n}$, the blending functions share some properties of the algebraic polynomials. In particular, the Lagrange interpolation by algebraic polynomials has an interesting extension to interpolation by blending functions.

Blending Grid

Let $\omega_{\mathbf{x_m},\mathbf{y_n}}(x,y) := \prod_{s=1}^{m}(x-x_s)\prod_{k=1}^{n}(y-y_k)$. We define the blending grid associated with the nodes $\mathbf{x_m} := \{x_s,\, 1 \le s \le m\}$ and $\mathbf{y_n} := \{\, y_k,\, 1 \le k \le n\,\}$ by

$$\mathbf{G}_{\mathbf{x_m},\mathbf{y_n}} = \{(x,y) \in [-1,1]^2 \,:\, \omega_{\mathbf{x_m},\mathbf{y_n}}(x,y) = 0\,\}.$$

Transfinite Lagrange Interpolation by Blending Functions

Let $f \in C(I^2)$. Let m, n be natural numbers and let $\{x_1, \ldots, x_m\}$ and $\{y_1, \ldots, y_n\}$ be sets of (in each case) distinct points in $[-1,1]$. Then there exists a uniquely determined blending function h^f from $B^{m,n}$, satisfying the following *interpolation (boundary value) conditions*:

$$h^f_{|\mathbf{G}_{\mathbf{x_m},\mathbf{y_n}}} = f_{|\mathbf{G}_{\mathbf{x_m},\mathbf{y_n}}} .$$

An explicit form of the transfinite Lagrange interpolant h^f and a Cauchy error representation are given by the next two formulas:

$$\begin{aligned} h^f(x,y) &= \sum_{s=1}^{m} f(x_s, y)\, l_{s,m}(x) + \sum_{k=1}^{n} f(x, y_k)\, l_{k,n}(y) \\ &\quad - \sum_{s=1}^{m} \sum_{k=1}^{n} f(x_s, y_k)\, l_{s,m}(x)\; l_{k,n}(y) \end{aligned}$$

and

$$f(x,y) - h^f(x,y) = \frac{D^{m,n} f(\xi,\eta)}{m!\,n!}\, \omega_{\mathbf{x_m},\mathbf{y_n}}(x,y),$$

where $(\xi,\eta) \in I^2$.

Note that $h_g \equiv g$ for each $g \in B^{m,n}$ and therefore *the Lagrange blending interpolant from* $B^{m,n}$ *with respect to any* (m,n) *blending grid exists and is unique.*

Transfinite Hermite Interpolation by Blending Functions

Let m, n be natural numbers and let

$$\omega_{\mathbf{x_m},\mathbf{y_n}}(x,y) := \prod_{\mu=1}^{m} (x - x_\mu) \prod_{\nu=1}^{n} (y - y_\nu),$$

where the nodes $\mathbf{x_m} := \{x_\mu,\, 1 \le \mu \le m\}$ and $\mathbf{y_n} := \{y_\nu,\, 1 \le k \le n\}$ are sets of (in each case) distinct points in $[-1,1]$. Let

$$\mathbf{G}_{\mathbf{x_m},\mathbf{y_n}} = \{(x,y) \in [-1,1]^2 : \omega_{\mathbf{x_m},\mathbf{y_n}}(x,y) = 0\}.$$

be the blending grid, associated with the nodes $\mathbf{x_m} := \{x_\mu,\, 1 \le \mu \le m\}$ and $\mathbf{y_n} := \{y_\nu,\, 1 \le \nu \le n\}$. Let us associate with each node x_μ a multiplicity s_μ and with each node y_ν a multiplicity j_ν and let $\sum_{\mu=1}^{m} s_\mu = M$

and $\sum_{\nu=1}^{n} j_\nu = N$. Then, for a given function $f \in C^{M,N}(I^2)$ there exists a unique Hermite blending interpolant H^f from $B^{M,N}$, satisfying the following *transfinite interpolation (boundary value) conditions*

$$D^{s,0}\, H^f(x_\mu, y) = D^{s,0}\, f(x_\mu, y),$$
$$\mu = 1, \ldots, m;\ s = 0, \ldots, s_\mu - 1;\ y \in [-1,1]$$

and

$$D^{0,j}\, H^f(x, y_\nu) = D^{0,j}\, f(x, y_\nu),$$
$$\nu = 1, \ldots, n;\ j = 0, \ldots, j_\nu - 1;\ x \in [-1,1]$$

with respect to the given blending grid $\mathbf{G}_{\mathbf{x_m},\mathbf{y_n}}$.

An explicit form of the transfinite Hermite interpolant H^f to f from $B^{m,n}$ is given by the following expression:

$$\begin{aligned} H^f(x,y) &= \sum_{\mu=1}^{m} \sum_{s=0}^{s_\mu-1} D^{s,0}\, f(x_\mu, y)\, l_{\mu,M}^{s}(x) \\ &\quad + \sum_{\nu=1}^{n} \sum_{j=0}^{j_\nu-1} D^{0,j}\, f(x, y_\nu)\, l_{\nu,N}^{j}(y) \\ &\quad - \sum_{\mu=1}^{m} \sum_{s=0}^{s_\mu-1} \sum_{\nu=1}^{n} \sum_{j=0}^{j_\nu-1} D^{s,j}\, f(x_\mu, y_\nu) l_{\mu,M}^{s}(x)\, l_{\nu,N}^{j}(y), \end{aligned}$$

where the fundamental Hermite polynomials $l_{\mu,M}^{s}(x)$, resp. $l_{\nu,N}^{j}(y)$ satisfy

$$D^{r,0}\, l_{\mu,M}^{s}(x_\kappa) = \delta_{\mu,\kappa}\, \delta_{s,r},\ \kappa = 1, \ldots, m;\ r = 0, \ldots, s_\mu - 1$$

and

$$D^{0,r}\, l_{\nu,M}^{j}(y_\kappa) = \delta_{\nu,\kappa}\, \delta_{j,r},\ \kappa = 1, \ldots, n;\ r = 0, \ldots, j_\nu - 1\,.$$

Error Representation in Terms of Normalized B–Splines

Let

$$\Omega_{\mathbf{x_M},\mathbf{y_N}}(x,y) := \prod_{\mu=1}^{m} (x - x_\mu)^{s_\mu} \prod_{\nu=1}^{n} (y - y_\nu)^{j_\nu}\,.$$

Then

$$\begin{aligned} f(x,y) - H^f(x,y) &= \Omega_{\mathbf{x_M},\mathbf{y_N}}(x,y) \\ \times \int_{-1}^{1}\!\int_{-1}^{1} & B_{M-1;(\mathbf{x_m},\mathbf{s_m}),x}(\xi)\, B_{N-1;(\mathbf{y_n},\mathbf{j_n}),y}(\eta)\ D^{M,N} f(\xi,\eta)\, d\xi\, d\eta \end{aligned}$$

where B_{M-1} is the normalized B–spline of degree $M-1$ with knots $(\mathbf{x_m}, x)$ with multiplicities $(\mathbf{s_m}, 1)$. For details about B–splines see [6], [8]. As a simple corollary from the above formula we obtain the following Cauchy type error representation for the blending Hermite interpolant :

$$f(x,y) - H^f(x,y) = \frac{D^{M,N} f(\xi,\eta)}{M!\,N!}\,\Omega_{\mathbf{x_M},\mathbf{y_N}}(x,y), \quad where\ (\xi,\eta) \in I^2 .$$

Note that $H^g(x,y) \equiv g(x,y)$ for each $g \in B^{M,N}$ and therefore any blending function from $B^{M,N}$ can be uniquely represented by its Hermite blending interpolant.

3 Newton–Cotes Type Cubature Formulae

Compound Trapezoidal Blending Cubature Formula

We construct compound trapezoidal cubature formula based on transfinite Lagrange interpolation by blending functions. Let $x_\mu = y_\mu = -1 + 2\mu/m$, $\mu = 0, 1, \ldots, m$ and let $\mathbf{G}_{\mathbf{x_{m+1}},\mathbf{y_{m+1}}}$ be the blending grid based on these points. Then (see [7]), for $f \in C^{2,2}(I^2)$, the following cubature holds :

$$\begin{aligned}\int_{I^2} f &= \left[\frac{2}{m}\int_{\mathbf{G}_{\mathbf{x_{m+1}},\mathbf{y_{n+1}}}} f - \frac{1}{m}\int_{\partial I^2} f\right] \\ &\quad - \frac{2}{m^2}\sum_{\mu=1}^{m-1}\left(f(-1,y_\mu) + f(1,y_\mu) + f(x_\mu,-1) + f(x_\mu,1)\right) \\ &\quad + \frac{4}{m^2}\sum_{\mu=1}^{m-1}\sum_{\nu=1}^{m-1} f(x_\mu,y_\nu) + \frac{1}{m^2}\left(f(-1,-1) + f(-1,1)\right. \\ &\quad \left. + f(1,-1) + f(1,1)\right) + \frac{4}{9\,m^4} D^{2,2} f(\xi,\eta) \quad with\ (\xi,\eta) \in I^2 .\end{aligned}$$

The cubature obtained is exact in $B^{2,2}(I^2)$. It has order of convergence $O\left(m^{-4}\right)$, $m \to \infty$, and yields approximate recovery of a bivariate integral by linear combinations of (*lower dimensional*) univariate integrals and functional values. Similar cubature holds in d–dimensional case ($d \geq 2$). It approximates a d–dimensional integral over I^d by finite linear combination of lower dimensional integrals on hyperplanes.

Compound Rectangular Blending Cubature Formula

Let $x_\mu = y_\mu = -1 + (2\mu + 1)/m,\ \mu = 0, \ldots, m-1$. Then,

$$\int_{I^2} f = \frac{2}{m} \int_{\mathbf{G}_{\mathbf{x_m},\mathbf{y_n}}} f - \frac{4}{m^2} \sum_{\mu=0}^{m-1} \sum_{\nu=0}^{m-1} f(x_\mu, y_\nu) + \frac{D^{2,2} f(\xi, \eta)}{9\, m^4}$$

with a suitable $(\xi, \eta) \in I^2$. For details we refer to [7], [23].

Cubature Formulae with Polyharmonic Degree of Precision

By a method which does not require any prior knowledge for the corresponding interpolation (that even may not exist) and on the basis of an appropriate multivariate extension of Bernoulli's functions we obtain a multivariate cubature of Euler–Maclaurin type that is exact in an infinite dimensional space of polyharmonic functions of a given order (see [15] and [16]).

4 Best L^1–Approximation

Let $(X, \|\cdot\|)$ be a normed vector space, let V be a subspace of X. Then $v^* \in V$ is called a best approximation to $x \in X$ from V if $\|x - v^*\| \le \|x - v\|$ for all $v \in V$.

Univariate Polynomial Case

One of the most exciting results in univariate Approximation Theory is a Theorem of A. Markov which characterizes the best L^1–approximant to some given function in terms of interpolation conditions with respect to a point set that consists of the zeros of the Chebyshev polynomial of the second kind $U_m(x) := (\sin(m+1) \arccos x) / \sqrt{1 - x^2}$ of degree m. First we state a result of S. Bernstein [11], [24].

Theorem A. *Let $f \in C^m[-1,1]$ and $f^{(m)} \ge 0$. The polynomial p^* of degree not exceeding $m-1$ is the unique best L^1–approximant from π_{m-1} to f on $[-1,1]$ if and only if*

$$p^*(\xi_k) = f(\xi_k) \qquad (1 \le k \le m),$$

where $\xi_k = \cos k\pi/(m+1)$ $(1 \le k \le m)$ are the zeros of the Chebyshev polynomial of the second kind of degree m.

The points ξ_k $(k = 1, \ldots, m)$ are called *canonical points.* Their collection is called *canonical point set*. They do not depend on the particular function f

under consideration provided it satisfies $f^{(m)} \geq 0$. *If the canonical point set exists, then the non-linear problem of best L^1-approximation becomes a linear one.* It is clear that a canonical point set may change with respect to the approximating space.

Corollary B. $2^{-m}U_m$ *is the unique monic polynomial of degree m which has minimal L^1-norm on $[-1,1]$.*

The Haar–Chebyshev Condition in the Univariate Case

A system of $m+1$ functions $\{f_0, f_1, \ldots, f_m\} \subset C[-1,1]$ is a Haar–Chebyshev system of order $m+1$ on $[-1,1]$ if each nontrivial linear combination of these functions possesses at most m zeros in $[-1,1]$. In other words, a system of $m+1$ functions is a Haar–Chebyshev system if it is *linearly independent with respect to an arbitrarily chosen set of $m+1$ distinct points.*

Here is a famous result of A. Markov [1], [11].

Theorem C (A. Markov). *Let $f \in C[-1,1]$ be such that the set*

$$\{1, x, x^2, \ldots, x^{m-1}, f\}$$

satisfies the Haar–Chebyshev condition of order $m+1$ on $[-1,1]$. Then the unique Lagrange interpolant p^* to f from π_{m-1} with respect to the zeros ξ_k, $k = 1, \ldots, m$ of the Chebyshev polynomial of the second kind of degree m is the unique best L^1-approximant to f from π_{m-1}.

Remark. In particular, if $f \in C^m(I)$ satisfies $D^m f > 0$, then $\{1, x, x^2, \ldots, x^{m-1}, f\}$ possesses the Haar–Chebyshev property. The Haar–Chebyshev property implies that any set of m distinct points in I is a maximal set of zeros of $f - p_{m-1}$, where p_{m-1} is a polynomial of degree at most $m-1$.

A Haar–Chebyshev Property in the Multivariate Case

The concept of a univariate Haar–Chebyshev system is based on counting zeros and it is essentially restricted to the one–dimensional case. It is known [9], [26] that there is no *universal* Haar–Chebyshev system in the multivariate case. However, on the basis of the Cauchy representation of the error for the Lagrange–Hermite blending interpolant, we can extend the Haar–Chebyshev concept *in a local way, i.e., with respect to some properly chosen set of grids*: Let f satisfy $D^{m,n} f > 0$ on I^2, and let h be an arbitrary blending function in $B^{m,n}(I^2)$. Then any (m,n) blending grid G is a maximal set of zeros for the function $f + h$, *i.e.*, if one adds one single point $(x_0, y_0) \in I^2 \setminus G$ to G, then $f + h$ cannot vanish on $G \cup (x_0, y_0)$, cf. the above Remark.

Best L^1–Approximation by Blending Functions

The above Haar–Chebyshev concept in the multivariate case is essential for the proof of the following theorem.

Theorem D (Haussmann, Zeller [23]). *Let $f \in C^{m,n}(I^2)$ satisfy $D^{m,n} f \geq 0$. Then f possesses a unique best L^1–approximant from $B^{m,n}$ and it is given by the unique Lagrange blending interpolant h^* with respect to the blending grid $\mathbf{G}(\xi_1, \ldots, \xi_m; \eta_1, \ldots, \eta_n)$, where the ξ_s's and the η_k's are the zeros of the Chebyshev polynomials of second kind U_m and U_n, respectively.*

The proof is based on the Cauchy error representation and the result holds in the d–dimensional case ($d \geq 2$) correspondingly. An immediate consequence of the above Theorem is the fact that the unique monic polynomial from the class

$$\pi_{m,n} := \left\{ \sum_{i=0}^{m} \sum_{j=0}^{n} a_{i,j}\, x^i\, y^j \ : \ a_{i,j} \in \mathbb{R}, \ a_{m,n} = 1 \right\}$$

of minimal L^1–deviation on I^2 is $2^{-m-n}\, U_m(x)\, U_n(y)$. Moreover,

$$\| U_m(x)\, U_n(y)/2^{m+n} \|_{L_1(I^2)} \leq \|x^m\, y^n - h(x,y)\|_{L_1(I^2)}, \quad h \in B^{m,n}(I^2).$$

Best L^1–Approximation by Harmonic Functions

Let $B(r)$ denote the open ball with center 0 and radius $r > 0$ in $\mathbb{R}^n$, and let $B := B(1)$ and $B_0 := B(2^{-1/n})$. Let $\mathcal{H}(\mathcal{B})$ and $\mathcal{S}(\mathcal{B})$ denote the sets of harmonic and subharmonic functions on B, respectively. Let f be continuous on the closed unit ball, i.e., $f \in C(\overline{B})$, and let $f \in \mathcal{S}(\mathcal{B})$. A function $h^* \in C(\bar{B}) \cap \mathcal{H}(B)$ which satisfies

$$\|f - h^*\|_{L_1} \leq \|f - h\|_{L_1}$$

for all $h \in C(\bar{B}) \cap \mathcal{H}(B)$ is called a best harmonic L^1–approximant to f. The following result is considered to be *a very essential achievement in the theory of multivariate canonical point sets*.

Theorem E (Armitage, Gardiner, Haussmann, Rogge [3]). *Let $f \in C(\bar{B}) \cap \mathcal{S}(B)$. Then h^* is a best harmonic L^1–approximant to f from $C(\bar{B}) \cap \mathcal{H}(B)$ if and only if*

(i) $h^* = f$ *on* ∂B_0

and

(ii) $h^* \leq f$ *on* $\bar{B} \backslash B_0$.

If a best harmonic approximant exists, it is unique. However, a best harmonic L^1–approximant to an arbitrary function $f \in C(\bar{B}) \cap \mathcal{S}(B)$ need not exist. For example, $f(x,y) = x^4 y^4$ is a subharmonic polynomial on the unit disc in $\mathbb{R}^2$ which fails to possess a best L^1–approximant from $C(\bar{B}) \cap H(B)$. However, $D^{2,2} x^4 y^4 \geq 0$ and the best $L^1(I^2)$ approximant from $B^{2,2}$ exists and it is unique.

The characterization in Theorem E is given in terms of a Dirichlet problem which can be considered as a transfinite interpolation by harmonic functions on the unit sphere. The canonical point set (canonical grid) is ∂B_0. Let us mention that Theorem E is a final result of more than 10 years of work in this direction (see [19], [20]).

5 Best One–Sided L^1–Approximation

Univariate Polynomial Case

Theorem F (Bojanić and DeVore, [10], [11]). *Let f be a differentiable, 2π–periodic function and let T_m denote the vector space of all real trigonometric polynomials of degree at most m. Then the best one–sided L^1–approximant from above, resp. from below to f from T_m exists and it is unique.*

The example $f(x) := |\cos m\theta|$ and $t^-(\theta) := \alpha \cos m\theta$ $(-1 \leq \alpha \leq 1)$ shows that continuity is not enough for uniqueness to be claimed.

Let $f \in C(I^2)$. A function $h^+ \in B^{m,n}(I^2)$ is defined to be a best one–sided L^1–approximant to f from above with respect to $B^{m,n}(I^2)$ if

$$h^+ - f \geq 0 \quad \text{and} \quad \int_{I^2} (h^+ - f) \leq \int_{I^2} (h - f)$$

for all $h \in \Omega$ which satisfy $h - f \geq 0$.

Unlike the situation of best L^1–approximation with respect to blending functions without constrains, we shall see that the best one–sided L^1–approximant from $B^{m,n}$ to a given function f satisfying $D^{m,n} f > 0$ is in fact *a special type of transfinite Hermite interpolant based on point sets which are not blending grids.*

First we state a general result showing that multivariate canonical point sets must satisfy certain conditions.

Proposition G (Armitage, Gardiner [2]). *Let U be a subspace of $C(\bar{\Omega})$ containing the constant functions. Let $u^* \in \mathcal{U}(f) := \{u : f \geq u \text{ on } \Omega\}$ and let $Z := Z(f - u^*)$ be the zero set of $f - u^*$. Then, the following are equivalent:*

(a) u^* is a best one–sided L^1–approximant to f from U;

(b) The Gaussian property of the zero set: $\int_{\bar{\Omega}} u \, d\lambda \geq 0$, whenever $u \in U$ and $u \geq 0$ on Z ;

(c) The domain under consideration is a quadrature domain: There is a positive measure μ with support in Z such that

$$\int_{\bar{\Omega}} u \, d\lambda = \int_Z u \, d\mu$$

for all u in U .

Let us mention that in the case of harmonic approximation a part of this proposition is essentially given in [22]. In the case of blending approximation we arrive also at a part of it (see [7]). In particular, Proposition G says that *the zero set (or the interpolation set)* in the case of one–sided L^1 approximation must satisfy the Gaussian property. By making use of this fact it has been shown in [7] that a blending grid of vertical and horizontal segments can not be a canonical point set for best one–sided L^1–approximation on I^2.

Theorem 1. *There exists a blending function $h \in B^{2m,2n}(I^2)$ such that it is negative on the Legendre (m,n)–blending grid defined as*

$$\mathbf{LG}_{\mathbf{x_m},\mathbf{y_n}} := \left\{(x, y) \in I^2 \ : \ L_m(x)\, L_n(y) \ = \ 0\right\} ,$$

where L_m and L_n are the Legendre polynomials of degree m, resp. n, i.e.

$$h_{|\mathbf{LG}_{\mathbf{x_m},\mathbf{y_n}}} < 0 ,$$

but its integral is positive

$$\int_{I^2} h > 0 .$$

Best One–Sided L^1–Approximation by $B^{1,1}$–Blending Functions

We give constructive characterization of the best one–sided L^1–approximant to a function $f \in C^{1,1}(I^2)$ satisfying $D^{1,1}f \geq 0$ on I^2 from the infinite dimensional vector space $B^{1,1}$. The characterization is in terms of matching (interpolation) conditions with respect to a diagonal of I^2 as a canonical point set (see [12]). *The occurrence of diagonals as a canonical point set (other than vertical–horizontal blending grids) is a phenomenon which to the best of our knowledge has not been observed so far.*

The proof [12] of existence and uniqueness of the best one–sided L^1–approximant from above and from below consists of three main steps.

A) *Transfinite Interpolation Formula with Remainder Expression:*

Theorem 2. *Let $f \in C^{1,1}(I^2)$ be given.*

a) The function

$$h^*(x,y) := f(-1,-1) + \int_{-1}^{x} D^{1,0} f(t,t)\,dt + \int_{-1}^{y} f(t,t)\,dt$$

is the unique $B^{1,1}$–function which satisfies the transfinite interpolation (boundary value) conditions

$$h^*{}_{|\Delta^*} = f_{|\Delta^*} \quad \text{and} \quad \operatorname{grad} h^*{}_{|\Delta^*} = \operatorname{grad} f_{|\Delta^*}$$

on the main diagonal Δ^*. Moreover, the following error representation holds true:

$$f(x,y) - h^*(x,y) = -\frac{(x-y)^2}{2} D^{1,1} f(\xi,\eta),$$

where $(\xi,\eta) \in I^2$.

b) Similarly, the function

$$h_*(x,y) := f(-1,1) + \int_{-1}^{x} D^{1,0} f(t,t)\,dt - \int_{-1}^{-y} f(t,-t)\,dt$$

is the unique $B^{1,1}$–function which satisfies the transfinite interpolation (boundary value) conditions

$$h_{*|\Delta_*} = f_{|\Delta_*} \quad \text{and} \quad \operatorname{grad} h_{*|\Delta_*} = \operatorname{grad} f_{|\Delta_*}$$

on the anti–diagonal Δ_*. The error representation in this case is given by:

$$f(x,y) - h_*(x,y) = \frac{(x+y)^2}{2} D^{1,1} f(\rho,\sigma),$$

where $(\rho,\sigma) \in I^2$.

B) *Cubature Formula Based on Transfinite Interpolation:*

Let $f \in C^{1,1}(I^2)$. Then

$$\int_{I^2} f \;-\; 2\int_I f(x,x)\,dx \;=\; -\frac{4}{3}\,D^{1,1} f(\alpha,\beta) \quad \Big((\alpha,\beta)\in I^2\Big)\,.$$

In particular we have

$$\int_{I^2} h \;=\; 2\int_I h(x,x)\,dx \quad \text{for} \quad h\in B^{1,1}(I^2)\,.$$

C) *Constructive Characterization of Best One–Sided L^1–Approximant:*

Theorem 3. *Let $f \in C^{1,1}(I^2)$ satisfy $D^{1,1} f \geq 0$ on I^2.*

a) Then f possesses a unique best one–sided L^1–approximant from above with respect to $B^{1,1}$. The best approximant h^* is characterized by the following simultaneous transfinite interpolation conditions:

$$i)\;\; {h^*}_{|\Delta^*} \;=\; f_{|\Delta^*} \quad \text{and} \quad ii)\;\; \operatorname{grad} {h^*}_{|\Delta^*} \;=\; \operatorname{grad} f_{|\Delta^*}\,.$$

b) The best one–sided L^1–approximant h_* from below also exists and is unique. It is characterized by

$$i)\;\; h_{*|\Delta_*} \;=\; f_{|\Delta_*} \quad \text{and} \quad ii)\;\; \operatorname{grad} h_{*|\Delta_*} \;=\; \operatorname{grad} f_{|\Delta_*}\,.$$

Proof of Theorem 3. By assumption $D^{1,1} f \geq 0$ hence we get from Theorem 2 that $h^* \geq f$ on I^2. We shall prove the assertion a) of Theorem 3; b) follows in a similar way. Let $h \in B^{1,1}$ and $h \geq f$ on I^2. We have

$$\int_{I^2}(h-f) \;=\; \int_{I^2}(h-h^*) \;+\; \int_{I^2}(h^* - f)\,.$$

By making use of the cubature formula we obtain

$$\int_{I^2}(h-h^*) \;=\; 2\int_{-1}^{1}(h-h^*)(x,x)\,dx\,.$$

This integral is ≥ 0, since $h \geq f$ on I^2 and ${h^*}_{|\Delta^*} = f_{|\Delta^*}$. From here

$$\int_{I^2}(h-f) \;\geq\; \int_{I^2}(h^* - f)$$

for each $h \in B^{1,1}$, satisfying $h - f \geq 0$. This means that h^* is a best one–sided L^1–approximant to f from above with respect to $B^{1,1}$.

If there exists a point $(x_0, x_0) \in \Delta^*$ such that $h(x_0, x_0) > f(x_0, x_0)$ then h can not be a best approximant. It follows that h is a best approximant only when

$$h_{|\Delta^*} = f_{|\Delta^*} \left(= {h^*}_{|\Delta^*}\right).$$

We have $h \geq f$ on I^2 and from here

$$\operatorname{grad} h_{|\Delta^*} = \operatorname{grad} f_{|\Delta^*} = \operatorname{grad} {h^*}_{|\Delta^*}$$

because $h \geq f$ on I^2 and by Theorem 2 we have $h \equiv h^*$. Hence, the best one–sided L^1–approximant from above by $B^{1,1}(I^2)$ to f is unique. □

Open problem. Let us consider the class Ω of functions $f \in C^{1,1,1}(I^3)$ ($I^3 := [-1,1]^3$) which satisfy

$$\frac{\partial^3}{\partial x \partial y \partial z} f \geq 0.$$

Find a canonical point set for best one–sided L^1–approximation from $B^{1,1,1}(I^3)$ to a function from Ω.

Remark. In contrast to Theorem D, Theorem E, and Theorem 3 on best L^1–approximation from $B^{m,n}(I^2)$, the best uniform (Chebyshev) approximant from $B^{1,1}(I^2)$ to $f \in C^{1,1}(I^2)$ satisfying $D^{1,1} f \geq 0$ is never unique unless $f \in B^{1,1}$ (see [18]).

The next theorem [13] gives a characterization for the best one–sided L^1–approximant from $B^{2,1}(I^2)$ in terms of transfinite interpolation on a canonical grid.

Theorem 4. *Let $f \in C^{2,1}(I^2)$ such that $D^{2,1} f \geq 0$ on I^2 and let*

$$V^* := \{ (x, 2|x| - 1) \ : \ x \in [-1,1] \}.$$

Then f possesses a unique best one–sided L^1–approximant from above h^* with respect to $B^{2,1}(I^2)$. The best approximant h^* is characterized by the following simultaneous Hermite type interpolation conditions:

$${h^*}_{|V^*} = f_{|V^*}, \qquad \operatorname{grad} {h^*}_{|V^*} = \operatorname{grad} f_{|V^*}.$$

An analogous result [13] holds for best one–sided L^1–approximation from below with respect to $B^{2,1}(I^2)$.

Example. Let us consider the polynomial $f_0(x, y) = x^2 y$ on I^2. Then, the best one–sided L^1–approximant from above with respect to $B^{2,1}(I^2)$ is

$$h^*(x, y) = (4/3)x^2|x| - x^2 + (1/12)(1+y)^3 .$$

Hence, the best one–sided L^1–approximant to the polynomial f_0 from $B^{2,1}(I^2)$ is not a polynomial. Moreover, it has a limited $C^{2,1}$ smoothness.

Best one–sided L^1–approximation by $B^{2,2}$ blending functions

Let $f \in C^{2,2}(I^2)$ be a real function satisfying $\partial^4 f/\partial x^2 \partial y^2 \geq 0$ on I^2. The unique best one–sided L^1–approximant from above to f with resprect to $B^{2,2}(I^2)$ is characterized [17] by transfinite Hermite interpolation on the canonical grid

$$\times^* := \{ (x, y) \in [-1, 1]^2 \; : \; |x| = |y| \} .$$

For f even with respect to one of its variables we characterize the unique best one–sided L^1–approximant to f from below by transfinite Hermite interpolation on the canonical grid $\{ (x, y) \in [-1, 1]^2 \; : \; |x| + |y| = 1 \}$. There is no canonical grid for the entire cone class of functions f with $\partial^4 f/\partial x^2 \partial y^2 \geq 0$ on $[-1, 1]^2$ when we approximate from below. The best one–sided L^1–approximant from above has the smoothness of f. The best one–sided L^1–approximant to f from below is a *blending–spline function with two line segment knots* $\{(x, 0) \; : \; -1 \leq x \leq 1\}$ and $\{(0, y) \; : \; -1 \leq y \leq 1\}$, i.e., the best one–sided approximation to f from below possesses a saturation effect with respect to the smoothness of f.

Example. Let $f(x, y) = x^2 y^2$. Then $h^*(x, y) = (x^4 + y^4)/2$. However,

$$h_*(x, y) = \frac{1}{2}(x^4 + y^4) - \frac{4}{3}(|x|^3 + |y|^3) + (x^2 + y^2) - \frac{1}{6}$$

and evidently $h_* \notin C^{3,0}(I^2) \cup C^{0,3}(I^2)$.

The following *maximum/ minimum principles* are useful in the proof of the $B^{2,2}(I^2)$ case. These principles are not maximum/ minimum principles in the proper classic sense. They refer to subsets on I^2 rather than to the boundary of the given set as it is in the case of the Laplace differential operator (subharmonic functions). However, these principles express the same idea in a more general form. Namely, under some assumptions for a given function on I^2 we can localize considerably the point set where the function attains its maximum/ minimum on I^2.

Maximum principle. *Let* $g \in C^{2,2}(I^2)$ *satisfy*

$$g_{\,|\times^*} = 0\,, \qquad \text{grad}\, g_{\,|\times^*} = \mathbf{0} \qquad \textit{and} \qquad D^{2,2}g > 0 \textit{ on } I^2.$$

Then g *can not attain a maximum on*

$$I^2\backslash\times^* := \Big\{(x,y) \in I^2 : |x| \neq |y|\Big\}\,.$$

In other words $g(x,y) < 0$ *on* $I^2\backslash\times^*$.

Minimum principle. *Let* $g \in C^{2,2}(I^2)$ *satisfy*

$$g_{\,|\diamond_*} = 0\,, \qquad \text{grad}\, g_{\,|\diamond_*} = \mathbf{0} \qquad \textit{and} \qquad D^{2,2}g > 0 \textit{ on } I^2,$$

where $\diamond_* = \{(x,y) : |x| + |y| = 1\}$. *Then* g *can not attain a minimum on* $\mathbf{int}\diamond_* := \{(x,y) \in I^2 : |x| + |y| < 1\}$. *In other words* $g(x,y) > 0$ *on* $\mathbf{int}\diamond_*$.

6 Cubature Formulae Exact for Blending Functions

Let us denote by $J_1 := \{(x,y) : a_1x + b_1y + c_1 = 0\}$ and $J_2 := \{(x,y) : a_2x + b_2y + c_2 = 0\}$ two line segments on $I^2 = [-1,1]^2$. It is easily proved by using a method of test functions that the unique two–line segments cubature formula of the form

$$\int_{I^2} f \approx \alpha_1 \int_{J_1} f + \alpha_2 \int_{J_2} f \quad (\alpha_1 \text{ and } \alpha_2 \text{ real})$$

which is exact in $B^{2,2}(I^2)$ is the cubature (see [17])

$$\int_{I^2} f = \int_I f(x,x)\,dx + \int_I f(x,-x)dx - \frac{4}{45} D^{2,2}f(\xi,\eta) \quad \Big((\xi,\eta) \in I^2\Big)\,.$$

This cubature formula plays an important role to obtain the characterization of the best one–sided L^1–approximant from above with respect to $B^{2,2}(I^2)$ by interpolation on the canonical grid [17]. From the error representation it is easily seen that the cubature is not exact in $B^{m,n}(I^2)$, $m \geq 3$ or $n \geq 3$. There is no *one–line segment* cubature on I^2 which is exact in $B^{2,2}(I^2)$ (see [14]).

Compound Cubature Formula for Approximate Integration

The above cubature formula gives rise to the following compound cubature formula:

$$\int_{I^2} f(x,y)\,dx\,dy = CF_{n\times n}(f) - \frac{4}{45\,n^4} D^{2,2} f(\xi,\eta) \quad \Big((\xi,\eta)\in I^2\Big),$$

where

$$CF_{n\times n}(f) := \frac{1}{n}\sum_{k=-n+1}^{n-1}\int_{\max\{-1,-1+2k/n\}}^{\min\{1,1+2k/n\}} f(x,-x+2k/n)dx$$

$$+\frac{1}{n}\sum_{k=-n+1}^{n-1}\int_{\max\{-1,-1+2k/n\}}^{\min\{1,1+2k/n\}} f(x,x-2k/n)dx\,.$$

It is exact in the infinite dimensional vector space $B^{2,2}(I^2)$, has order of convergence $O\left(n^{-4}\right)$, $n\to\infty$, and allows to calculate double integrals by using line integrals, i.e. bivariate integrals are approximated by univariate ones. The line integral involved are taken over the diagonals of I^2 as well as over lines parallel to the diagonals. The union of the diagonals in I^2 constitute the canonical point set for the best one–sided L^1–approximation from above with respect to $B^{2,2}(I^2)$.

Blending–Type Hermite Interpolation Formula

Let $f\in C^{2,2}(I^2)$ and $\Delta^* := \{(x,y)\in I^2 : x=y\}$. We construct a unique blending interpolant

$$h(x,y) = x\,b_1(y) + y\,a_1(x) + b_0(y) + a_0(x)$$

from $B^{2,2}(I^2)$ satisfying the following blending–type interpolation conditions

$$\frac{\partial^{i+j}}{\partial x^i \partial y^j}(f-h)_{|\Delta^*} = 0 \qquad (i+j\le 3).$$

From $(f-h)^{(2,1)}(x,x) = 0$ and $(f-h)^{(1,2)}(x,x) = 0$ we obtain $a_1''(x) = f^{(2,1)}(x,x)$ and $b_1''(y) = f^{(1,2)}(y,y)$. Therefore

$$a_1(x) = a_1(-1) + a_1'(-1)(x+1) + \int_{-1}^{x}(x-t)f^{(2,1)}(t,t)dt$$

and

$$b_1(y) = b_1(-1) + b_1'(-1)(y+1) + \int_{-1}^{y}(y-t)f^{(1,2)}(t,t)dt\,.$$

On the other hand $(f - h)^{(1,1)}(x,x) = 0$, thus $a_1'(x) + b_1'(x) = f^{(1,1)}(x,x)$. From here

$$\begin{aligned} a_1(x)y + b_1(y)x &= a_1(-1)\,y + b_1(-1)\,x + f^{(1,1)}(-1,-1)\,xy \\ &\quad + a_1'(-1)y + b_1'(-1)x \\ &\quad + y\int_{-1}^{x}(x-t)f^{(2,1)}(t,t)\,dt + x\int_{-1}^{y}(y-t)f^{(1,2)}(t,t)\,dt. \end{aligned}$$

Finally, for $g(x,y) = f(x,y) - a_1(x)\,y - b_1(y)\,x$ we determine $a_0(x) + b_0(y)$ such that (see [12, Theorem 2 (a)])

$$(\,g(x,y) - a_0(x) - b_0(y))^{(i,j)}_{|x=y} = 0 \quad (\,i+j=1\,)$$

to obtain

$$\begin{aligned} h(x,y) &= f(-1,-1) - f^{(1,1)}(-1,-1)\frac{(x-y)^2}{2} + \int_{-1}^{x} f^{(1,0)}(t,t)\,dt \\ &\quad + \int_{-1}^{y} f^{(0,1)}(t,t)\,dt + \int_{-1}^{x}\frac{(x-t)^2}{2}(f^{(2,1)}(t,t) - f^{(1,2)}(t,t))\,dt \\ &\quad + \int_{-1}^{y}\frac{(y-t)^2}{2}(f^{(1,2)}(t,t) - f^{(2,1)}(t,t))\,dt \\ &\quad + \int_{-1}^{x}(x-t)f^{(2,1)}(t,t)dt\,(y-x) + \int_{-1}^{y}(y-t)f^{(1,2)}(t,t)\,dt\,(x-y)\,. \end{aligned}$$

A technique, similar to that in [12], can be applied to obtain the following error representation formula:

$$f(x,y) = h(x,y) + \frac{f^{(2,2)}(\xi,\eta)}{4!}(x-y)^4, \quad \text{where } (\xi,\eta) \in I^2\,.$$

Integration over I^2 yields the cubature formula

$$\int_{I^2} f = 2\int_{-1}^{1} f(t,t)dt - \int_{-1}^{1}(1-t)^2 f^{(1,1)}(t,t)\,dt + \frac{8}{45}D^{2,2}f(\tilde{\xi},\tilde{\eta})\,,$$

$(\tilde{\xi},\tilde{\eta}) \in I^2$, that gives rise (see [14] for a method) to a compound Hermite type cubature formula with order of convergence $O(n^{-4})$, $n \to \infty$, which is similar to that given in the previous subsection.

References

[1] N. I. Achieser: *Theory of Approximation,* Frederic Ungar, New York 1956.

[2] D. H. Armitage, S. J. Gardiner: *Best one–sided L^1–approximation by harmonic and subharmonic functions,* In: *Advances in Multivariate Approximation*, Mathematical Research Vol. **107**, 43–56, Wiley–VCH, Berlin 1999.

[3] D. H. Armitage, S. J. Gardiner, W. Haussmann, L. Rogge: *Characterization of best harmonic and superharmonic L^1-approximants,* J. Reine Angew. Math. **478** (1996), 1–15.

[4] R. E. Barnhill, R. F. Riesenfeld: *Computer Aided Geometric Design,* Proceedings of a Conference held at the University of Utah, Academic Press, New York 1974.

[5] G. Birkhoff, R. E. Lynch: *Numerical Solution of Elliptic Problems,* SIAM Studies in Applied Mathematics Vol. **6**, Society for Industrial and Applied Mathematics (SIAM), Philadelphia 1984.

[6] B. D. Bojanov, H. A. Hakopian, A. A. Sahakian: *Spline Functions and Multivariate Interpolations,* Mathematics and its Applications Vol. **248**, Kluwer, Dordrecht 1993.

[7] B. D. Bojanov, D. P. Dryanov, W. Haussmann, G. P. Nikolov: *Best one–sided L^1–approximation by blending functions,* In: *Advances in Multivariate Approximation,* Mathematical Research Vol. **107**, 85–106, Wiley–VCH, Berlin 1999.

[8] C. de Boor: *A Practical Guide to Splines,* Applied Mathematical Sciences Vol. **27**, Springer, New York–Berlin 1978.

[9] P. J. Davis: *Interpolation and Approximation,* Blaisdell Publ. Co. Ginn and Co., New York 1963.

[10] R. A. DeVore: *One–sided approximation of functions,* J. Approx. Theory **1** (1968), 11–25.

[11] R. A. DeVore, G. G. Lorentz: *Constructive Approximation,* Springer, Berlin 1993.

[12] D. Dryanov, W. Haussmann, P. Petrov: *Best one-sided L^1-approximation of bivariate functions by sums of univariate ones,* Arch. Math. (Basel) **75** (2000), 125–131.

[13] D. Dryanov, W. Haussmann, P. Petrov: *Best one-sided L^1-approximation by $B^{2,1}$-blending functions,* In: *Recent Progress in Multivariate Approximation,* International Series of Numerical Mathematics Vol. **137**, 115–134, Birkhäuser, Basel 2001.

[14] D. Dryanov, W. Haussmann, P. Petrov: *Best one-sided L^1-approximation by blending functions and approximate cubature,* In: *Approximation Theory X,* Innovations in Applied Mathematics Vol. **X**, 145–152, Vanderbilt Univ. Press, Nashville 2002.

[15] D. Dryanov, O. Kounchev: *Polyharmonically exact formula of Euler–Maclaurin, multivariate Bernoulli functions, and Poisson type formula,* C. R. Acad. Sci. Paris Sr. I Math. **327** (1998), 515–520.

[16] D. Dryanov, O. Kounchev: *Multivariate Bernoulli functions and polyharmonically exact cubature of Euler–Maclaurin,* Math. Nachr. **226** (2001), 65–83.

[17] D. Dryanov, P. Petrov: *Best one-sided L^1-approximation by blending functions of order* $(2,2)$, J. Approx. Theory **115** (2002), 72–99.

[18] L. Flatto: *The approximation of certain functions of several variables by sums of functions of fewer variables,* Amer. Math. Monthly **73** (1966), 131–132.

[19] M. Goldstein, W. Haussmann, K. Jetter: *Best harmonic L^1-approximation to subharmonic functions,* J. London Math. Soc. **30** (1984), 257–264.

[20] M. Goldstein, W. Haussmann, L. Rogge: *On the mean value property of harmonic functions and best harmonic L^1-approximation,* Trans. Amer. Math. Soc. **305** (1988), 505–515.

[21] W. J. Gordon, C. A. Hall: *Transfinite element methods: blending-function interpolation over arbitrary curved element domains,* Numer. Math. **21** (1973/74), 109–129.

[22] B. Gustafsson, M. Sakai, H. S. Shapiro: *On domains in which harmonic functions satisfy generalized mean value properties,* Potential Anal. **7** (1997), 467–484.

[23] W. Haussmann, K. Zeller: *Blending interpolation and best L^1–approximation,* Arch. Math. (Basel) **40** (1983), 545–552.

[24] W. Haussmann, K. Zeller: *Canonical point sets in multivariate constructive function theory,* In: *Proceedings of the Ashkelon Workshop on Complex Function Theory* (1996), Israel Math. Conf. Proc. **11**, 91–104, Bar–Ilan University, Ramat Gan 1997.

[25] S. Y. Khavinson: *Best Approximation by Linear Superpositions,* Amer. Math. Soc., Providence 1997.

[26] J. C. Mairhuber: *On Haar's Theorem concerning Chebyshev approximation problems having unique solution,* Proc. Amer. Math. Soc. **7** (1956), 609–615.

[27] V. M. Mordashev: *Approximation of functions of several variables by a sum of functions of fewer variables,* Doklad. Akad. Nauk SSSR **183** (1968), 778–779; English transl., Soviet Math. Dokl. 9 (1968), 1462–1463.

[28] A. Pinkus: *On L^1–Approximation,* Cambridge Univ. Press, Cambridge 1989.

[29] T. J. Rivlin, R. J. Sibner: *The degree of approximation of certain functions of two variables by a sum of functions of one variable,* Amer. Math. Monthly **72** (1965), 1101–1103.

Address:

Dimiter P. Dryanov
Départment de Mathématiques et de Statistique
Université de Montréal
Pavillon André–Aisenstadt, C.P. 6128, Succursaile Centre–Ville
Montréal (QC) H3C 3J7
Canada

International Series of Numerical Mathematics
Vol. 145,

Refining Oscillatory Signals by Non–Stationary Subdivision Schemes

Nira Dyn, David Levin, and Ariel Luzzatto

Abstract

The paper presents a method for refining real highly oscillatory signals. The method is based upon interpolation by a finite set of trigonometric basis functions. The set of trigonometric functions is chosen (identified) by minimizing a natural error norm in the Fourier domain. Both the identification and the refining processes are computed by linear operations. Unlike the Yule–Walker approach, and related algorithms, the identification of the approximating trigonometric space is not repeated for every new input signal. It is rather computed off–line for a family of signals with the same support of their Fourier transform, while the refinement calculations are done in real–time. Statistical estimates of the point–wise errors are derived, and numerical examples are presented.

Introduction

Given a series of equidistant samples of a band limited real signal we would like to evaluate samples of the signal at intermediate points. I.e., we look for an approximation to a series of samples with a double sampling rate. We approach this problem as an interpolation problem, and the basic issue is to find an appropriate space of basis functions for interpolation. A naive local polynomial interpolation is not appropriate here since the signals may be highly oscillatory. It seems rather natural to try to interpolate by using trigonometric basis functions, and the main issue is to identify the optimal set of trigonometric functions which are suitable for local interpolation of the given signal at the given sampling rate. Based upon a previous work by the authors [3], we look for an 'orthogonal scheme', which is a scheme with the property of annihilating any signal taken from a related space of exponentials. Finding such an orthogonal scheme, following [3], is of course enough for identifying the space of exponentials. Moreover, by solving a linear system, this scheme also provides us with the rule for local interpolation by functions in the identified space.

Identifying the optimal orthogonal scheme for a given signal at a certain sampling rate is the same as identifying the optimal linear predictor. However, unlike the Yule–Walker approach to finding a linear predictor, which involves the solution of a linear system which varies with each portion of a signal, we are looking for an orthogonal scheme which will suit a class of signals with the same support Ω of their Fourier transforms. The computation of such a scheme may thus be done off–line, once and for all the signals in the prescribed class.

Both the orthogonal scheme and the related interpolating scheme are represented by associated Laurent polynomials. Assuming that the signals share the same support of their Fourier transforms, the actual algorithm for finding the optimal orthogonal scheme involves minimizing a certain weighted L_2–norm of the associated Laurent polynomial. It is shown that there exists always a minimizing Laurent polynomial with a full set of simple roots. This implies that the related space of exponentials is non–degenerate, and defines a space of trigonometric functions with frequencies in Ω.

In the rest of the paper we present estimates for the error resulting in refining signals by the new interpolating scheme, and we conclude with numerical experiments.

1 Preliminaries

In this paper we are interested in approximating signals by functions in finite dimensional spaces of exponential polynomials of the form

$$E_\ell^c = \{f : \mathbb{R} \to \mathbb{C},\ f \in C^\ell(\mathbb{R}),\ \sum_{n=0}^{\ell} c_n D^n f = 0\}\ , \tag{1.1}$$

where $c = (c_0, \ldots, c_\ell) \in \mathbb{R}^{\ell+1}$, and where D^n denotes differentiation of order n.

We can write explicitly the space E_ℓ^c in terms of the distinct roots of the polynomial $c(z) = \sum_{n=0}^{\ell} c_n z^n$ with their multiplicities $\{\gamma_n, \mu_n\}_{n=1}^{\nu}$, such that $\sum_{n=1}^{\nu} \mu_n = \ell$, as

$$E_\ell^c = \text{span}\{x^r \exp(\gamma_n x),\ r = 0, 1, \ldots, \mu_n - 1,\ n = 1, \ldots, \nu\}\ . \tag{1.2}$$

In the following we use the notation $\gamma_n = \alpha_n + i\beta_n$, $n = 1, \ldots \nu$. For oscillatory signals the relevant spaces are defined in terms of the parameters

$$\ell = 2m,\ \nu = \ell,\ \gamma_j = i\beta_j,\ \gamma_{j+m} = \bar{\gamma}_j = -i\beta_j,\ j = 1, \ldots, m\ . \tag{1.3}$$

Such spaces are termed here *strictly oscillatory.*

For a given space E_ℓ^c, there exists a reproducing subdivision scheme of minimal rank [3]. Such a scheme is an interpolatory subdivision scheme, which consists of reproducing refinement rules of minimal rank $\{R_k\}_{k\geq k_0}$. The rule R_k maps samples of any signal f from the space E_ℓ^c at refinement level k, $f^k = \{f_j^k = f(2^{-k}j) : j \in \mathbb{Z}\}$, to its samples at the next refinement level, namely at the points $2^{-(k+1)}\mathbb{Z}$ in the following way,

$$f_{2j}^{k+1} = f_j^k, \quad f_{2j+1}^{k+1} = \sum_{n=-M}^{N-1} a_{2n+1}^{[k]} f_{j-n}^k, \; j \in \mathbb{Z}, \tag{1.4}$$

where $M = [\frac{\ell}{2}]$, $N = [\frac{\ell+1}{2}]$. This rule is of minimal rank, since any other rule reproducing E_ℓ^c is based on at least ℓ values at level k for the generation of a new value at level $k+1$. The operation in equation (1.4) is written formally as $f^{k+1} = R_k f^k$. The symbol of the rule R_k is defined as $a^{[k]}(z) = \sum_{n=-M}^{N-1} a_{2n+1}^{[k]} z^{2n+1} + 1$. The following theorem characterizes the symbol of the rule R_k [2]. It is used later for the derivation of a linear system defining this rule.

Theorem 1.1. *The symbol $a^{[k]}(z)$ is the symbol of a reproducing refinement rule of minimal rank at level k for a space E_ℓ^c of the form (1.2), if and only if*

$$D^r a^{[k]}(z_{\gamma_n,k+1}) = 2\delta_{r,0}, \quad D^r a^{[k]}(-z_{\gamma_n,k+1}) = 0, \quad r = 0,\dots,\mu_n - 1, \; n = 1,\dots,\nu, \tag{1.5}$$

with $z_{\gamma_n,k} = \exp(-2^{-k}\gamma_n)$, $n = 1,\dots\nu$.

There might be values of k for which a reproducing refinement rule of minimal rank (of the form (1.4)) does not exist. Yet, such rules exist if k is large enough [3]. There is a sufficient condition for the existence and uniqueness of the reproducing refinement rule of minimal rank at level k.

Theorem 1.2. *For a space E_ℓ^c of the form (1.2), with $\gamma_n = \alpha_n + i\,\beta_n$, if for any pair $m, n \in \{1,\dots,\nu\}$ such that $m \neq n$ and $\alpha_n = \alpha_m$, we have*

$$\beta_n - \beta_m \neq 2^{k+1} s\pi, \quad s \in \mathbb{Z} \tag{1.6}$$

then a unique reproducing refinement rule of minimal rank at level k exists.

Remark. If

$$\beta_n < 2^{k_0}\pi, \; n = 1,\dots,\nu, \tag{1.7}$$

then by Theorem 1.2 there exists at each level $k \geq k_0$ a unique reproducing refinement rule of minimal rank for the space E_ℓ^c, and therefore there exists a unique minimal rank reproducing scheme for $k \geq k_0$.

In the case of a strictly oscillatory space E^c_{2m}, conditions (1.7) on $\{\beta_n : n = 1\dots, 2m\}$ guarantee that the sampling rate of 2^k samples per unit at level k is above the Nyquist rate [5].

The minimal rank reproducing scheme of a space E^c_ℓ, can refine exactly samples of a signal from that space. Yet, for a given signal in a space E^c_ℓ , the space or its minimal rank reproducing scheme have first to be identified. The tool for a suboptimal identification of an E^c_ℓ space, given a signal from this space with or without added small noise, is the notion of orthogonal schemes [3, 4]. A minimal rank orthogonal scheme for E^c_ℓ consists of orthogonal refinement rules of minimal rank $\{O_k : k \geq 0\}$ such that O_k refines and maps to zero the samples of any signal from E^c_ℓ at level k, in the following way,

$$f^{k+1}_{2j+1} = 0, \quad f^{k+1}_{2j} = \sum_{n=-N}^{M} p^{[k]}_{2n} f^k_{j-n} = 0, \tag{1.8}$$

where $N = [\frac{\ell}{2}]$, $M = [\frac{\ell+1}{2}]$ as before. Thus, the symbol of an orthogonal refinement rule of minimal rank at level k has the form

$$p^{[k]}(z) = \sum_{n=-N}^{M} p^{[k]}_{2n} z^{2n}. \tag{1.9}$$

For an orthogonal refinement rule of minimal rank there exists a characterization of its symbol [3].

Theorem 1.3. *The symbol $p^{[k]}(z)$ is the symbol of an orthogonal refinement rule of minimal rank at level k for the space E^c_ℓ of the form* (1.2), *if and only if*

$$D^r p^{[k]}(\pm z_{\gamma_n,k+1}) = 0, \quad r = 0, \dots, \mu_n - 1, \ n = 1, \dots, \nu, \tag{1.10}$$

with $z_{\gamma_n,k} = \exp(-2^{-k}\gamma_n)$, $n = 1, \dots \nu$.

Note that in contrast to Theorem 1.1, from which the explicit form of the symbol of the reproducing refinement rule of minimal rank cannot be deduced, this theorem implies that up to a multiplicative constant (normalization), the symbol of O_k has the form,

$$z^{-2N} \prod_{n=1}^{\nu} (1 - \exp(-2^{-k}\gamma_n) z^2)^{\mu_n}. \tag{1.11}$$

When the multiplicative constant is real, we see from (1.11) that the coefficients of the symbol of O_k are real.

Hereafter we term a refinement rule with real coefficients in its symbol as a real refinement rule.

The existence of an orthogonal refinement rule of minimal rank at any level k is a consequence of Theorem 1.3. The following Theorem establishes its uniqueness and the existence, up to normalization, under conditions of the type (1.6), see [3].

Theorem 1.4. *An orthogonal refinement rule of minimal rank at level k (of the form* (1.8)) *exists and is unique up to normalization, if the conditions in Theorem* 1.2 *hold for* $k-1$.

Remark. Note that by the last theorem, the conditions of Theorem 1.2 on the structure of an E_ℓ^c space, guarantee the existence and uniqueness of a reproducing refinement rule of minimal rank at level k, and of a unique orthogonal refinement rule of minimal rank at level $k+1$, up to normalization factor. We assume hereafter that the free coefficient of the symbol of the orthogonal refinement rule is not zero, and thus it may be normalized so that $p_0^{[k]} = 1$.

For a given space E_ℓ^c, there is a simple relation between the symbols of the reproducing refinement rule of minimal rank at level k and the orthogonal refinement rule of minimal rank at level $k+1$ with free coefficient 1, as can be deduced from Theorems 1.1, 1.3, cf. [3, 4].

Theorem 1.5. *Let* $a^{[k]}$ *be the symbol of a reproducing refinement rule of minimal rank at level k for a space* E_ℓ^c, *and let* $p^{[k+1]}$ *with* $p_0^{[k+1]} = 1$ *be the symbol of an orthogonal refinement rule of minimal rank at level* $k+1$ *for the same* E_ℓ^c *space. Then*

$$a^{[k]}(z^2) - 2 = p^{[k+1]}(z)q^{[k]}(z), \tag{1.12}$$

where $q^{[k]}(z) = \sum_{n=-(N-1)}^{M-1} q_{2n}^{[k]} z^{2n}$. *The coefficients of* $q^{[k]}$ *are determined uniquely as the solution of the linear system*

$$\sum_{n=-(N-1)}^{M-1} p_{2(2j-n)}^{i[k+1]} q_{2n}^{[k]} = -\delta_{j,0}, \quad j = -(N-1), \ldots, M-1. \tag{1.13}$$

It follows from the last theorem that once the symbol of the orthogonal refinement rule of minimal rank with free coefficient 1 at level $k+1$ is known for an E_ℓ^c space, then the reproducing refinement rule of minimal rank at level k can be obtained by solving a non–singular linear system, and that this scheme is real, since the coefficients of the given symbol are real.

In the next section we discuss how to choose the orthogonal refinement rule of minimal rank at level k, based on some knowledge about the Fourier transform of the signal.

2 Choosing the Approximating Exponential Space

We now turn to one of the main issues of this paper, namely the identification of an approximating space E_ℓ^c for a given band–limited signal.

To be able to make a valuable choice, we assume prior knowledge about the Fourier transform of the signal f, namely, knowledge of $|\widehat{f}|$, or at least the support of $\widehat{f}$. Both cases may appear in real applications, and we shall start with the first, namely, we assume that $|\widehat{f}|$ is known.

We shall identify the space E_ℓ^c through identifying the orthogonal scheme $p^{[k]}$, annihilating k–level samples of functions in E_ℓ^c. With the special normalization $p_0^{[k]} = 1$, the quantity

$$e_j = \sum_{n=-N}^{M} p_{2n}^{[k]} f_{j-n}^k \ , \tag{2.1}$$

may be viewed as the approximation error in predicting the value f_j^k by interpolation based on a function in E_ℓ^c, interpolating the values $\{f_{j-n}^k \ : \ 0 < |n| \leq \ell\}$, at the points $\{2^{-k}(j-n) \ : \ 0 < |n| \leq \ell\}$. Thus, we search for the orthogonal scheme $p^{[k]}$ that minimizes $\|\{e_j\}\|_\infty$.

Using the equality

$$e_j = \sum_{n=-N}^{M} p_{2n}^{[k]} f_{j-n}^k = \frac{1}{2\pi}\int_{-\infty}^{\infty} p^{[k]}(e^{-i2^{-k-1}\omega})\widehat{f}(\omega)e^{i\omega 2^{-k}j}d\omega \ , \tag{2.2}$$

and denoting by Ω the support of $\widehat{f}$, we have

$$\|\{e_j\}\|_\infty \leq \frac{1}{2\pi}\int_\Omega |p^{[k]}(e^{-i2^{-k-1}\omega})||\widehat{f}(\omega)|d\omega \ . \tag{2.3}$$

Note that Ω is symmetric since f is real.

Let us identify the r.h.s. of (2.3) as a weighted L_1 norm of the polynomial $p^{[k]}$, and denote it by $\|p^{[k]}\|_{1,\widehat{f}}$. We also have

$$\|p^{[k]}\|_{1,\widehat{f}} \leq |\Omega|^{\frac{1}{2}}\|p^{[k]}\|_{2,\widehat{f}} \equiv \frac{|\Omega|^{\frac{1}{2}}}{2\pi}\{\int_\Omega |p^{[k]}(e^{-i2^{-k-1}\omega})|^2|\widehat{f}(\omega)|^2 d\omega\}^{\frac{1}{2}} \ , \tag{2.4}$$

where $|\Omega|$ is the measure of Ω.

The relation (2.3) implies that, in the case of prior knowledge about $\widehat{f}$, one would like to define the orthogonal scheme $p^{[k]}$ by minimizing $\|p^{[k]}\|_{1,\widehat{f}}$. Yet, since it is much simpler to minimize $\|p^{[k]}\|_{2,\widehat{f}}$, we define $p^{[k]}$ to be the minimizer of the $|\widehat{f}(\omega)|^2$ weighted L_2 norm. Let us further assume that $\ell = 2N$, that Ω contains at least ℓ distinct points and that $\Omega \subset 2^{k-1}(-\pi, \pi)$.

The following results are central in the characterization of the zeros of $p^{[k]}$ which minimizes $\|p^{[k]}\|_{2,\widehat{f}}$.

Lemma 2.1. *The polynomial $p^{[k]}$, of the form (1.9) with $p_0^{[k]} = 1$, minimizing $\|p^{[k]}\|_{2,\widehat{f}}$, is a symmetric polynomial. Namely, $p_{2n}^{[k]} = p_{-2n}^{[k]}$, $n = 1, .., N$.*

Proof. Setting

$$c_n = p_{2n}^{[k]} + p_{-2n}^{[k]}\ , \quad d_n = p_{2n}^{[k]} - p_{-2n}^{[k]}\ , \tag{2.5}$$

we get

$$p^{[k]}(e^{-i2^{-k-1}\omega}) = \eta(\omega) - i\phi(\omega)\ , \tag{2.6}$$

where

$$\eta(\omega) = 1 + \sum_{n=-N}^{N} c_n \cos(2^{-k}\omega n)\ , \quad \phi(\omega) = \sum_{n=-N}^{N} d_n \sin(2^{-k}\omega n)\ . \tag{2.7}$$

Consequently we have

$$\|p^{[k]}\|_{2,\widehat{f}}^2 = \frac{1}{2\pi}\int_\Omega (|\eta(\omega)|^2 + |\phi(\omega)|^2)|\widehat{f}(\omega)|^2 d\omega\ . \tag{2.8}$$

It thus follows that for the minimizing polynomial $\phi(\omega) \equiv 0$, which is equivalent to the claim of the Lemma. □

Remark. Under the conditions of Lemma 2.1, it is possible to show by its method of proof that if $\ell = 2N+1$ then $p_{2N+2}^{[k]} = 0$.

Thus the minimizer of $\|p^{[k]}\|_{2,\widehat{f}}$ has the form

$$p^{[k]}(z) = 1 + \sum_{n=1}^{N} p_{2n}^{[k]}(z^{2n} + z^{-2n})\ , \quad N = [\frac{\ell}{2}]\ .$$

With the substitution $z = \exp(-i2^{-k-1}\omega)$, the above polynomial becomes

$$p^{[k]}(\exp(-i2^{-k-1}\omega) = \tilde{p}^{[k]}(\omega) = 1 + \sum_{n=1}^{N} p_{2n}^{[k]} \cos(2^{-k}n\omega)\ . \tag{2.9}$$

Lemma 2.2. *The system* $\{\cos(j\cdot)\}_{j=1}^N$ *is a Haar system on* $[0, \frac{\pi}{2})$.

By the substitution $x = \cos t$, this lemma is equivalent to the result,

Lemma 2.3. *The system of Chebyshev polynomials* $\{T_j\}_{j=1}^N$ *is a Haar system on* $(0, 1]$.

Proof. It is sufficient to show that any linear combination of the elements of the system $\{T_j\}_{j=1}^N$ can have at most $N-1$ zeros in $(0, 1]$. To prove this we will show that any algebraic polynomial of degree at most N with N zeros in $(0, 1]$, has the representation $\sum_{j=0}^N t_j T_j(x)$, with $t_0 \neq 0$.

Let p be a polynomial of degree N with N zeros in $(0, 1]$. Then $p(x) = C\pi_{j=1}^N(x - \xi_j)$, with $\xi_j \in (0, 1]$, $j = 1, \ldots, N$, and C a constant. It is easy to conclude from this form of p that the sign of $p(x)$, $x \in (-1, 0)$ is fixed and that $|p(x)| > |p(-x)|$, $x \in (-1, 0)$. Thus $|\int_{-1}^0 p(x)(1 - x^2)^{-\frac{1}{2}} dx| > |\int_0^1 p(x)(1 - x^2)^{-\frac{1}{2}} dx|$, and

$$t_0 = \int_{-1}^1 p(x)(1 - x^2)^{-\frac{1}{2}} dx \neq 0.$$

□

With Lemma 2.2 proved, we can use ideas from the theory of orthogonal algebraic polynomials, in order to conclude that,

Theorem 2.4. *There exists a unique polynomial* $p^{[k]}$ *of the form* (1.9) *with* $p_0^{[k]} = 1$, *minimizing* $\|p^{[k]}\|_{2,\widehat{f}}$. *The roots of* $p^{[k]}$ *are simple and are of the form* $\{\pm \exp(\pm i 2^{-k-1}\beta_q)\}_{q=1}^N$ *with* $\{\beta_q\}_{q=1}^N \subset \Omega$, $\Omega = \text{supp}\ \widehat{f}$.

Proof. First we observe that the minimizing polynomial should be symmetric. Next, we define the following inner product on the space of symmetric trigonometric polynomials.

$$< p, q >_{\widehat{f}} = \frac{1}{2\pi} \int_\Omega p(\omega) q(\omega) |\widehat{f}(\omega)|^2 d\omega \ . \tag{2.10}$$

If $p^{[k]}$ with $p_0^{[k]} = 1$, minimizes $\|p^{[k]}\|_{2,\widehat{f}}$, then $\tilde{p}^{[k]}$ must satisfy the orthogonality conditions

$$< \tilde{p}^{[k]}, \cos(2^{-k} j\cdot) >_{\widehat{f}} = 0 \ , \quad j = 1, ..., N \ . \tag{2.11}$$

Using Lemma 2.2, it follows from (2.11) that $\tilde{p}^{[k]}(\omega)$ must have exactly n simple roots in Ω. Otherwise, since $\Omega \subset 2^k[0, \frac{\pi}{2})$, we can find a linear combination of the functions $\{\cos(2^{-k} j\omega)\}_{j=1}^N$ which has the same sign structure as $\tilde{p}^{[k]}(\omega)$

for $\omega \in \Omega$. This would imply a contradiction to the orthogonality conditions (2.11).

Now, since $\tilde{p}^{[k]}$ is a symmetric trigonometric polynomial its roots are of the form

$$\pm\beta_q \ , \ \beta_q \in \Omega \ , \ q = 1, \ldots, N \ .$$

Hence by (2.9) the roots of $p^{[k]}(z)i$, which is an even polynomial, are of the form $\{\pm\exp(\pm i 2^{-k-1}\beta_q)\}_{q=1}^{N}$ with $\{\beta_q\}_{q=1}^{N} \subset \Omega$.
□

A direct consequence of Theorems 2.4 and 1.3 is:

Corollary 2.5. *The space E_ℓ^c determined by the $2N$ roots of the polynomial $p^{[k]}$ minimizing $\|p^{[k]}\|_{2,\widehat{f}}$ is strictly oscillatory.*

The space in the last Corollary is denoted hereafter by $E_\ell^{c^*}$ with $c^* = c^*(|\widehat{f}|, k, \ell)$.

In the case that only the support Ω of the spectrum is known, we are restating the problem as follows:

Consider the set F_Ω of all signals f having a Fourier transform with support contained in Ω, and such that $|\widehat{f}| \leq 1$. We would like to find an optimal E_ℓ^c space, corresponding to a minimal rank orthogonal rule at level k, which is as closely orthogonal as possible to the k–level samples of all signals in F_Ω. Using (2.2) we observe that for any given polynomial $p^{[k]}$ we can find a function in F_Ω such that

$$\|\{e_j\}\|_\infty = \frac{1}{2\pi}\int_\Omega |p^{[k]}(e^{-i2^{-k-1}\omega})|d\omega \ . \tag{2.12}$$

This implies that

$$\inf_{p^{[k]}}\{\sup_{f\in F_\Omega}\|\{e_j\}\|_\infty\} = \inf_{p^{[k]}}\{\frac{1}{2\pi}\int_\Omega |p^{[k]}(e^{-i2^{-k-1}\omega})|d\omega\} = \inf_{p^{[k]}}\{\|p^{[k]}\|_{1,\chi_\Omega}\} \ , \tag{2.13}$$

where χ_Ω is the characteristic function of Ω. Therefore, the optimal solution to the above stated problem is to find the polynomial $p^{[k]}$ minimizing $\|p^{[k]}\|_{1,\chi_\Omega}$. Here again, for practical reasons, we shall be content with a sub–optimal solution defined by minimizing $\|p^{[k]}\|_{2,\chi_\Omega}$.

3 Refining Oscillatory Signals

The problem we solve with the tools presented in the previous sections is:

Given in real–time consecutive samples at level k of a zero–mean real signal $f \in L^2(\mathbb{R})$

$$f^k = \{f_j^k = f(2^{-k}j) \ : \ j \in J\}, \tag{3.14}$$

and a compact set $\Omega \in \mathbb{R}$, which is the support of $\hat{f}$,

compute in real–time values

$$\tilde{f}^{k+1} = \{\tilde{f}_j^{k+1} \ : \ j \in I\}$$

approximating

$$f_j^{k+1} = \{f(2^{k+1}j) \ : \ j \in I\},$$

with I a finite set of consecutive indices in $\mathbb{Z}$, which is almost the set $2J$. This problem is termed hereafter *the spectral refinement problem.*

In view of the results in the previous sections the procedure we suggest for solving the spectral refinement problem is:

1. Choose $\ell = 2N$ (the dimension of the approximating exponentials space).
2. Compute $p^{[k+1]}$ by minimizing $\|p^{[k+1]}\|_{2,\chi_\Omega}$.
3. Compute the symbol of R_k, $a^{[k]}$, from $p^{[k+1]}$, by solving a linear system.
4. Compute in real–time $\tilde{f}^{k+1} = R_k f^k$.

The procedure above is termed *spectral refinement* or *spectral interpolation.* Next, we analyze the error in our procedure, under a reasonable statistical assumption, and bound the variance of the point–wise error.

4 Statistical Error Analysis

In this section we present an error analysis based upon a statistical assumption on the signal, which is valid in many communication applications. The statistical assumption at level k on $f \in L^2(\mathbb{R})$ with zero mean and with Fourier transform of compact support Ω, is:

Any set of consecutive samples of f at level k, w.l.o.g.

$$f_j^k = f(2^{-k}j) \ , \quad j = -N, \ \dots \ , N+L \ ,$$

for k such that $2^{-k}\Omega \subset [-\frac{\pi}{2}, \frac{\pi}{2}]$, and for large L, can be written as

$$f_j^k = g(2^{-k}j) + \nu_j^k \ , \quad j = -N, \ \dots \ , N+L \ , \tag{4.1}$$

with $g \in E_\ell^{c^*}$, $c^* = c^*(\chi_\Omega, k, \ell)$, $\ell = 2N$, and where $\{\nu_j^k : j = -N, \dots, N+L\}$ are samples of an ergodic random process which are independent Gaussian random variables with zero mean and variance σ^2.

Under this statistical assumption we can bound σ.

Lemma 4.1. *Under the statistical assumptions at level $k+1$,*

$$\sigma \leq \|p^{[k+1]}\|_{2,\chi_\Omega} \|f\|_2 + o(1) \ , \ \text{as } L \to \infty \ , \tag{4.2}$$

where $p^{[k+1]}(z)$ is the symbol of the orthogonal rule of minimal rank for $E_\ell^{c^*}$, $c^* = c^*(\chi_\Omega, k+1, \ell)$, with a free coefficient 1.

Proof. Denote $p^{[k+1]}(z) = 1 + \sum_{n=-N}^{N} p_{2n}^{[k+1]} z^{2n}$. Then for $j \in \{0, 1, \dots, L\}$, by (4.1) and since $g \in E_\ell^{c^*}$,

$$\sum_{n=-N}^{N} p_{2n}^{[k+1]} f_{j-n}^{k+1} = \sum_{n=-N}^{N} p_{2n}^{[k+1]} \nu_{j-n}^{k+1} \ . \tag{4.3}$$

Defining the random variables $\xi_j = \sum_{n=-N}^{N} p_{2n}^{[k+1]} \nu_{j-n}^{k+1}$, $j \in \{0, 1, \dots, L\}$, we observe that each ξ_j has zero mean, and

$$var(\xi_j) = \sigma^2 \sum_{n=-N}^{N} (p_{2n}^{[k+1]})^2 \ . \tag{4.4}$$

By the ergodic assumption, and since L is large,

$$var(\xi_j) = \frac{1}{L+1} \sum_{j=0}^{L} \xi_j^2 + o(1) \ , \tag{4.5}$$

while by (2.2) and (4.3)

$$\xi_j = \frac{1}{2\pi} \int_\Omega p^{[k+1]}(e^{-i2^{-k-2}\omega}) \widehat{f}(\omega) e^{i\omega 2^{-k-1} j} d\omega \ , \tag{4.6}$$

and therefore

$$|\xi_j|^2 \leq \|p^{[k+1]}\|^2_{2,\chi_\Omega} \|\widehat{f}\|^2_2 \, . \tag{4.7}$$

Then, by (4.5)

$$var(\xi_j) \leq \|p^{[k+1]}\|^2_{2,\chi_\Omega} \|\widehat{f}\|^2_2 + o(1) \, , \tag{4.8}$$

and since $\sum_{n=-N}^{N}(p_{2n}^{[k+1]})^2 \geq 1$, we conclude (4.2) from (4.4) and (4.8). □

Without loss of generality we bound the error incurred by step 4 of our procedure, namely the variance of the point–wise error

$$\epsilon_j^{k+1} = \tilde{f}_j^{k+1} - f_j^{k+1} \;, \quad j \in I = \{0, 1, \ldots, 2L\}.$$

The same bound applies everywhere. Observe that under the statistical assumption at level $k+1$

$$f_j^{k+1} = g(2^{-k-1}j) + \nu_j^{k+1} \;, \quad j = -N, \ldots, N+L \;, \tag{4.9}$$

with $g \in E_\ell^{c^*}$, $c^* = c^*(\chi_\Omega, \ell, k)$, and with $var(\nu_{2j+1}^{k+1}) = \sigma^2$, where σ is bounded by (4.2). Since the operation $\tilde{f}^{k+1} = R_k f^k$ is described by (1.4), step 4 generates samples at level $k+1$ of the form,

$$\tilde{f}_{2j}^{k+1} = f_j^k, \quad \tilde{f}_{2j+1}^{k+1} = \sum_{n=-N}^{N-1} a_{2n+1}^{[k]} f_{j-n}^k, \; 2j, 2j+1 \in I \;, \tag{4.10}$$

we have $\epsilon_{2j}^{k+1} = 0$, $2j \in I$, and by the statistical assumption at level $k+1$,

$$\epsilon_{2j+1}^{k+1} = \sum_{n=-N}^{N-1} a_{2n+1}^{[k]} f_{j-n}^k - f_{2j+1}^{k+1} = \sum_{n=-N}^{N-1} a_{2n+1}^{[k]} \nu_{2(j-n)}^{k+1} - \nu_{2j+1}^{k+1} \, .$$

Recalling the statistical properties of $\{\nu_j^{k+1}\}$ we get,

$$var(\epsilon_{2j+1}^{k+1}) = \sum_{n=-N}^{N-1} (a_{2n+1}^{[k]})^2 + 1)\sigma^2 \;, \quad 2j+1 \in I \, . \tag{4.11}$$

Finally, using Lemma 4.1, we obtain the bound

$$(var(\epsilon_j^{k+1}))^{\frac{1}{2}} \leq (\sum_{n=-N}^{N-1} (a_{2n+1}^{[k]})^2 - 1)^{\frac{1}{2}} \|p^{[k+1]}\|_{2,\chi_\Omega} \|f\|_2 + o(1) \;, \quad j \in I. \tag{4.12}$$

with $o(1)$ small relative to L. Thus, in order to bound the variance of the point–wise error, we derive in the next section a bound on $\|p^{[k+1]}\|_{2,\chi_\Omega}$ in terms of Ω, ℓ, k.

5 Bounding $\|p^{[k]}\|_{2,\chi_\Omega}$

In this section we derive a bound on $\|p^{[k]}\|_{2,\chi_\Omega}$, in terms of Ω, ℓ, and k. This bound is then used in bounding the refinement error analyzed in the previous section.

Let us assume that $\Omega = [-b,-a] \cup [a,b]$, and that $\ell = 2N$. Since $p^{[k]}$ is a symmetric trigonometric polynomial, we may write

$$p^{[k]}(e^{-i2^{-k-1}\omega}) = 1 + 2\sum_{n=1}^{N} p_{2n}^{[k]} \cos(2^{-k}n\omega) \ . \tag{5.1}$$

Substiuting $x = \cos(2^{-k}\omega)$ we obtain an algebraic polynomial representation

$$p^{[k]}(e^{-i2^{-k-1}\omega}) = 1 + 2\sum_{n=1}^{N} p_{2n}^{[k]} T_n(x) \ , \ x = \cos(2^{-k}\omega) \ , \tag{5.2}$$

where T_n is the nth degree Chebyshev polynomial of the first kind. As ω varies in Ω, $x \in [c_k, d_k] = [\cos(2^{-k}b), \cos(2^{-k}a)]$, and for large enough k, this is an interval of length $O(2^{-2k})$ near $x = 1$.

In order to bound $\|p^{[k]}\|_{2,\chi_\Omega}$, we recall its optimal property as the minimizer of $\|q\|_{2,\chi_\Omega}$, among all trigonometric polynomials q of the form (5.1), and bound $\|p^{[k]}\|_{2,\chi_\Omega}$ by $\|h^{[k]}\|_{2,\chi_\Omega}$, where $h^{[k]}$ is of the form (5.1). To obtain a trigonometric polynomial for which we can estimate its $\|\cdot\|_{2,\chi_\Omega}$-norm, we consider the Chebyshev polynomial of degree N transformed to the interval $[c_k, d_k]$,

$$g(x) = T_N\left(\frac{x - \frac{c_k+d_k}{2}}{\frac{d_k-c_k}{2}}\right) . \tag{5.3}$$

and normalize it to be a polynomial of the form (5.2), namely with a free coefficient 1 in its Chebyshev expansion,

$$\overline{g}(x) = g(x) / \int_{-1}^{1} (1-t^2)^{-\frac{1}{2}} g(t) dt \equiv 1 + 2\sum_{n=1}^{N} g_{2n} T_n(x) \ . \tag{5.4}$$

Analysing the integral in (5.4), we have

$$\int_{-1}^{1} (1-t^2)^{-\frac{1}{2}} g(t) dt \geq g(0)(1 + O(2^{-2k})) = \frac{2^{N-1}}{(\frac{d_k-c_k}{2})^N}(1 + O(2^{-2k})) \ . \tag{5.5}$$

Recalling that g is bounded by 1 in $[c_k, d_k]$, we obtain,

$$\|\overline{g}\|_{[c_k,d_k],\infty} = O((b^2 - a^2)^N 2^{-(2k+3)N}) , \tag{5.6}$$

as k and $\ell = 2N$ increase. Defining $h^{[k]}(x) = 1 + \sum_{n=1}^{N} g_{2n}(z^n + z^{-n})$, we have

$$\|p^{[k]}\|_{2,\chi_\Omega} \le \|h^{[k]}\|_{2,\chi_\Omega} = O((b^2 - a^2)^N 2^{-(2k+3)N}) . \tag{5.7}$$

6 Numerical Examples

Here we present numerical experiments for a family of signals with an almost compactly supported Fourier transform

$$f(t) = \cos(2\pi F t + \beta \sin(2\pi F_s t)) . \tag{6.1}$$

This is a widely used family of frequency modulated signals [1]. A function of the form (6.1) is essentially of a band limited support, centered at $\omega = 2\pi F$, with a bandwidth $\Delta\omega = 2\pi(\beta + 1)F_s$. We applied the spectral interpolation algorithm to refine equidistant data values measured from such functions, and we compared the results achieved with local $\ell - 1$ degree polynomial interpolation. In the following figures we present in the upper part the spectral density of the signal, in the middle part the percentage of the error in local polynomial interpolation, and in the lower part the percentage of the error by the spectral interpolation algorithm. In Figure 1 $F = 0.1$, $F_s = 0.0062$ and $\beta = 5.75$, in Fugure 2 $F = 0.3$, $F_s = 0.0062$ and $\beta = 5.75$, and in Figure 3 $F = 0.3$, $F_s = 0.0062$ and $\beta = 2.375$. In all cases the spectral interpolation gives superior results, and the advantage is larger when the bandwidth is smaller, and when the center of the support is larger. This is in accordance with the error estimates developed above and the bound (5.7).

FIGURE 1: $F = 0.1,\ F_S = 0.0062\ ,\ \beta = 5.75$

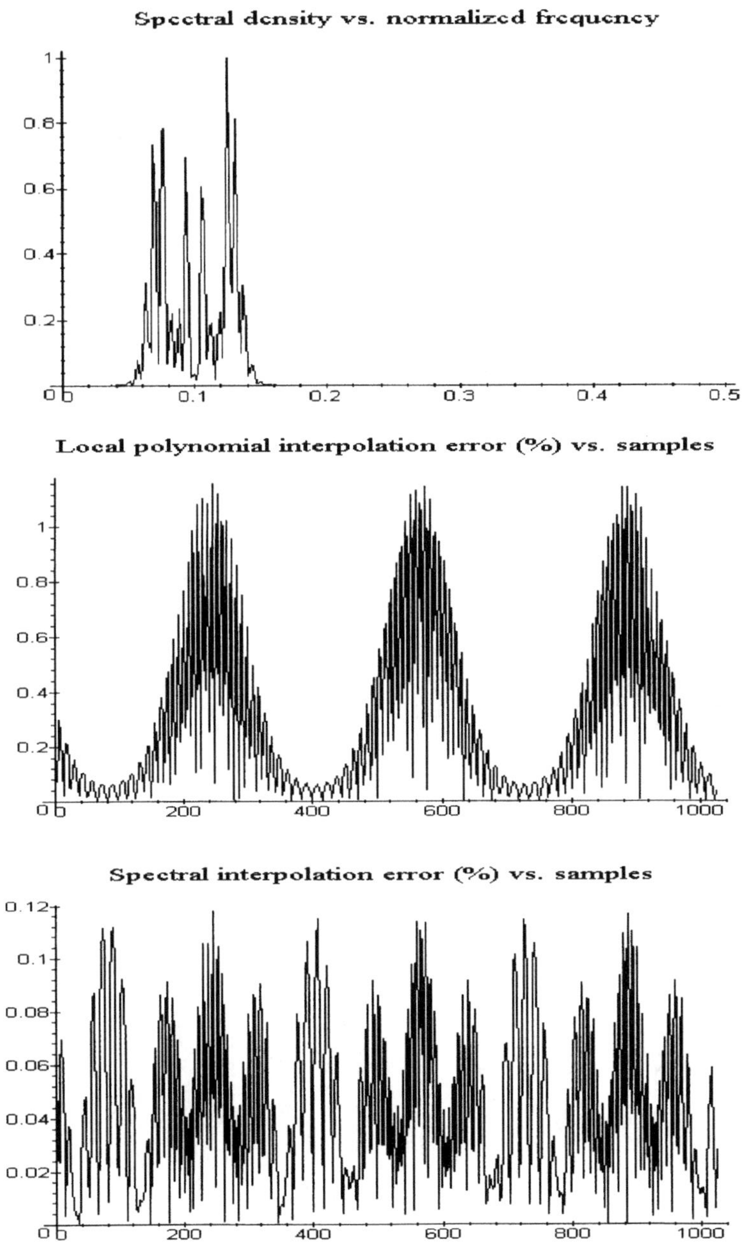

FIGURE 2: $F = 0.3$, $F_s = 0.0062$, $\beta = 5.75$

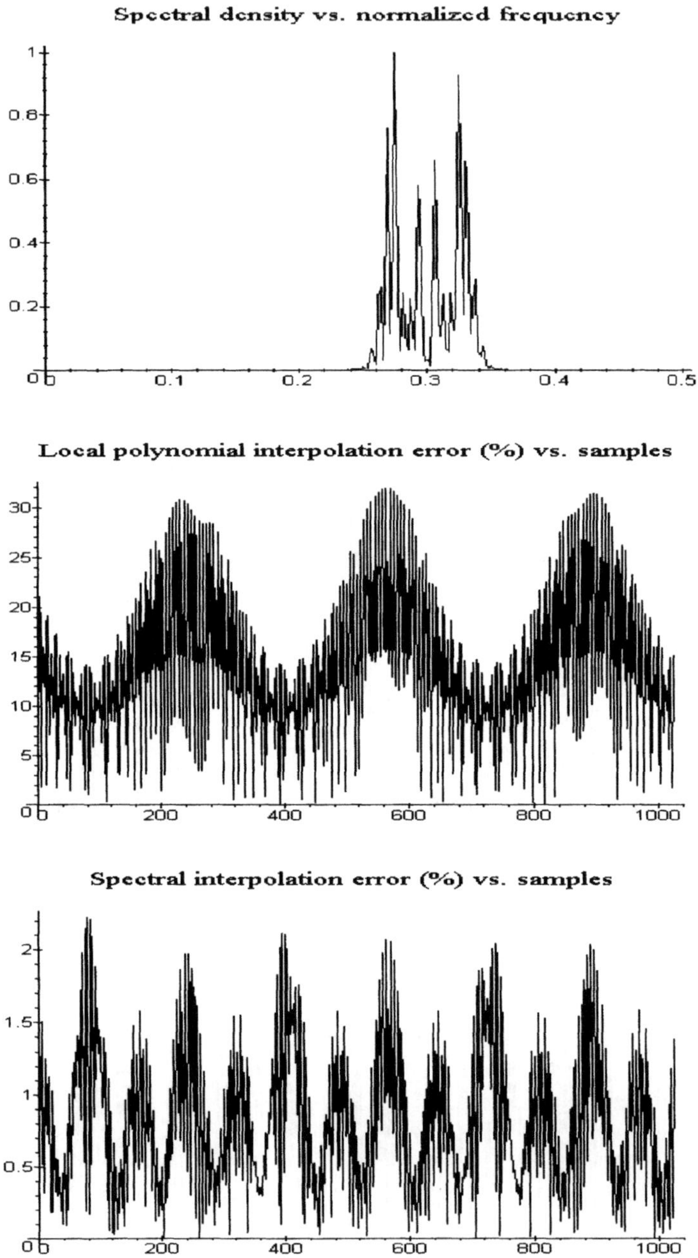

FIGURE 3: $F = 0.3, \; F_S = 0.0062 \;, \; \beta = 2.375$

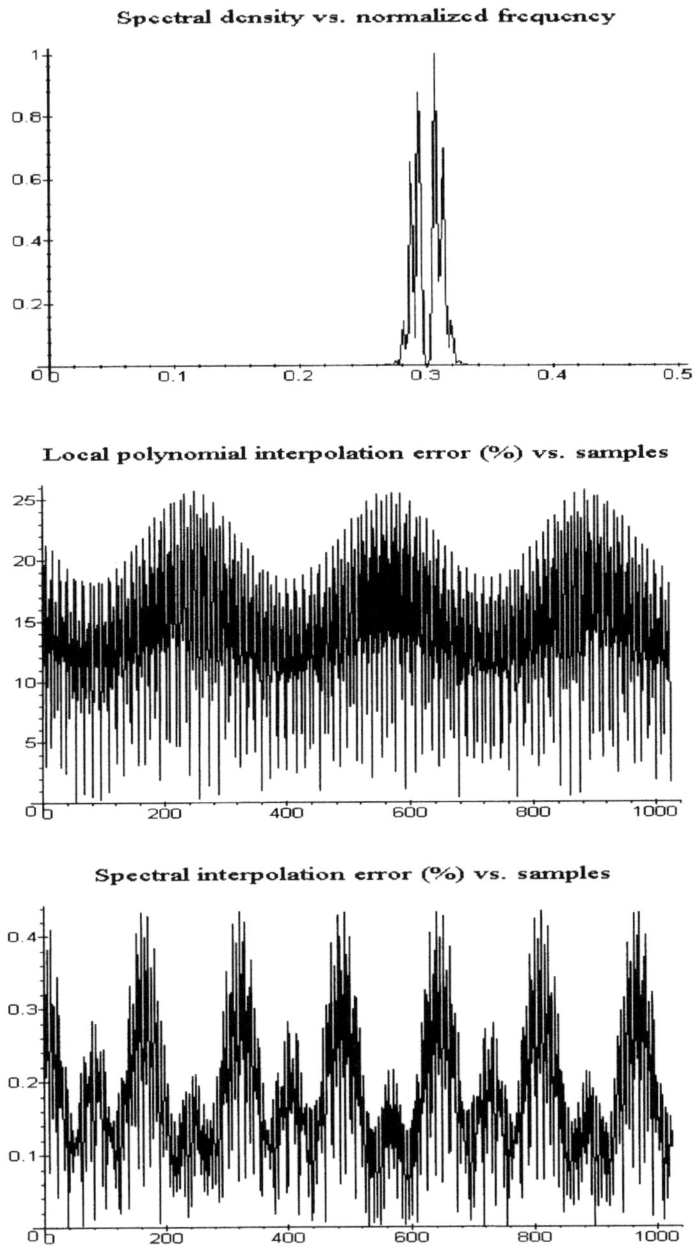

References

[1] K. Clarke, D. Hess: *Communication Circuits, Analysis and Design*, Addison Wesley, 1978.

[2] N. Dyn, D. Levin: *Subdivision schemes in geometric modeling*, Acta Numerica (2002), 73–144.

[3] N. Dyn, D. Levin, A. Luzzatto: *Non-stationary interpolatory subdivision schemes reproducing spaces of exponential polynomials*, To appear.

[4] A. Luzzatto: *Multi-Scale Signal Processing based on Non-Stationary Subdivision*, PhD Thesis, Tel–Aviv University, 2000.

[5] M. Schwartz: *Information, Transmission, Modulation and Noise*, McGraw–Hill, 1990.

Addresses:

Nira Dyn, David Levin
Tel–Aviv University
School of Mathematical Sciences
Tel–Aviv 69978
Israel

Ariel Luzzatto
Motorola Communications Israel
3 Kremenetzky st.
Tel–Aviv 67899
Israel

International Series of Numerical Mathematics
Vol. 145, ©2003 Birkhäuser Verlag Basel/Switzerland

Asymptotic Error Expansions for Schoenberg Type Operators

Karol Dziedziul and Kurt Jetter

Dedicated to Professor Manfred Reimer on his seventieth birthday

Abstract

We derive an asymptotic expansion of the L_2–error of scaled Schoenberg operators in terms of powers of the scale parameter. The class of operators includes cardinal interpolation operators, in one or several variables. The polyphase case is also considered. Connections to multiwavelet expansions are inherent, but not worked out here.

1 Introduction

We consider a continuous function $\phi : \mathbb{R}^d \to \mathbb{R}$ satisfying a decay condition of type

$$|\phi(x)| \quad \lesssim \quad q^{|x|} \quad \text{for some} \quad 0 < q = e^{-a} < 1 . \tag{1.1}$$

Here, $A(x) \lesssim B(x)$ is short for saying that $A(x)/B(x)$ is bounded, and

$$|x| := \|x\|_1 = |x_1| + \ldots + |x_d| \quad \text{for} \quad x = (x_1, \ldots, x_d) \in \mathbb{R}^d .$$

The associated *Schoenberg operator* is defined as

$$I(f) := I_\phi(f) := \sum_{\alpha \in \mathbb{Z}^d} f(\alpha)\, \phi(x - \alpha) \tag{1.2}$$

where $f : \mathbb{R}^d \to \mathbb{R}$ is allowed to be any continuous function growing at most polynomially at infinity. In the sequel, $\sum_\alpha$ indicates that the sum is taken with respect to the entire lattice $\mathbb{Z}^d$. The scaled operator, for some scale parameter $h > 0$, is then defined by

$$I_h(f) := (I(f_h))_{1/h} \tag{1.3}$$

where the index of a function refers to dilation, $g_\tau(x) := g(\tau x)$.

A special type of Schoenberg operators are the so called *cardinal interpolation operators.* In this case the function ϕ is *fundamental with respect to the lattice* $\mathbb{Z}^d$, *i.e.*,

$$\phi(\alpha) = \delta_\alpha \,, \quad \alpha \in \mathbb{Z}^d \,,$$

with δ_α denoting the usual Kronecker delta. This type of operators was often studied in the literature (*e.g.* [8]), where also the connection of cardinal interpolation with L_2–projection onto shift–invariant spaces is described. This L_2–projection plays a prominent role as the starting point for wavelet expansions, and this explains how our error expansions can be applied to wavelet and multiwavelet theory as well.

We say that the Schoenberg operator has *polynomial order* $r \in \mathbb{N}$, if $I(p) = p$ for all polynomials p of degree at most $r-1$. Under this condition it is well known [9] that the $L_2(\mathbb{R}^d)$–error can be bounded as

$$\|f - I_h(f)\|_2 \le \text{const}\; h^r \; |f|_{r,2} \tag{1.4}$$

for functions f in the Sobolev space $W_2^r(\mathbb{R}^d)$. Here, we have to assume that the space is continuously embedded into the space of continuous functions, *i.e.*, we have to assume that $r > d/2$, and the Sobolev seminorm of order r is defined as

$$|f|_{r,2} := \Big(\int_{\mathbb{R}^d} (\xi_1^2 + \ldots + \xi_d^2)^r \; |f^\wedge(\xi)|^2 \; d\xi \Big)^{1/2} .$$

It refers to the Fourier transform of f, which we normalize as follows:

$$f^\wedge(\xi) := \int_{\mathbb{R}^d} f(x) \; e^{-ix\cdot\xi} \; dx \;.$$

More detailed estimates of type (1.4) can be given in various directions. By assuming higher order smoothness of f, one can expand the error in terms of powers of h. This idea was first developed for wavelet expansions, by Sweldens and Piessens in [11, 12], and then pursued and generalized by Blu and Unser in [4 – 6]. Daubechies and Unser [7] gave interesting examples by specifying the leading constant in these expansions, while Beśka and Dziedziul [1 – 3] gave an independent approach, and pointed to the resulting saturation phenomena for the underlying approximation process.

In this paper we revisit error expansions by looking at special types of operators which were studied in a more general setting by Zhou and Jetter in [9, 10]. Using their results, the kernel of all operators studied so far in the literature is readily available (and explicitly used in [10]). This will be elaborated for operators of type (1.2) in the following Section 2, while Section 3 is devoted to the more general case where the underlying shift–invariant space is generated by more than just one basis function.

We restrict our analysis to the case where f is band–limited, *i.e.*, it is assumed to be an entire function with

$$\text{supp} f^\wedge \subset C := [-\pi, +\pi]^d \,. \tag{1.5}$$

However, using typical cut–off procedures and denseness properties of function spaces, our main results (Theorem 2.3 and Theorem 3.1 below) could be used to find error expansions for functions from higher order Sobolev spaces as well.

2 Error Expansion for Schoenberg Operators

We start with the definition of the kernel

$$\kappa(\xi) := |\xi|^{2r} K(\xi) := |1 - \phi^\wedge(\xi)|^2 + \sum_{\alpha \neq 0} |\phi^\wedge(\xi + 2\pi\alpha)|^2 \tag{2.1}$$

associated with the operator (1.2), where $|\xi|^2 := \|\xi\|^2 := \xi_1^2 + \cdots + \xi_d^2$. According to [9, Theorem 3] the function K is bounded in case I has polynomial order r (this result applies even for non–integer r). Also, for $f \in W_2^r(\mathbb{R}^d)$,

$$(f - I(f))^\wedge = (1 - \phi^\wedge)\, f^\wedge - \phi^\wedge \sum_{\alpha \neq 0} f^\wedge(\cdot + 2\pi\alpha)$$

(see [8, Section 3]), and the terms $f^\wedge(\cdot + 2\pi\alpha)$, $\alpha \in \mathbb{Z}^d$, are orthogonal in case (1.5) holds. This shows that

$$\|(f - I(f))^\wedge\|_2^2 = \int_C \kappa(\xi)\, |f^\wedge(\xi)|^2 \, d\xi \,.$$

After rescaling according to (1.3), we arrive at the basic

Theorem 2.1. *If the Schoenberg operator* (1.2) *has polynomial order* $r \in \mathbb{N}$ *with* $r > d/2$, *then for* $f \in W_2^r(\mathbb{R}^d)$ *satisfying* (1.5) *we have*

$$\|(f - I_h(f))^\wedge\|_2^2 = \int_C \kappa(h\xi)\, |f^\wedge(\xi)|^2 \, d\xi = h^{2r} \int_C K(h\xi)\, |\xi|^{2r}\, |f^\wedge(\xi)|^2 \, d\xi$$

for $0 < h < 1$.

For the next result we write eq. (2.1) as

$$\kappa(\xi) = 1 - 2\,\Re\phi^\wedge(\xi) + \Phi^\sim(\xi) \tag{2.2}$$

with

$$\Phi^\sim(\xi) := \sum_\alpha |\phi^\wedge(\xi + 2\pi\alpha)|^2 \tag{2.3}$$

the 2π–periodization of the Fourier transform of $\Phi^\wedge := |\phi^\wedge|^2$. Here $\Phi = \phi * \phi^*$, with $\phi^*(x) := \overline{\phi(-x)}$, is the autocorrelation function of ϕ,

$$\Phi(x) := \int_{\mathbb{R}^d} \phi(t)\,\overline{\phi(t-x)}\,dt\,. \tag{2.4}$$

According to the Poisson summation formula (which applies in the classical sense due to the decay conditions assumed) we have

$$\Phi^\sim(\xi) = \sum_\alpha \Phi(\alpha)\,e^{-i\alpha\cdot\xi}\,. \tag{2.5}$$

Now the coefficients in this Fourier series decay exponentially,

$$|\Phi(\alpha)| \;\lesssim\; q^{|\alpha|}\,, \quad \alpha \in \mathbb{Z}^d\,,$$

with the same $q = e^{-a}$ as in (1.1). Hence, the Fourier series is uniformly convergent, as is the series of all partial derivatives. We have

$$D^{\mathbf{k}}\Phi^\sim(\xi) = \sum_\alpha \Phi(\alpha)\,(-i\alpha)^{\mathbf{k}}\,e^{-i\alpha\cdot\xi}\,, \quad \mathbf{k} = (k_1,\ldots,k_d) \in \mathbb{N}_0^d\,,$$

such that

$$|D^{\mathbf{k}}\Phi^\sim(\xi)| \;\lesssim\; \sum_\alpha e^{-a|\alpha|}\prod_{i=1}^d |\alpha_i|^{k_i} \;\lesssim\; \prod_{i=1}^d a^{-(k_i+1)}\,\Gamma(k_i+1)\,. \tag{2.6}$$

Also,

$$\begin{aligned} |D^{\mathbf{k}}\phi^\wedge(\xi)| &= \left|\textstyle\int_{\mathbb{R}^d}(-ix)^{\mathbf{k}}\,\phi(x)\,e^{-ix\cdot\xi}\,dx\right| \\ &\lesssim \textstyle\int_{\mathbb{R}^d} e^{-a|x|}\prod_{i=1}^d |x_i|^{k_i}\,dx \;\lesssim\; \prod_{i=1}^d a^{-(k_i+1)}\,\Gamma(k_i+1)\,, \end{aligned} \tag{2.7}$$

and it should be noted that the constants appearing in these two estimates do not depend on $\mathbf{k}$. Using (2.2), (2.6) and (2.7), we can bound the derivatives of the kernel.

Lemma 2.2. *Given the fundamental function ϕ satisfying (1.1), the kernel (2.1) allows the estimates*

$$|D^{\mathbf{k}}\kappa(\xi)| \quad \lesssim \quad \prod_{i=1}^{d} k_i!\, a^{-k_i} \quad \text{for} \quad \mathbf{k} = (k_1, \ldots, k_d) \in \mathbb{N}_0^d\,,$$

where the bound is uniform with respect to $\mathbf{k}$. *Whence,* $\kappa(\xi)$ *can be written in terms of its Taylor expansion at the origin:*

$$\kappa(\xi) = \sum_{\mathbf{k}\in\mathbb{N}_0^d} \frac{D^{\mathbf{k}}\kappa(0)}{\mathbf{k}!}\, \xi^{\mathbf{k}}$$

for all $\xi = (\xi_1, \ldots, \xi_d)$ *satisfying* $|\xi_i| < a$ *for* $i = 1, \ldots, d$.

Proof: It remains to show that the kernel is represented by its Taylor series. Clearly, the series converges absolutely with the given assumption. For given $n \in \mathbb{N}$, consider the error term

$$R_n(\xi) := \kappa(\xi) - \sum_{|\mathbf{k}|\le n} \frac{D^{\mathbf{k}}\kappa(0)}{\mathbf{k}!}\, \xi^{\mathbf{k}} = \sum_{|\mathbf{k}|=n+1} \frac{D^{\mathbf{k}}\kappa(\theta\xi)}{\mathbf{k}!}\, \xi^{\mathbf{k}}\,,$$

with $0 < \theta < 1$ depending on ξ. Then

$$|R_n(\xi)| \quad \lesssim \quad \sum_{|\mathbf{k}|=n+1} \prod_{i=1}^{d} \left|\frac{\xi_i}{a}\right|^{k_i} \quad \to \quad 0 \quad \text{for} \quad n \to \infty\,. \quad \square$$

Combining this with Theorem 2.1 we arrive at the following error expansion formula. It should be noted that $\kappa(\xi)/|\xi|^{2r}$ is bounded in a neighborhood of the origin, whence

$$D^{\mathbf{k}}\kappa(0) = 0 \quad \text{for} \quad |\mathbf{k}| = k_1 + \ldots + k_d < 2r\,.$$

Theorem 2.3. *With the assumptions of Theorem 2.1 and Lemma 2.2 we have*

$$\|(f - I_h(f))^\wedge\|_2^2 = \sum_{n=2r}^{\infty} \frac{h^n}{n!} \sum_{|\mathbf{k}|=n} \binom{n}{\mathbf{k}} D^{\mathbf{k}}\kappa(0) \int_C \xi^{\mathbf{k}}\, |f^\wedge(\xi)|^2\, d\xi$$

for $0 < h < \min\{1, a/\pi\}$.

We point to the special case where ϕ leads to a symmetric kernel, *i.e.*, $\kappa(\xi) = \kappa(-\xi)$:

Corollary 2.4. *If, in addition to the assumptions of Theorem 2.3, $d = 1$ and ϕ is real–valued and symmetric, then*

$$\|(f - I_h(f))^\wedge\|_2^2 = \sum_{n=r}^{\infty} h^{2n} \, \frac{D^{2n}\kappa(0)}{(2n)!} \, |f|_{n,2}^2$$

for $0 < h < \min\{1, a/\pi\}$, and f satisfying (1.5).

3 The Polyphase Case

In this section we consider more general operators of the type

$$I(f) := \sum_{\alpha} \sum_{j=1}^{n} f(\alpha + \alpha_j) \, \phi_j(x - \alpha) \tag{3.1}$$

for given pairwise different shifts

$$\alpha_j \in [0, 1[^d \,, \quad j = 1, \dots, n \,,$$

and for *continuous* basis functions

$$\phi_j \, : \, \mathbb{R}^d \to \mathbb{R} \,, \quad j = 1, \dots, n \,,$$

satisfying the same decay condition as before,

$$|\phi_j(x)| \quad \lesssim \quad q^{|x|} \quad \text{for some} \quad 0 < q = e^{-a} < 1 \,, \quad j = 1, \dots, n \,. \tag{3.2}$$

For $f \in W_2^r(\mathbb{R}^d)$, we find

$$\begin{aligned} &(f - I(f))^\wedge(\xi) \\ &= f^\wedge(\xi)\Big(1 - \sum_{j=1}^{n} e^{i\alpha_j\cdot\xi} \, \phi_j^\wedge(\xi)\Big) - \sum_{\alpha\neq 0} f^\wedge(\xi + 2\pi\alpha) \sum_{j=1}^{n} e^{i\alpha_j\cdot(\xi+2\pi\alpha)} \, \phi_j^\wedge(\xi) \,. \end{aligned} \tag{3.3}$$

This motivates us to define the kernel now as

$$\kappa(\xi) := |\xi|^{2r} K(\xi) := \Big|1 - \sum_{j=1}^{n} e^{i\alpha_j\cdot\xi} \, \phi_j^\wedge(\xi)\Big|^2 + \sum_{\alpha\neq 0} \Big|\sum_{j=1}^{n} e^{i\alpha_j\cdot\xi} \, \phi_j^\wedge(\xi + 2\pi\alpha)\Big|^2 \tag{3.4}$$

in order to see that Theorem 2.1 holds true literally for scaled versions of the operator (3.1) as well. It should be noted that in case $I(p) = p$ for polynomials p of total degree at most $r-1$, the kernel

$$K(\xi) = |\xi|^{-2r}\kappa(\xi)$$

is again bounded (see [10, Section 4.3]) if $r > d/2$.

In order to extend our error expansion to the present case, we again write the kernel as

$$\kappa(\xi) = 1 - 2\,\Re\Big(\sum_{j=1}^{n} e^{i\alpha_j\cdot\xi}\,\phi_j^\wedge(\xi)\Big) + \mathbf{\Phi}^\sim(\xi)$$

with

$$\begin{aligned}\mathbf{\Phi}^\sim(\xi) &:= \textstyle\sum_\alpha |\sum_{j=1}^n e^{i\alpha_j\cdot\xi}\,\phi_j^\wedge(\xi+2\pi\alpha)|^2 \\ &= \textstyle\sum_{j,\ell=1}^n e^{i(\alpha_j-\alpha_\ell)\cdot\xi}\sum_\alpha \phi_j^\wedge(\xi+2\pi\alpha)\,\overline{\phi_\ell^\wedge(\xi+2\pi\alpha)}\,.\end{aligned} \tag{3.5}$$

By the Poisson summation formula we have

$$\sum_\alpha \phi_j^\wedge(\xi+2\pi\alpha)\,\overline{\phi_\ell^\wedge(\xi+2\pi\alpha)} = \sum_\alpha (\phi_j * \phi_\ell^*)(\alpha)\,e^{-i\alpha\cdot\xi} \tag{3.6}$$

for $j,\ell = 1,\dots,n$, and this shows that the Fourier coefficients of $\mathbf{\Phi}^\sim$ and of all its partial derivatives decay exponentially again. In addition, for $\mathbf{k} = (k_1,\dots,k_d) \in \mathbb{N}_0^d$,

$$\begin{aligned}|D^{\mathbf{k}} e^{i\alpha_j\cdot\xi}\,\phi_j^\wedge(\xi)| &= \Big|\int_{\mathbb{R}^d}(-ix)^{\mathbf{k}}\,\phi_j(x+\alpha_j)\,e^{-ix\cdot\xi}\,dx\Big| \\ &\lesssim \int_{\mathbb{R}^d} e^{-a|x+\alpha_j|}\prod_{i=1}^d |x_i|^{k_i}\,dx \;\lesssim\; \prod_{i=1}^d a^{-(k_i+1)}\,\Gamma(k_i+1)\,,\end{aligned}$$

for $j = 1,\dots,n$. This shows that also Lemma 2.2 extends to the present case. We thus arrive at the corresponding extension of Theorem 2.3:

Theorem 3.1. *Given the polyphase Schoenberg operator* (3.1) *with basis functions ϕ_j satisfying the decay condition* (3.2)*, let the kernel κ be defined by* (3.4)*. If the operator has polynomial order $r > d/2$, then the error expansion*

$$\|(f - I_h(f))^\wedge\|_2^2 = \sum_{n=2r}^{\infty}\frac{h^n}{n!}\sum_{|\mathbf{k}|=n}\binom{n}{\mathbf{k}} D^{\mathbf{k}}\kappa(0)\int_C \xi^{\mathbf{k}}\,|f^\wedge(\xi)|^2\,d\xi$$

holds true for $0 < h < \min\{1, a/\pi\}$, *subject that* f *is an entire function satisfying* (1.5).

As a corollary we have the following estimate which is a proper extension of Corollary 2.4: *If the kernel* (3.4) *is symmetric, i.e.,* $\kappa(\xi) = \kappa(-\xi)$, *then*

$$\|(f - I_h(f))^\wedge\|_2^2 \lesssim \sum_{n=r}^{\infty} h^{2n} \frac{\kappa_{2n}}{(2n)!} |f|_{n,2}^2$$

with $\kappa_{2n} := \max_{|\mathbf{k}|=2n} |D^{\mathbf{k}}\kappa(0)|$.

In closing we point to the special case (the true polyphase case) where we start with one basis function ϕ by putting

$$\phi_j(x) = \frac{1}{n}\, \phi(x - \alpha_j)\,, \quad j = 1, \ldots, n\,,$$

in formula (3.1). Here, the kernel takes the form

$$\kappa(\xi) = |1 - \phi^\wedge(\xi)|^2 + \sum_{\alpha \neq 0} |\phi^\wedge(\xi + 2\pi\alpha)|^2 \cdot \Big|\frac{1}{n}\sum_{j=1}^{n} e^{2\pi i \alpha_j \cdot \alpha}\Big|^2 . \tag{3.7}$$

In some cases we can interpret this as the kernel of type (2.1), associated with a rescaled function ϕ. To this end, let $M \in \mathbb{Z}^{d\times d}$ be an integer matrix with $|\det M| = n$, and let α_j, $j = 1, \ldots, n$, be the representers of the cosets of $(M^T)^{-1}\mathbb{Z}^d/\mathbb{Z}^d$ chosen in the way that

$$\{\alpha_1, \ldots, \alpha_n\} = (M^T)^{-1}\mathbb{Z}^d \cap [0, 1[^d .$$

In this situation, we have

$$\frac{1}{n}\sum_{j=1}^{n} e^{2\pi i \alpha_j \cdot \alpha} = \begin{cases} 0, & \text{if } \alpha \in \mathbb{Z}^d \setminus M\mathbb{Z}^d, \\ 1, & \text{if } \alpha \in M\mathbb{Z}^d, \end{cases}$$

and we see that the kernel (3.7) reduces to

$$\kappa(\xi) = |1 - \phi^\wedge(\xi)|^2 + \sum_{0 \neq \alpha \in M\mathbb{Z}^d} |\phi^\wedge(\xi + 2\pi\alpha)|^2 .$$

This is the kernel (2.1) modulo the substitution $\phi^\wedge \to \phi^\wedge \circ M$.

References

[1] M. Beśka, K. Dziedziul: *Asymptotic formula for the error in cardinal interpolation*, Numer. Math. **89** (2001), 445–456.

[2] M. Beśka, K. Dziedziul: *Asymptotic formulas in cardinal interpolation and orthogonal projection*, in: *Recent Progress in Multivariate Approximation* (W. Haußmann, K. Jetter, M. Reimer, eds.), 139–157, Birkhäuser, Basel, 2001.

[3] M. Beśka, K. Dziedziul: *The asymptotic formula for the error in orthogonal projection*, Math. Nachrichten **233–234** (2002), 47–53.

[4] T. Blu, M. Unser: *Approximation error for quasi–interpolators and (multi–)wavelet expansions*, Appl. Comp. Harm. Analysis **6** (1999), 219–251.

[5] T. Blu, M. Unser: *Quantitative Fourier analysis of approximation techniques: Part I – Interpolators and projectors*, IEEE Trans. Signal Processing **47** (1999), 2783–2795.

[6] T. Blu, M. Unser: *Quantitative Fourier analysis of approximation techniques: Part II – Wavelets*, IEEE Trans. Signal Processing **47** (1999), 2796–2806.

[7] I. Daubechies, M. Unser: *On the approximation power of convolution–based least squares versus interpolation*, IEEE Trans. Signal Processing **45** (1997), 1697–1711.

[8] K. Jetter: *Multivariate approximation from the cardinal interpolation point of view*, in: *Approximation Theory VII* (E. W. Cheney, C. K. Chui, L. L. Schumaker, eds.), 131–161, Academic Press, New York, 1993.

[9] K. Jetter, D.–X. Zhou: *Order of linear approximation from shift invariant spaces*, Constr. Approx. **11** (1995), 423–438.

[10] K. Jetter, D.–X. Zhou: *Approximation order of linear operators onto finitely generated shift–invariant spaces*, manuscript, 28 pp. (July 1998).

[11] W. Sweldens, R. Piessens: *Quadrature formulae and asymptotic error expansions for wavelet approximations of smooth functions*, SIAM J. Numer. Anal. **31** (1994), 1240–1264.

[12] W. Sweldens, R. Piessens: *Asymptotic error expansion of wavelet approximations of smooth functions II*, Numer. Math. **68** (1994), 377–401.

Addresses:

Karol Dziedziul
Faculty of Applied Mathematics
Technical University of Gdańsk
Naturowicza 11/12
80–952 Gdańsk
Poland

Kurt Jetter
Institut für Angewandte Mathematik und Statistik
Universität Hohenheim
D–70593 Stuttgart
Germany

International Series of Numerical Mathematics
Vol. 145,

On the Approximation Order and Numerical Stability of Local Lagrange Interpolation by Polyharmonic Splines

Armin Iske

Dedicated to Professor Manfred Reimer on the occasion of his 70th birthday

Abstract

This paper proves convergence rates for local scattered data interpolation by polyharmonic splines. To this end, it is shown that the Lagrange basis functions of polyharmonic spline interpolation are invariant under uniform scalings. Consequences of this important result for the numerical stability of the local interpolation scheme are discussed. A stable algorithm for the evaluation of polyharmonic spline interpolants is proposed.

1 Introduction

Polyharmonic splines, also often referred to as *surface splines*, are powerful tools for multivariate scattered data interpolation (see [5, 10, 11] for surveys). In this problem, a data vector $f|_X = (f(x_1), \dots, f(x_n))^T \in \mathbb{R}^n$ of function values, sampled from an unknown function $f : \mathbb{R}^d \to \mathbb{R}$ at a *scattered* point set $X = \{x_1, \dots, x_n\} \subset \mathbb{R}^d$, $d \geq 1$, is assumed to be given. Scattered data interpolation requires computing a *suitable* interpolant $s : \mathbb{R}^d \to \mathbb{R}$ satisfying $s|_X = f|_X$, i.e.,

$$s(x_j) = f(x_j), \qquad \text{for all } 1 \leq j \leq n. \tag{1.1}$$

To this end, the polyharmonic spline interpolation scheme works with a fixed *radial* function,

$$\phi_{d,k}(r) = \begin{cases} r^{2k-d} \log(r), & \text{for } d \text{ even}, \\ r^{2k-d}, & \text{for } d \text{ odd}, \end{cases}$$

where $2k > d$, and with $k \in \mathbb{N}$. The interpolant s in (1.1) is required to have the form

$$s(x) = \sum_{j=1}^{n} c_j \phi_{d,k}(\|x - x_j\|) + p(x), \qquad p \in \mathcal{P}_m^d, \tag{1.2}$$

where $\|\cdot\|$ is the Euclidean norm on $\mathbb{R}^d$. Moreover, $\mathcal{P}_m^d$ denotes the linear space containing all real–valued polynomials in d variables of degree at most $m-1$, where we let $m = k - \lceil d/2 \rceil + 1$ for the *order* of the conditionally positive (negative) definite function $\phi_{d,k}$. For more details concerning conditionally positive definite functions, see [19, 24].

By the n linear equations in the interpolation conditions (1.1), and by additionally requiring the q *vanishing moment conditions*

$$\sum_{j=1}^{n} c_j p(x_j) = 0, \qquad \text{for all } p \in \mathcal{P}_m^d,$$

with $q = \binom{m-1+d}{d}$ being the dimension of $\mathcal{P}_m^d$, the $n+q$ unknown coefficients of s in (1.2) can be computed by solving the resulting linear system of size $(n+q) \times (n+q)$. This linear system has always a solution, which is unique, provided that the point set X is $\mathcal{P}_m^d$*–unisolvent* [19]. The latter is equivalent to requiring that there is no nontrivial polynomial in $\mathcal{P}_m^d$ that vanishes on all points in X, i.e., for $p \in \mathcal{P}_m^d$ we have the implication

$$p(x_j) = 0 \quad \text{for } 1 \leq j \leq n \qquad \Longrightarrow \qquad p \equiv 0.$$

Much of the ground–breaking work on the theory of polyharmonic spline interpolation was done by Duchon [7, 8, 9] and by Meinguet [17, 18] in the late 70s. Apart from fundamental questions concerning the solvability and uniqueness of polyharmonic spline interpolation, Duchon [7, 8, 9] did also address the central aspect of *convergence* and *convergence rates*, albeit for special cases, such as for *thin plate spline* interpolation in the plane, where $k = d = 2$, and therefore $\phi_{2,2} = r^2 \log(r)$. Since then, several improvements on various different aspects concerning the error analysis of scattered data interpolation by polyharmonic splines have been made [14, 15, 16, 21, 25, 26].

In the setting of these papers, the resulting convergence rates are, for a fixed bounded and open domain $\Omega \subset \mathbb{R}^d$ comprising X, $X \subset \Omega$, proven in terms of the *fill distance*

$$h_{X,\Omega} = \sup_{y \in \Omega} \min_{x \in X} \|y - x\|$$

of X in Ω. So were e.g. pointwise error estimates, due to Wu and Schaback [26], used in [25] in order to obtain, for functions f in the *native space* $\mathcal{F}_{\phi_{d,k}}$ of $\phi_{d,k}$, uniform error bounds of the form

$$\|f - s\|_{L_\infty(\Omega)} \leq C \cdot |f|_{\mathcal{F}_{\phi_{d,k}}} \cdot h_{X,\Omega}^{k-d/2}, \qquad \text{for } f \in \mathcal{F}_{\phi_{d,k}}, \tag{1.3}$$

where Ω is required to satisfy an *interior cone condition.*

In this sense, the *approximation order* of polyharmonic spline interpolation is $p = k - d/2$ for functions f in the native function space $\mathcal{F}_{\phi_{d,k}}$. We remark that this approximation order can, due to Schaback [22], be doubled, but with requiring more regularity for f. Moreover, *saturation theorems* concerning special cases of polyharmonic spline interpolation are proven in [25].

This paper proves convergence rates for *local* Lagrange interpolation by polyharmonic splines. To this end, we consider for any fixed point x_0 and a $\mathcal{P}_m^d$-unisolvent point set $X = \{x_1, \dots, x_n\} \subset \mathbb{R}^d$ the interpolation problem

$$s^h(x_0 + hx_j) = f(x_0 + hx_j), \qquad 1 \leq j \leq n, \tag{1.4}$$

in a local neighbourhood $U_h(x_0)$ of x_0, where h is a positive scaling parameter. Due to our assumptions on X, there is for any $h > 0$ a unique polyharmonic spline interpolant s^h, using $\phi_{d,k}$, satisfying (1.4). The following Section 2 discusses further details concerning local polyharmonic spline interpolation.

Then, in Section 3 the convergence rate of polyharmonic spline interpolation is determined in terms of asymptotic bounds of the form

$$|s^h(x_0 + hx) - f(x_0 + hx)| = \mathcal{O}(h^p), \quad h \to 0, \tag{1.5}$$

in which case p is said to be the *approximation order* at x_0. A precise definition of this term is given in Section 3, Definition 3.1. Note that this concept for a *local* convergence rate is in contrast to the *global* one above. Indeed, the bound in (1.3) is obtained for a small fill distance $h_{X,\Omega}$ of X in the fixed domain Ω. This requires a high density of X in the global domain Ω, whereas the asymptotic bound (1.5) is obtained by locally scaling the interpolation points relative to the center x_0.

It is shown in Section 3 that the convergence rate p in (1.5) is, for functions f which are C^m locally around x_0, given by the abovementioned order m of $\Phi_{d,k}$, i.e., $p = k - \lceil d/2 \rceil + 1$. Note that this is in contrast to the global case in (1.3), where we only obtain $p = k - d/2$ for functions in the native space $\mathcal{F}_{\phi_{d,k}}$.

In order to establish approximation orders for local interpolation, it is shown that the Lagrange basis functions of the polyharmonic spline interpolant s^h are invariant under scalings of h. This key observation, to be proven in Section 3, Lemma 3.2, is due to the homogeneity of the radial basis functions $\phi_{d,k}$. For a discussion on further consequences from this essential property of polyharmonic splines we refer to the related work [1].

The scale–invariance of the Lagrange basis allows us to draw further conclusions concerning the condition number of the interpolation problem (1.1) and the numerical stability of polyharmonic spline interpolation. These important aspects are subject of the discussion in Section 4, where a stable algorithm for evaluating the interpolant s^h is proposed for situations where the point set X is of *moderate* size, i.e., for *small* n.

We finally remark that the motivation for analyzing the approximation order and numerical stability of local polyharmonic spline interpolation comes from applications concerning the numerical simulation of multiscale phenomena in transport processes. To this end, an adaptive and meshfree method of backward characteristics for solving both linear and nonlinear transport equations is proposed in the previous papers [2, 4]. This particle–based Lagrangian advection scheme works with local Lagrange interpolation by polyharmonic splines, at a *small* number of scattered nodes, each of which corresponds (at a time t) to one flow particle. In this method, the local interpolation at the nodes is combined in order to obtain global approximations to the solution of a hyperbolic equation. The performance of this advection scheme essentially relies on both the approximation order and the numerical stability of local Lagrange interpolation. For further details on this, see [2, 3, 4].

2 Local Interpolation by Polyharmonic Splines

In this section, we recall selected basic features of polyharmonic spline interpolation, which are relevant for the following discussion in this text. Moreover, we introduce some notations. Starting point of the subsequent discussion is the *scaled* interpolation problem (1.4). Due to the shift–invariance of polyharmonic spline interpolation, we assume from now $x_0 = 0$ without loss of generality, so that (1.4) becomes

$$s^h(hx_j) = f(hx_j), \qquad 1 \le j \le n. \tag{2.1}$$

For any $h > 0$, and a fixed $\mathcal{P}_m^d$–unisolvent point set X, the interpolation problem (2.1) has under constraints

$$\sum_{j=1}^{n} c_j^h p(hx_j) = 0, \qquad \text{for all } p \in \mathcal{P}_m^d, \tag{2.2}$$

a unique solution s^h of the form

$$s^h(hx) = \sum_{j=1}^{n} c_j^h \phi_{d,k}(\|hx - hx_j\|) + \sum_{|\alpha|<m} d_\alpha^h (hx)^\alpha, \tag{2.3}$$

whose coefficients $c^h = (c_1^h, \ldots, c_n^h)^T \in \mathbb{R}^n$ and $d^h = (d_\alpha^h)_{|\alpha|<m} \in \mathbb{R}^q$ can be computed by solving the linear system

$$\begin{bmatrix} \Phi_h & \Pi_h \\ \Pi_h^T & 0 \end{bmatrix} \cdot \begin{bmatrix} c^h \\ d^h \end{bmatrix} = \begin{bmatrix} f|_{hX} \\ 0 \end{bmatrix}, \tag{2.4}$$

where we let

$$\begin{aligned} \Phi_h &= (\phi_{d,k}(\|hx_i - hx_j\|))_{1\le i,j\le n} \in \mathbb{R}^{n\times n}, \\ \Pi_h &= ((hx_i)^\alpha)_{1\le i\le n; |\alpha|<m} \in \mathbb{R}^{n\times q}, \\ f|_{hX} &= (f(hx_i))_{1\le i\le n} \in \mathbb{R}^n. \end{aligned}$$

In the following discussion it will be useful to abbreviate the above linear system (2.4) as

$$A_h \cdot b^h = f_h, \tag{2.5}$$

i.e., for the sake of notational brevity, we let

$$A_h = \begin{bmatrix} \Phi_h & \Pi_h \\ \Pi_h^T & 0 \end{bmatrix}, \quad b^h = \begin{bmatrix} c^h \\ d^h \end{bmatrix}, \quad \text{and} \quad f_h = \begin{bmatrix} f|_{hX} \\ 0 \end{bmatrix}.$$

We recall the *Lagrange representation* of s^h in (2.3), given by

$$s^h(hx) = \sum_{i=1}^{n} \lambda_i^h(hx) f(hx_i), \tag{2.6}$$

with the *Lagrange basis functions* λ_i^h satisfying

$$\lambda_i^h(hx_j) = \begin{cases} 1 & \text{for } i = j, \\ 0 & \text{for } i \ne j. \end{cases}$$

Moreover, due to the reproduction of polynomials from $\mathcal{P}_m^d$, we have

$$\sum_{i=1}^{n} \lambda_i^h(hx) p(hx_i) = p(hx), \qquad \text{for all } p \in \mathcal{P}_m^d. \tag{2.7}$$

The Lagrange functions can pointwise, at any hx, be evaluated by solving the linear system

$$\begin{bmatrix} \Phi_h & \Pi_h \\ \Pi_h^T & 0 \end{bmatrix} \cdot \begin{bmatrix} \lambda^h(hx) \\ \mu^h(hx) \end{bmatrix} = \begin{bmatrix} \varphi_h(hx) \\ \pi_h(hx) \end{bmatrix}, \tag{2.8}$$

where we let $\lambda^h(hx) = (\lambda_i^h(hx))_{1\leq i\leq n} \in \mathbb{R}^n$, and $\mu^h(hx) = (\mu_\alpha^h(hx))_{|\alpha|<m} \in \mathbb{R}^q$ for the vector of the *Lagrange multipliers*. Moreover, the right hand side

$$\begin{aligned} \varphi_h(hx) &= (\phi_{d,k}(\|hx - hx_j\|))_{1\leq j\leq n} \in \mathbb{R}^n, \\ \pi_h(hx) &= ((hx)^\alpha)_{|\alpha|<m} \in \mathbb{R}^q, \end{aligned}$$

in (2.8) contains the point evaluations of the basis functions at hx. Again, it is convenient to abbreviate the above system (2.8) by

$$A_h \cdot \nu^h(hx) = \beta_h(hx),$$

where we let

$$\nu^h(hx) = \begin{bmatrix} \lambda^h(hx) \\ \mu^h(hx) \end{bmatrix} \quad \text{and} \quad \beta_h(hx) = \begin{bmatrix} \varphi_h(hx) \\ \pi_h(hx) \end{bmatrix}.$$

Starting with the Lagrange representation of s^h in (2.6), we obtain

$$\begin{aligned} s^h(hx) &= < \lambda^h(hx), f|_{hX} > \\ &= < \nu^h(hx), f_h > \\ &= < A_h^{-1} \cdot \beta_h(hx), f_h > \\ &= < \beta_h(hx), A_h^{-1} \cdot f_h > \\ &= < \beta_h(hx), b_h >, \end{aligned} \tag{2.9}$$

where $< \cdot, \cdot >$ denotes the inner product of the Euclidean space $\mathbb{R}^d$. This in particular combines the two alternative representations for s^h in (2.6) and (2.3).

3 Approximation Order

The discussion in this section is dominated by the following definition.

Definition 3.1. *Let s^h denote the polyharmonic spline interpolant, using $\phi_{d,k}$, satisfying* (2.1). *We say that the* approximation order *of local polyharmonic spline interpolation with respect to the function space $\mathcal{F}$ is p, iff for any $f \in \mathcal{F}$ the asymptotic bound*

$$|f(hx) - s^h(hx)| = \mathcal{O}(h^p), \quad h \to 0,$$

holds for any $x \in \mathbb{R}^d$, and any finite $\mathcal{P}_m^d$–unisolvent point set $X \subset \mathbb{R}^d$.

Recall from the discussion in the previous Section 2 that any interpolant s^h satisfying (2.1) has a unique Lagrange representation of the form (2.6). The following lemma concerning the scale–invariance of the Lagrange basis plays a key role in the following discussion of this paper.

Lemma 3.2. *The Lagrange basis functions of polyharmonic spline interpolation are invariant under uniform scalings, i.e., for any $h > 0$, we have*

$$\lambda^h(hx) = \lambda^1(x), \qquad \textit{for every } x \in \mathbb{R}^d.$$

Proof. For $h > 0$, let

$$\mathcal{S}_h = \left\{ \sum_{j=1}^n c_j \phi_{d,k}(\| \cdot - hx_j \|) + p \, : p \in \mathcal{P}_m^d, \sum_{j=1}^n c_j q(x_j) = 0 \text{ for all } q \in \mathcal{P}_m^d \right\}$$

denote the space of all possible polyharmonic spline interpolants of the form (2.3) satisfying (2.2). In what follows, we show that $\mathcal{S}_h$ is a scaled version of $\mathcal{S}_1$, so that $\mathcal{S}_h = \{\sigma_h(s) : s \in \mathcal{S}_1\}$, where the dilatation operator σ_h is given by $\sigma_h(s) = s(\cdot/h)$. This then implies that, due to the unicity of the interpolation in either space, $\mathcal{S}_h$ or $\mathcal{S}_1$, their Lagrange basis functions satisfy $\lambda^h = \sigma_h(\lambda^1)$, as stated above.

In order to show that $\mathcal{S}_h = \sigma_h(\mathcal{S}_1)$, we distinguish the special case where d is even from the one where d is odd. If the space dimension d is odd, then $\mathcal{S}_h = \sigma_h(\mathcal{S}_1)$ follows immediately from the homogeneity of $\phi_{d,k}$, where $\phi_{d,k}(hr) = h^{2k-d}\phi_{d,k}(r)$.

Now suppose that d is even. In this case we have

$$\phi_{d,k}(hr) = h^{2k-d}\left(\phi_{d,k}(r) + r^{2k-d}\log(h)\right).$$

Therefore, any function $s^h \in \mathcal{S}_h$ has, for some $p \in \mathcal{P}_m^d$, the form

$$s^h(hx) = h^{2k-d}\left(\sum_{j=1}^{n} c_j\phi_{d,k}(\|x - x_j\|) + \log(h)r(x)\right) + p(x),$$

where we let

$$r(x) = \sum_{j=1}^{n} c_j\|x - x_j\|^{2k-d}.$$

In order to see that s^h is contained in $\sigma_h(\mathcal{S}_1)$, it remains to show that the degree of the polynomial r is at most $m-1$. To this end, we rewrite r as

$$r(x) = \sum_{j=1}^{n} c_j \sum_{|\alpha|+|\beta|=2k-d} c_{\alpha,\beta} \cdot x^\alpha (x_j)^\beta = \sum_{|\alpha|+|\beta|=2k-d} c_{\alpha,\beta} \cdot x^\alpha \sum_{j=1}^{n} c_j (x_j)^\beta,$$

for some coefficients $c_{\alpha,\beta} \in \mathbb{R}$ with $|\alpha| + |\beta| = 2k - d$. Due to the vanishing moment conditions (2.2) for the coefficients $c_1, \ldots, c_n$, this implies that the degree of r is at most $2k - d - m = k - d/2 - 1 < m$. Therefore, $s^h \in \sigma_h(\mathcal{S}_1)$, and so $\mathcal{S}_h \subset \sigma_h(\mathcal{S}_1)$. The inclusion $\mathcal{S}_1 \subset \sigma_h^{-1}(\mathcal{S}_h)$ can be proven accordingly.

Altogether, we find that $\mathcal{S}_h = \sigma_h(\mathcal{S}_1)$ for any d, which completes our proof. □

We are now in a position to discuss the *approximation order* of local polyharmonic spline interpolation, according to Definition 3.1, for functions f in C^m.

To this end, regard for fixed $h > 0$ and $x \in \mathbb{R}^d$ the m–th order Taylor polynomial

$$T_{f,hx}^m(y) = \sum_{|\alpha|<m} \frac{1}{\alpha!} D^\alpha f(hx)(y - hx)^\alpha,$$

of $f \in C^m$ around hx, which yields the identity

$$f(hx) = T_{f,hx}^m(hx_i) - \sum_{0<|\alpha|<m} \frac{1}{\alpha!} D^\alpha f(hx)(hx_i - hx)^\alpha, \qquad \text{for all } 1 \le i \le n.$$

By using the representation (2.6) for s^h and the polynomial reproduction property (2.7), this immediately implies

$$f(hx) - s^h(hx) = \sum_{i=1}^{n} \lambda_i^h(hx)\left[T_{f,hx}^m(hx_i) - f(hx_i)\right].$$

Now due to Lemma 3.2, the *Lebesgue function*

$$\Lambda(x) = \sum_{i=1}^{n} |\lambda_i^h(hx)| = \sum_{i=1}^{n} |\lambda_i^1(x)|$$

is uniformly bounded in any local neighbourhood of the origin. Since we have

$$T_{f,hx}^m(hx_i) - f(hx_i) = \mathcal{O}(h^m), \quad h \to 0, \qquad \text{for all } 1 \le i \le n,$$

this implies

$$|f(hx) - s^h(hx)| = \mathcal{O}(h^m), \quad h \to 0.$$

Altogether, this proves the following

Theorem 3.3. *The approximation order of local polyharmonic spline interpolation, using $\phi_{d,k}$, with respect to C^m is $m = k - \lceil d/2 \rceil + 1$.* □

We remark that the above Theorem 3.3 generalizes a previous result in [12] concerning the approximation order of local thin plate spline interpolation in the plane.

Corollary 3.4. *The approximation order of local thin plate spline interpolation, using $\phi_{2,2} = r^2 \log(r)$, with respect to C^2 is $m = 2$.* □

4 Numerical Stability

This section is devoted to the construction of a numerically stable algorithm for the evaluation of polyharmonic spline interpolants. Recall that the stability of an algorithm always depends on the conditioning of the given problem. For a more general discussion on the relevant principles and concepts from error analysis, especially the *condition number* of a given *problem* versus the *stability* of a *numerical algorithm*, we recommend the textbook [13].

In order to briefly explain the conditioning of polyharmonic spline interpolation, let $U \subset \mathbb{R}^d$ denote a compact domain comprising $X = \{x_1, \ldots, x_n\} \subset U$, the $\mathcal{P}_m^d$–unisolvent set of interpolation points. The condition number of polyharmonic spline interpolation w.r.t. the L_∞–norm $\|\cdot\|_{L_\infty(U)}$ is given by the *Lebesgue constant*

$$\Lambda = \max_{x \in U} \sum_{i=1}^{n} |\lambda_i^1(x)|.$$

Note that a corresponding result for (univariate) polynomial interpolation is well–known from numerical analysis, see e.g. the textbook [6], where the arguments from there directly carry over to the situation of polyharmonic spline interpolation. Due to Lemma 3.2 the Lebesgue constant $\Lambda = \Lambda(U, X)$ is invariant under uniform scalings. But $\Lambda(U, X)$ is also invariant under rotations and translations. We summarize these observations as follows.

Theorem 4.1. *The condition number of polyharmonic spline interpolation is invariant under rotations, translations and uniform scalings.* □

Now let us turn to the construction of a numerically stable algorithm for evaluating the polyharmonic spline interpolant s^h satisfying (2.1). To this end, we require that the given interpolation problem (2.1) is well–conditioned. Note that according to Theorem 4.5, this requirement depends on the geometry of the interpolation points X, but not on the scale h.

However, the spectral condition number of the matrix A_h depends on h. The following rescaling can be viewed as a simple way of preconditioning the matrix A_h for very small h. To this end, in order to evaluate the polyharmonic spline interpolant s^h satisfying (2.1), we prefer to work with the representation

$$s^h(hx) = < \beta_1(x), A_1^{-1} \cdot f_h >, \tag{4.1}$$

which immediately follows from the identity (2.9) and the scale–invariance of the Lagrange basis, Lemma 3.2. Due to (4.1) we can evaluate s^h at hx by solving the linear system

$$A_1 \cdot \mathbf{b} = f_h. \tag{4.2}$$

The solution $\mathbf{b} \in \mathbb{R}^{n+q}$ in (4.2) then yields the coefficients of $s^h(hx)$ w.r.t. the basis functions in $\beta_1(x)$.

By working with the representation (4.1) for s^h instead of the one in (2.3), we can avoid solving the linear system (2.5). This is useful insofar as the linear system (2.5) is ill–conditioned for very small h, but well–conditioned for sufficiently large h. The latter relies on earlier results due to Narcowich and Ward [20], where it is shown that the spectral norm of the matrix Φ_h^{-1} is bounded above by a monotonically decreasing function of the minimal Euclidean distance between the points in hX. This in turns implies that one should, for the sake of numerical stability, avoid solving the system (2.5) directly for very small h. For further details on this, see [20] and the more general discussion provided by the recent paper [23] of Schaback.

Acknowledgment

I wish to thank the anonymous referees for their useful comments and suggestions. This work was partly supported by the European Union within the project NetAGES (Network for Automated Geometry Extraction from Seismic), contract no. IST–1999–29034.

References

[1] R. K. Beatson, W. A. Light, S. Billings: *Fast solution of the radial basis function interpolation equations: domain decomposition methods,* SIAM J. Sci. Comput. **22** (2000), 1717–1740.

[2] J. Behrens, A. Iske: *Grid–free adaptive semi–Lagrangian advection using radial basis functions,* Computers and Mathematics with Applications, **43**:3–5 (2002), 319–327.

[3] J. Behrens, A. Iske, S. Pöhn: *Effective node adaption for grid–free semi–Lagrangian advection,* Discrete Modelling and Discrete Algorithms in Continuum Mechanics, Th. Sonar, I. Thomas (eds.), Logos Verlag, Berlin, 2001, 110–119.

[4] J. Behrens, A. Iske, M. Käser: *Adaptive meshfree method of backward characteristics for nonlinear transport equations,* Meshfree Methods for Partial Differential Equations, M. Griebel, M.A. Schweitzer (eds.), Springer, Heidelberg, 2002, 21–36.

[5] M. D. Buhmann: *Radial basis functions,* Acta Numerica, 2000, 1–38.

[6] P. Deuflhard, A. Hohmann: *Numerical Analysis in Modern Scientific Computing: An Introduction.* Second edition, Springer, New York, 2003.

[7] J. Duchon: *Interpolation des Fonctions de deux variables suivant le principe de la flexion des plaques minces,* R.A.I.R.O. Analyse Numeriques **10** (1976), 5–12.

[8] J. Duchon: *Splines minimizing rotation–invariant semi–norms in Sobolev spaces,* in Constructive Theory of Functions of Several Variables, W. Schempp, K. Zeller (eds.), Springer, Berlin, 1977, 85–100.

[9] J. Duchon: *Sur l'erreur d'interpolation des fonctions de plusieurs variables par les D^m-splines,* R.A.I.R.O. Analyse Numeriques **12** (1978), 325–334.

[10] N. Dyn: *Interpolation of scattered data by radial functions,* in *Topics in Multivariate Approximation,* C. K. Chui, L. L. Schumaker, F. I. Utreras (eds.), Academic Press, 1987, 47–61.

[11] N. Dyn: *Interpolation and approximation by radial and related functions,* in *Approximation Theory VI,* C. K. Chui, L. L. Schumaker, J. D. Ward (eds.), Academic Press, New York, 1989, 211–234.

[12] T. Gutzmer, A. Iske: *Detection of discontinuities in scattered data approximation,* Numer. Algorithms **16** (1997), 155–170.

[13] N. J. Higham: *Accuracy and Stability of Numerical Algorithms.* SIAM, Philadelphia, 1996.

[14] W. Light, H. Wayne: *On power functions and error estimates for radial basis function interpolation,* J. Approx. Theory **92**, 1998, 245–266.

[15] W. R. Madych, S. A. Nelson: *Multivariate interpolation and conditionally positive definite functions,* Approx. Theory Appl. **4** (1988), 77–89.

[16] W. R. Madych, S. A. Nelson: *Multivariate interpolation and conditionally positive definite functions II,* Math. Comp. **54** (1990), 211–230.

[17] J. Meinguet: *Multivariate interpolation at arbitrary points made simple,* Z. Angew. Math. Phys. **30** (1979), 292–304.

[18] J. Meinguet: *An intrinsic approach to multivariate spline interpolation at arbitrary points,* in *Polynomial and Spline Approximations,* N.B. Sahney (ed.), Reidel, Dordrecht, 1979, 163–190.

[19] C. A. Micchelli: *Interpolation of scattered data: distance matrices and conditionally positive definite functions,* Constr. Approx. **2**, 1986, 11–22.

[20] F. J. Narcowich, J. D. Ward: *Norm estimates for the inverses of a general class of scattered-data radial-function interpolation matrices,* J. Approx. Theory **69**, 1992, 84–109.

[21] M. J. D. Powell: *The uniform convergence of thin plate spline interpolation in two dimensions,* Numer. Math. **68**, 1994, 107–128.

[22] R. Schaback: *Improved error bounds for scattered data interpolation by radial basis functions,* Math. Comp. **68** (1999), 201–216.

[23] R. Schaback: *Stability of radial basis function interpolants,* in *Approximation Theory X: Wavelets, Splines, and Applications,* C.K. Chui, L.L. Schumaker, J. Stöckler (eds.), Vanderbilt Univ. Press, Nashville, 2002, 433–440.

[24] R. Schaback, H. Wendland: *Characterization and construction of radial basis functions,* in *Multivariate Approximation and Applications,* N. Dyn, D. Leviatan, D. Levin, A. Pinkus (eds.), Cambridge University Press, Cambridge, 2001, 1–24.

[25] R. Schaback, H. Wendland: *Inverse and saturation theorems for radial basis function interpolation,* Math. Comp. **71** (2002), 669–681.

[26] Z. Wu, R. Schaback: *Local error estimates for radial basis function interpolation of scattered data,* IMA J. Numer. Anal. **13** (1993), 13–27.

Address:

Armin Iske
Zentrum Mathematik
Technische Universität München
D–85747 Garching
Germany

International Series of Numerical Mathematics
Vol. 145, ©2003 Birkhäuser Verlag Basel/Switzerland

Isolated Singularities of Harmonic Functions

Lavi Karp and Harold S. Shapiro

Abstract

The main result of this paper gives a sufficient condition for removability of an isolated singularity of a harmonic function. The condition is given in terms of Newtonian capacity. In addition, an application to an approximation problem is presented.

Introduction

This note deals with a problem of removable isolated singularities of harmonic functions. Suppose u is harmonic and of at most polynomial growth on a punctured neighborhood of a point x_0, and its gradient ∇u satisfies $|\nabla u(x)| \leq C|x - x_0|^{-1}$ on a set K, where x_0 is an accumulation point of K. Under these hypotheses, what size restrictions on K ensure that x_0 is a removable singularity of u? This type of growth constraint is rather natural in the case of harmonic functions, while for analytic functions, a bound on the function is often used, and the answer in this case is very well known (see e.g. [1], Proposition 2.4.4):

Theorem A. *Assume f is holomorphic and single valued on the punctured disk $\{z : 0 < |z - z_0| < r\}$ and bounded on a sequence of points clustering at z_0. Then f cannot have a polar singularity at z_0.*

The two dimensional case of the removable singularities problem for harmonic functions mentioned above, and with a bounded gradient on the set K, is settled by Theorem A. However, in higher dimensions the situation is more complicated. We present here a generalization of the above for harmonic functions in $\mathbb{R}^n$, $n \geq 3$. The Newtonian capacity will measure the "*size*" of K.

In order to state our result, we introduce some notations. We let $B_r(x_0) = \{x : |x - x_0| < r\}$ be the ball of radius r centered at x_0, $B_r = B_r(0)$, $S(a, b, x_0) = \{x : a \leq |x - x_0| \leq b\}$ and $S(a, b)$ shall denote $S(a, b, 0)$. The gradient of a function is denoted by ∇, Δ is the Laplace operator and cap (K) denotes the Newtonian capacity of the set K (see [4]).

Theorem 0.1. *Let m be a positive integer and assume that a function u satisfies*

(*a*) $\Delta u = 0$ *in* $B_1(x_0) \setminus \{x_0\}$,

(*b*) $u(x) = O(|x - x_0|^{-m})$ *as* $x \to x_0$,

(*c*) $|\nabla u(x)| \leq \frac{C}{|x-x_0|}$ *on* $K \cap (B_1(x_0) \setminus \{x_0\})$.

If

$$\limsup_{r\to 0} \frac{\text{cap}\left(K \cap S(\tfrac{1}{2}r, r, x_0)\right)}{\text{cap}\left(S(\tfrac{1}{2}r, r)\right)} > 0, \tag{0.1}$$

then u extends to a harmonic function on all of $B_1(x_0)$.

Remark. The condition (0.1) is called a capacity density condition. It has appeared in various papers on potential theory, see [2] and the references there.

The plan of the paper is as follows: In Section 1 we prove Theorem 0.1 and we will also provide an example showing that the result is not true, if the capacity density condition (0.1) is abandoned. In Section 2 we will apply Theorem 0.1 to an approximation problem.

1 Proof of the Main Result

The proof relies on certain results from [2]. Let F be a closed bounded set of $\mathbb{R}^n$, and $C^1(F)$ the set of restriction to F of continuously differentiable functions on $\mathbb{R}^n$.

Definition 1.1 (Λ–stable). *A family $\mathcal{E}$ of subsets of F is Λ–stable, if for every sequence $\{E_j\} \subset \mathcal{E}$, the set*

$$\bigcap_{m=1}^{\infty} \left(\overline{\bigcup_{j=m}^{\infty} E_j} \right)$$

belongs to the family $\mathcal{E}$. Here $\overline{(\,\cdot\,)}$ denotes the closure.

Definition 1.2 (Sets of uniqueness). *Let P be a closed subspace of $C^1(F)$. A closed set K of F is a set of uniqueness for P, if the only p in P with gradient vanishing on all of K are constants.*

Lemma 1.3 ([2]). *Suppose that P is a finite–dimensional subspace of $C^1(F)$ and that $\mathcal{E}$ is a Λ–stable family of subsets of F, each of which is a set of uniqueness for P. Then there is a constant $C_{\mathcal{E}}$ such that for each K in $\mathcal{E}$*

$$\max_{x\in F}|p(x)| \le C_{\mathcal{E}} \max_{x\in K}|\nabla p(x)|$$

for every p in P with $p(x_0) = 0$, where x_0 is a fixed point in F.

Remark. The requirement $p(x_0) = 0$ is not necessary in case the subspace P does not contain a nonzero constant function.

Set $F = S(\frac{1}{2}, 1) = \{\frac{1}{2} \le |x| \le 1\}$ and let $\mathcal{E}$ be the family of all Borel subsets E of F satisfying

$$\text{cap}\,(E) \ge \gamma, \tag{1.1}$$

where γ is a positive constant.

To verify that $\mathcal{E}$ is Λ–stable one has to check the following three conditions:

(i) $\mathcal{E}$ is closed under formation of countable unions.

(ii) If $E \in \mathcal{E}$, then $\overline{E} \in \mathcal{E}$.

(iii) The intersection of a countable, monotone decreasing sequence of closed nonempty sets from $\mathcal{E}$ is again in $\mathcal{E}$.

It is known that the family defined by (1.1) has these properties. We refer to [4] and [2] for further details. We now introduce the finite dimensional space $\mathcal{H}_m$, consisting of harmonic functions of the following form:

$$h \in \mathcal{H}_m \Leftrightarrow h(x) = \sum_{j=0}^{m} \frac{p_j(x)}{|x|^{2j+n-2}},$$

where p_j is a homogeneous harmonic polynomial of degree j.

The set $\{x : \ |\nabla u(x)| = 0\}$ is called the critical set of the function u. Kuran proved that the critical set of a non–constant harmonic function has zero Newtonian capacity (see [3] Lemma 1). That is, a compact subset K of $S(\frac{1}{2}, 1)$ of positive capacity is a set of uniqueness for the space $\mathcal{H}_m$.

Applying Lemma 1.3 with $\mathcal{E}$ as above and with $P = \mathcal{H}_m$ we get

Corollary 1.4. *Given a positive constant γ and a positive integer m, there is a positive constant $c(\gamma, m)$ such that*

$$\max_{x \in S(\frac{1}{2},1)} |h(x)| \le c(\gamma, m) \max_{x \in K} |\nabla h(x)| \tag{1.2}$$

holds for every h in $\mathcal{H}_m$ and for every compact set $K \subset S(\frac{1}{2}, 1)$ with cap $(K) \ge \gamma$.

Proof of Theorem 0.1. First, we may assume that $x_0 = 0$. Conditions (a) and (b) of Theorem 0.1 imply that $u = u_0 + h$, where u_0 is harmonic in B_1 and h in $\mathcal{H}_m$. By the capacity density condition (0.1), there is a sequence of positive numbers $\{r_j\}$, $r_j \searrow 0$, and there is a constant $\gamma > 0$ such that

$$\frac{\operatorname{cap}\Big(K \cap S(\frac{1}{2}r_j, r_j)\Big)}{\operatorname{cap}\Big(S(\frac{1}{2}r_j, r_j)\Big)} \ge \gamma, \qquad j = 1, 2, \ldots \tag{1.3}$$

Let

$$h_j(x) = \frac{h(r_j x)}{r_j^2}$$

and

$$K_j = \{x : \ r_j x \in K\}.$$

Then, (1.3) becomes

$$\frac{\operatorname{cap}\Big(K_j \cap S(\frac{1}{2}, 1)\Big)}{\operatorname{cap}\Big(S(\frac{1}{2}, 1)\Big)} \ge \gamma, \qquad j = 1, 2, \ldots$$

and by condition (c),

$$|\nabla h_j(x)| = \frac{|\nabla h_j(r_j x)|}{r_j} \le \frac{C}{r_j^2 |x|} \le \frac{2C}{r_j^2} \qquad \text{for } x \in K_j \cap S(\frac{1}{2}, 1).$$

Since $h_j \in \mathcal{H}_m$, we may apply Corollary 1.4 and (1.2) and obtain

$$\max_{S(\frac{1}{2},1)} |h_j(x)| \le \frac{2Cc(\gamma, m)}{r_j^2},$$

or

$$\max_{S(\frac{1}{2}r_j,r_j)} |h(x)| \le 2Cc(\gamma, m),$$

where $r_j \to 0$. This, together with h belonging to $\mathcal{H}_m$, implies $h \equiv 0$. Hence $u = u_0$ is harmonic in B_1. □

Example. We give an example showing that if (0.1) is not assumed, then the conclusion of Theorem 0.1 no longer holds. Let $n = 3$, and take

$$K = \{x \in \mathbb{R}^3 : \sqrt{x_1^2 + x_2^2} \le x_3^4\}.$$

We verify that (0.1) does not hold. For each $0 < r < 1$, the set $S(\frac{1}{2}r, r) \cap K = \{\frac{1}{2}r \le |x| \le r\} \cap K$ is included in a cylinder of height r and base radius r^4, and hence it is contained in an ellipsoid with semi–axes r^4, r^4 and r. The Newtonian capacity of an elongated ellipsoid

$$\{x \in \mathbb{R}^3 : \frac{x_1^2 + x_2^2}{a^2} + \frac{x_3^2}{b^2}, \quad 0 < a < b\}$$

is equal to

$$\frac{2\sqrt{b^2 - a^2}}{\pi \log\left(\frac{b+\sqrt{b^2-a^2}}{b-\sqrt{b^2-a^2}}\right)},$$

see [4] page 165. Setting $a = r^4$ and $b = r$, we have

$$\text{cap}\left(K \cap S(\frac{1}{2}r, r)\right) \le \frac{2r\sqrt{1-r^6}}{\pi\left(\log\left(\frac{1+\sqrt{1-r^6}}{1-\sqrt{1-r^6}}\right)\right)} = \frac{2r\sqrt{1-r^6}}{\pi\left(\log\left(1 + \frac{2\sqrt{1-r^6}}{1-\sqrt{1-r^6}}\right)\right)}.$$

Since the shell $S(\frac{1}{2}r, r)$ contains a ball of radius $\frac{r}{2}$,

$$\text{cap}\,(S(\frac{1}{2}r, r)) \ge \text{cap}\,(B_{\frac{r}{2}}) = \frac{r}{2\pi}.$$

Having these two estimates, we see that,

$$\frac{\text{cap}\left(S(\frac{1}{2}r, r) \cap K\right)}{\text{cap}\left(S(\frac{1}{2}r, r)\right)} \le \frac{4\sqrt{1-r^6}}{\log\left(1 + \frac{2\sqrt{1-r^6}}{1-\sqrt{1-r^6}}\right)} \to 0 \quad \text{as } r \to 0.$$

Thus (0.1) fails.

Let

$$u(x) = \frac{x_1 x_2}{|x|^5}.$$

It is easy to check that $|\nabla u(x)| \le \frac{C}{|x|}$ on K, for a positive constant C. Thus u satisfies the conditions (a), (b) and (c) of Theorem 0.1, and the origin is not a removable singularity of u.

2 An Approximation Problem

Let Ω be a bounded domain in $\mathbb{R}^n$, $n \geq 3$. We denote by $L^p(\Omega)$ the usual Lebesgue space, and by $HL^p(\Omega)$ its subspace of harmonic functions in Ω. We consider the subspace $\Phi_{x_0}(\Omega)$, of $HL^1(\Omega)$ consisting of harmonic functions whose partial derivatives of order less than or equal m vanish at the point x_0. We address the question: Under which conditions, is the subspace $\Phi_{x_0}(\Omega)$ dense in $HL^1(\Omega)$ in the L^1–norm. One should note that if x_0 is an interior point of Ω, then it is impossible that $\Phi_{x_0}(\Omega)$ will be dense in $HL^1(\Omega)$, because convergence in L^1–norm for harmonic functions implies the convergence in the supremum–norm in a neighborhood of x_0 and hence, for example, the function $h \equiv 1$ cannot be approximated by functions from $\Phi_{x_0}(\Omega)$. However, in case x_0 is a boundary point of Ω, we will show that the capacity density condition (2.2) suffices for such approximation.

We recall first a result of Sakai, [5]. Let $N(x) = c_n|x|^{2-n}$ be the Newtonian kernel, and $U^\mu = N * \mu$ the Newtonian potential of a distribution μ with a compact support. The constant c_n is chosen so that

$$\Delta U^\mu = \mu. \tag{2.1}$$

We set

$$\Phi(\Omega) = \text{span}\,\{N(y-x),\ \partial_i N(y-x),\ x \in \mathbb{R}^n \setminus \Omega, i = 1, ..., n\}.$$

Theorem B. (Sakai [5]). *For any bounded set Ω of $\mathbb{R}^n$, $\Phi(\Omega)$ is dense in $HL^1(\Omega)$.*

For a point $x_0 \in \partial\Omega$ and a positive integer m, we set

$$\Phi_{x_0,m}(\Omega) = \text{span}\Big\{ N(x-y) \ - \sum_{|\alpha|\leq m} \frac{(y-x_0)^\alpha}{\alpha!}(\partial^\alpha N)(x_0 - x),$$

$$\partial_i N(y-x) - \sum_{|\alpha|\leq m} \frac{(y-x_0)^\alpha}{\alpha!}(\partial^\alpha \partial_i N)(x_0 - x),$$

$$x \in \mathbb{R}^n \setminus \Omega, x \neq x_0, i = 1, ..., n \ \Big\}.$$

Theorem 2.1. *Let Ω be a bounded domain, m a positive integer and $x_0 \in \partial\Omega$. If*

$$\limsup_{r\to 0} \frac{\operatorname{cap}\left((\mathbb{R}^n \setminus \Omega) \cap S(\frac{1}{2}r, r, x_0)\right)}{\operatorname{cap}\left(S(\frac{1}{2}r, r)\right)} > 0, \tag{2.2}$$

then $\Phi_{x_0,m}(\Omega)$ is dense in $HL^1(\Omega)$ in the topology of the L^1–norm.

Corollary 2.2. *Let Ω be a bounded domain such that Ω is the interior of its closure, m a positive integer and $x_0 \in \partial\Omega$. If (2.2) holds, then the harmonic functions on a neighborhood of the closure of Ω such that their partial derivatives of order less or equal to m vanish at the point x_0, are dense in $HL^1(\Omega)$.*

Proof of Theorem 2.1. By the the Hahn–Banach theorem, we have to show that if $f \in L^\infty(\Omega)$ annihilates $\Phi_{x_0,m}(\Omega)$, then f annihilates $HL^1(\Omega)$ too. Let

$$v(x) = \int_\Omega \left(N(y-x) - \sum_{|\alpha|\le m} \frac{(y-x_0)^\alpha}{\alpha!} (\partial^\alpha N)(x_0 - x) \right) f(y)dy.$$

Then

$$\partial_i v(x) = \int_\Omega \left(\partial_i N(y-x) - \sum_{|\alpha|\le m} \frac{(y-x_0)^\alpha}{\alpha!} (\partial^\alpha \partial_i N)(x_0 - x) \right) f(y)dy$$

and by (2.1),

$$\Delta v = \chi_\Omega f \qquad \text{in } \mathbb{R}^n \setminus \{x_0\},$$

where χ_Ω denotes the characteristic function of Ω. Also, $v(x)$ and $\nabla v(x)$ vanish on $\mathbb{R}^n \setminus (\Omega \cup \{x_0\})$ since f annihilates $\Phi_{x_0,m}(\Omega)$. Setting

$$u(x) = U^{(\chi_\Omega f)}(x) - v(x),$$

we have that u satisfies:

(a) $\Delta u = 0$ in $\mathbb{R}^n \setminus \{x_0\}$,

(b) $u(x) = O(|x - x_0|^{(-n+2-m)})$ as $x \to x_0$,
and since $U^{(\chi_\Omega f)}$ is C^1 and $\nabla v(x)$ vanishes on $\mathbb{R}^n \setminus \Omega$,

(c) $|\nabla u(x)| \le C$ on $\mathbb{R}^n \setminus \Omega$.

So we are now in the situation where Theorem 0.1 can be applied. By the hypothesis (2.2), we conclude that u is harmonic in a neighborhood of x_0. This means that

$$\int_\Omega \left(\sum_{|\alpha| \le m} \frac{(y - x_0)^\alpha}{\alpha!} (\partial^\alpha N)(x_0 - x) \right) f(y) dy = 0$$

for all x. Therefore, $U^{(\chi_\Omega f)}$ and $\nabla U^{(\chi_\Omega f)}$ vanish on $\mathbb{R}^n \setminus \Omega$. That is, f annihilates the class $\Phi(\Omega)$ and by Sakai's theorem, f annihilates $HL^1(\Omega)$. □

References

[1] C. A. Berenstein, R. Gay: *Complex Variables*, Springer, New York, 1997.

[2] W. K. Hayman, L. Karp, H. S. Shapiro: *Newtonian capacity and quasi balayage*, Rend. Mat. Appl. **20**, no. 7 (2000), 93–129.

[3] Ü. Kuran: *n–Dimensional extensions of theorems on conjugate functions*, Proc. London Math. Soc. **15**, no. 3 (1965), 713–730.

[4] N. S. Landkof: *Foundation of Modern Potential Theory*, Springer, Berlin, 1972.

[5] M. Sakai: *Solution to the obstacle problem as a Green potential*, J. Analyse Math. **44** (1984/85), 97–116.

Addresses:

Lavi Karp
Department of Mathematics
ORT Braude College
P.O. Box 78, Karmiel 21982, Israel

Harold S. Shapiro
Department of Mathematics
Royal Institue of Technology
S–100 44 Stockholm, Sweden

International Series of Numerical Mathematics
Vol. 145,

Two Applications of Adaptive Wavelet Methods

Angela Kunoth

Dedicated to Manfred Reimer on the occasion of his 70th birthday

Abstract

In this survey two different applications are discussed where the particular features of wavelet bases, namely, the Riesz basis property together with their locality are exploited. The first example concerns *explicitly* given information and discusses the problem of fitting nonuniformly distributed data to approximate surfaces. The second application, in which the information is contained *implicitly*, is concerned with an optimal control problem with constraints appearing in form of a linear elliptic partial differential equation. Both applications have in common that the formulation of the solution method is based on minimizing a quadratic functional and that the concept of adaptivity in a coarse–to–fine fashion plays a central role. An appropriate refinement criterion together with a thresholding technique guarantees to avoid a large number of small and unnecessary expansion coefficients of the target objects.

1 Wavelet Bases

Both applications I would like to present here are based on appropriate wavelet bases for a function space $H \subseteq L_2(\Omega)$ living on a bounded domain $\Omega \subset \mathbb{R}^d$. Wavelets are indexed by a parameter $\lambda = (j, k, \mathbf{e})$ which encodes information such as the *resolution level* or *scale* $|\lambda| := j$, the *spatial location* k, and possibly the type $\mathbf{e}$ of wavelet in the multivariate case. In view of the finite domain Ω, there is a coarsest level j_0. The infinite set of all possible indices will be denoted by $I\!I$. What is called a *wavelet basis* is a collection of functions

$$\Psi := \{\psi_\lambda : \lambda \in I\!I\} \subset H \tag{1.1}$$

with the following properties:

(R) Ψ constitutes a *Riesz basis* for H. This means that every $v \in H$ has a unique expansion in terms of Ψ,

$$v = \sum_{\lambda \in \mathit{II}} v_\lambda \, \psi_\lambda =: \mathbf{v}^T \, \Psi, \qquad \mathbf{v} := (v_\lambda)_{\lambda \in \mathit{II}}, \tag{1.2}$$

and its expansion coefficients satisfy a *norm equivalence*, that is, there exist constants $0 < c_H \leq C_H < \infty$ such that

$$c_H \|\mathbf{v}\|_{\ell_2(\mathit{II})} \leq \|\mathbf{v}^T \Psi\|_H \leq C_H \|\mathbf{v}\|_{\ell_2(\mathit{II})}, \qquad \mathbf{v} \in \ell_2(\mathit{II}). \tag{1.3}$$

Thus, wavelet expansions induce isomorphisms between the space of functions H and the sequence space $\ell_2(\mathit{II})$.

(L) The functions ψ_λ are local. That is, for each $\lambda \in \mathit{II}$ one has

$$\operatorname{diam}(\operatorname{supp} \psi_\lambda) \;\sim\; 2^{-|\lambda|}. \tag{1.4}$$

(CP) There exists an integer $\tilde{m}$ such that

$$\langle v, \psi_\lambda \rangle \;\lesssim\; 2^{-|\lambda|(d/2+\tilde{m})} \, |v|_{W^{\tilde{m}}_\infty(\operatorname{supp} \psi_\lambda)}, \tag{1.5}$$

where $\langle \cdot, \cdot \rangle$ is the dual form between H and its topological dual H'.

Throughout this paper, the relation $a \sim b$ is to express $a \lesssim b$ and $a \gtrsim b$, where the latter relation means that b can be estimated from above by a constant multiple of a independent of all parameters on which a or b may depend.

Property (CP) means that integration against a wavelet acts like taking an $\tilde{m}$th order difference which annihilates the smooth part of v. This feature called *cancellation property* comes into play for the efficient treatment of the control problem in that it entails quasi–sparse representations of a wide class of operators. (CP) is guaranteed to hold for Ψ if Ψ has a *dual* or *biorthogonal* basis $\tilde{\Psi} \subset H'$, i.e.,

$$\langle \psi_\lambda, \tilde{\psi}_\mu \rangle = \delta_{\lambda,\mu}, \qquad \lambda, \mu \in \mathit{II}, \tag{1.6}$$

which is a Riesz basis for H' and has members $\tilde{\psi}_\lambda$ such that the multiresolution spaces $\operatorname{span}\{\tilde{\psi}_\lambda : |\lambda| < j\}$ contain all polynomials of order $\tilde{m}$. By duality arguments, one can show that (1.3) implies

$$C_H^{-1} \|\tilde{\mathbf{v}}\|_{\ell_2(\mathit{II})} \leq \|\tilde{\mathbf{v}}^T \tilde{\Psi}\|_{H'} \leq c_H^{-1} \|\tilde{\mathbf{v}}\|_{\ell_2(\mathit{II})} \tag{1.7}$$

for any $\tilde{v} = \tilde{\mathbf{v}}^T \, \tilde{\Psi} \in H'$, see e.g. [8, 13–15, 25].

Properties (R), (L) and (CP) are strong enough to allow for the proof of theoretical statements such as (asymptotically) optimal condition number estimates or convergence results of adaptive methods for linear and nonlinear variational problems [9–12, 14, 16–18]. At the same time, they still allow to work computationally with piecewise polynomials. Concrete constructions of wavelet bases on bounded Euclidean tensor product domains based on B–splines can be found in [19]. These wavelets will be employed for the class of control problems discussed later in Section 3. For the data fitting approach in Section 2, tensor products of the piecewise linear boundary–adapted B–spline pre–wavelets from [27] are used which are a special case of the pre–wavelets on the interval constructed in [7]. These bases have the additional property that they are

(O) *semi–orthogonal* with respect to $L_2(\Omega)$, i.e., for $|\lambda| \neq |\mu|$ one has

$$\int_\Omega \psi_\lambda(x)\,\psi_\mu(x)\,dx = 0. \tag{1.8}$$

(Note that then the dual basis $\tilde{\Psi}$ will have global support which will, however, not affect the computations since we actually never employ the dual basis.) The semi–orthogonality will be very useful for the fast iterative solution of the linear systems appearing in this context.

In the sequel, we will always view Ψ both as a *collection* of functions as in (1.1) as well as a (possibly infinite) column *vector* which contains all functions always assembled in some fixed order. For a countable collection of functions Θ and some single function σ, the terms $\langle\Theta,\sigma\rangle$ and $\langle\sigma,\Theta\rangle$ are to be understood as the column (row) vector with entries $\langle\theta,\sigma\rangle$ (or $\langle\sigma,\theta\rangle$), $\theta\in\Theta$. For two collections Θ,Σ, the expression $\langle\Theta,\Sigma\rangle$ is then a (possibly infinite) matrix with entries $(\langle\theta,\sigma\rangle)_{\theta\in\Theta,\sigma\in\Sigma}$ for which $\langle\Theta,\Sigma\rangle = \langle\Sigma,\Theta\rangle^T$. This also implies for a (possibly infinite) matrix $\mathbf{C}$ that $\langle\mathbf{C}\Theta,\Sigma\rangle = \mathbf{C}\langle\Theta,\Sigma\rangle$ and $\langle\Theta,\mathbf{C}\Sigma\rangle = \langle\Theta,\Sigma\rangle\mathbf{C}^T$. In this notation, the expansion coefficients in (1.2) and (1.7) can explicitly be expressed as $\mathbf{v}^T = \langle v,\tilde{\Psi}\rangle$ and $\tilde{\mathbf{v}} = \langle\Psi,\tilde{v}\rangle$. Furthermore, the *biorthogonality* or *duality conditions* (1.6) can be reexpressed as $\langle\Psi,\tilde{\Psi}\rangle = \mathbf{I}$ with the infinite identity matrix $\mathbf{I}$.

The purpose of this paper is to discuss two different applications where the particular features of wavelet bases, the Riesz basis property and the locality, are exploited. The main goal is to solve the problems up to a certain accuracy with a minimal number of degrees of freedom, once for explicitly, once for implicitly given data. This leads us to adaptive 'coarse–to–fine' algorithms, where in both cases an appropriate refinement criterion together with a thresholding technique guarantees to avoid a large number of small and unnecessary expansion coefficients of the target objects. In addition, we strive for algorithms that are asymptotically optimal, that is, their overall complexity shall be proportional to the number of unknowns employed to resolve the objects. Common in both applications is that the solution methods are based on minimizing a quadratic functional.

The first application concerns *data fitting*, that is, fitting nonuniformly distributed data stemming from functions or surfaces. The second example discusses an *optimal control problem* with constraints appearing in form of a linear elliptic partial differential equation.

The remainder of this paper is structured as follows. In the next section, we will introduce the problem of fitting nonuniformly distributed surface data. A fully adaptive wavelet algorithm will be presented which is based on an appropriate refinement procedure that also takes into account a thresholding procedure to keep only relevant coefficients. In Section 3, recent theoretical results on an adaptive algorithm based on wavelets for the fast numerical solution of control problems governed by an elliptic boundary value problem are presented.

2 Fitting Irregularly Distributed Data

2.1 Problem Description

Let for some fixed $N \in \mathbb{N}$ a set $X = \{x_i\}_{i=1,\ldots,N}$ consist of irregularly spaced and pairwise distinct points $x_i \in \overline{\Omega} := [0,1]^d$ where $d \in \{1,2\}$. For each i we denote by $z_i \in \mathbb{R}$ corresponding data which is catenated in the set Z. The problem to *fit scattered data* is the following. Given the cloud of points $P := (X, Z) = \{(x_i, z_i),\ i = 1, \ldots, N\}$, find a function $f : \overline{\Omega} \to \mathbb{R}$ that approximates P in a least squares sense, that is, $f \in L_2(\Omega) \cap C^0(\overline{\Omega})$ minimizes the functional

$$J(f) := \sum_{i=1}^{N} (z_i - f(x_i))^2 . \tag{2.1}$$

In particular, we wish to construct a wavelet expansion of f of the form

$$f(x) = \sum_{\lambda \in \Lambda} d_\lambda \, \psi_\lambda(x), \quad x \in \Omega, \tag{2.2}$$

where the $\{\psi_\lambda\}_{\lambda\in\Lambda}$ are (tensor products of) the above mentioned boundary adapted B–spline–(pre)wavelets, shortly called wavelets in this section, which satisfy in addition property (O). Moreover, Λ is an appropriately determined lacunary set of indices which shall result from an adaptive coarse–to–fine procedure. We do *not* want to assume here that the data lies on a particular grid with uniform grid spacing, and we do not want to associate an (often artificially chosen) 'finest' grid Ω_J with the data. Therefore, any 'adaptive' method that requires in a first step to apply the Fast Wavelet Transform on such a uniform grid Ω_J followed by some thresholding is not considered to be an appropriate strategy. In such a case the full complexity $\#\Omega_J$ would enter into the algorithm independent of the cardinality or distribution of the data points.

On the contrary, in order to be able to handle very large data sets, our goal is to work with as few wavelets as possible, taking into account only those basis functions which are necessary to resolve the given data. Furthermore, the representation developed here allows for a fast update of the surface in case of time–dependent data. The underlying application for the motivation of this approach is real–time analysis of images stemming from reconstructions of stereoscopic images. Since only finitely many data are to be approximated, one expects that only finitely many indices from $I\!I$ are essential for the representation of f. For given points P, the determination of an appropriate index set $\Lambda \subset I\!I$, $\#\Lambda < \infty$, is therefore the central issue. On the one hand, Λ should be rich enough to be able to represent all features of P. On the other hand, if Λ is too large, one carries the risk of *overfitting* the data as it happens also with a full uniform grid and very irregularly distributed data, see e.g. [6]. Such overfitting effects usually cannot be remedied by adding a smoothing semi–norm to the functional (2.1) since this would result in a smoothing everywhere. A localized smoothing may, however, be possible [5]. Moreover, processing 'redundant' information that stems from a very fine sampling of a very smooth function also leads to an unnecessary increase in work.

In the sequel, we assume that the data is not noisy and that it does not contain outliers. The latter problem would require additional statistical methods to detect and handle these effects, see [5] for first results in this direction. Also we dispense here with taking into account additional regularization by adding, e.g., a Besov semi–norm to the functional (2.1) [28]. First experiments on bivariate domains regarding this issue can be found in [5].

2.2 An Adaptive Coarse–To–Fine Algorithm

We present here the adaptive coarse–to–fine algorithm developed in [5, 6] which is based on wavelets and has the following features.

(I) The choice of relevant indices in the index set Λ is determined progressively adapted to the given data;

(II) at no point a 'finest' artificial full grid is introduced or a finest resolution prescribed;

(III) an appropriate thresholding guarantees that only relevant wavelet coefficients are kept;

(IV) only very few iterations on the normal equations derived from (2.1) are necessary;

(V) which combined with a nested iteration strategy yields a faster solver than for hierarchical B–splines.

The scale localization property (L) of wavelets allows for a constructive algorithm which introduces degrees of freedom depending on the particular data set P. We progressively construct 'from coarse to fine' a sequence of (nested) index sets $\Lambda_j \subset \mathbb{I}$, $j \geq j_0$, while circumventing the effects discussed in the previous subsection as follows. On the coarsest level j_0, all indices are selected. On each finer level, we only consider those wavelets whose supports contain enough data points. That is, we fix q as the minimal number of data points that have to lie in the support of a wavelet in order to be considered a candidate for Λ_{j+1}. Recall from [27] that a univariate wavelet ψ_λ with $\lambda = (j,k)$ is said to be a *child* of a wavelet of index $\lambda' = (j',k')$ if $j = j'+1$ and $k = 2k'$ or $k = 2k'+1$, and correspondingly in two dimensions. (Note that the support of a child is always contained in the support of its parent which suffices for a coarse–to–fine algorithm to generate a well–defined tree.) From a set of indices Λ_j we construct a new set Λ_{j+1} by considering only children of previously chosen wavelets, with the additional requirement that those children have to have at least q points in their support, generating a binary tree in one dimension and a quadtree for two spatial variables. We denote by $\delta(\Lambda_j)$ the indices of newly allowable candidates and by $S_q(\delta(\Lambda_j)) \subset \delta(\Lambda_j)$ their subset of indices for wavelets having at least q points in their support. After an appropriate index set Λ_J has been found, the minimum of the least squares functional

$$J(f_J) := \sum_{i=1}^{N} \Big(z_i - \sum_{\lambda \in \Lambda_J} d_\lambda^J \, \psi_\lambda(x_i) \Big)^2 \tag{2.3}$$

can be computed by a standard method as discussed later.

In order to take into account possibly 'redundant' information, the data Z contained in P must come into play. As often exploited in wavelet expansions, the magnitude of the coefficients d_λ^J of the solution f_J of (2.3) can be interpreted as a *local smoothness indicator:* Large wavelet coefficients indicate the need for further local refinement. This in turn suggests a coarse–to–fine *thresholding* procedure to discard those wavelet coefficients which are smaller than a certain threshold. Combining these ideas, the algorithm may be formulated as follows.

ALGORITHM: FIX q AND A THRESHOLD ε AND DO FOR EACH $j = j_0, \ldots$

(I) DETERMINE $\delta(\Lambda_j)$,

(II) SELECT $S_q(\delta(\Lambda_j))$,

(III) IF $S_q(\delta(\Lambda_j)) = \emptyset$ STOP, ELSE SET $\tilde{\Lambda}_{j+1} := \Lambda_j \cup S_q(\delta(\Lambda_j))$,

(IV) COMPUTE $\{d_\lambda^{j+1}\}_{\lambda \in \tilde{\Lambda}_{j+1}}$ BY MINIMIZING

$$\sum_{i=1}^{N} \Big(z_i - \sum_{\lambda \in \tilde{\Lambda}_{j+1}} d_\lambda^{j+1} \, \psi_\lambda(x_i) \Big)^2,$$

(V) DETERMINE $\Lambda_{j+1}^\varepsilon := \{\lambda \in \tilde{\Lambda}_{j+1} \setminus \Lambda_j : \ |d_\lambda^{j+1}| \geq \varepsilon\}$,

(VI) IF $\Lambda_{j+1}^\varepsilon = \emptyset$ STOP, ELSE SET $\Lambda_{j+1} = \Lambda_j \cup \Lambda_{j+1}^\varepsilon$ AND $j = j + 1$.

The algorithm terminates after finitely many steps, stopping at a *finest level* J with an index set Λ_J which only depends on the density and the distribution of the data. Thus, the finest level is neither user–specified nor previously chosen. Since by (1.4) the supports of the wavelets shrink as the refinement level increases, for given finitely many data X the algorithm always terminates since either $S_q(\delta(\Lambda_j))$ or $\Lambda_{j+1}^\varepsilon$ are at some point empty. Moreover, as the algorithm grows exactly one level in each step until termination, the highest resolution level contained in the index set Λ_J is J. Note that this algorithm can be efficiently implemented by only considering the points from X which have already been considered in Λ_j. The numerical solution of the least squares problem in Step (IV) will be discussed below. Of course, we could formulate the algorithm also with level–dependent thresholds ε_j which we have dispensed with here. By construction, the final index set Λ_J has tree structure, and the representation is efficient in the sense that it provides a least squares approximation to the data with minimal degrees of freedom in comparison to working with the full grid of all wavelets up to level J. Since one usually is given only discrete data and no underlying continuous surface, it is not clear how to define 'accuracy' of the approximation as a measure for quality. The method is essentially 'zoom–

invariant' in the sense that it depends to a large extent on the relative position of the data points X and not on the absolute distances Z.

Once an index set Λ in (2.2) is fixed, the necessary conditions for minimizing J in (2.1) yield the *normal equations*

$$A^T A\mathbf{d} = A^T\mathbf{z} \quad \Longleftrightarrow: \quad M\mathbf{d} = \mathbf{b}, \tag{2.4}$$

where $\mathbf{d}$ and $\mathbf{z}$ are the column vectors for the coefficients and the given values in Z. Since the *observation matrix* $A \in \mathbb{R}^{N\times(\#\Lambda)}$ has entries $A_{i,\lambda} = \psi_\lambda(x_i)$, the *cross product matrix* $M = A^T A$ and b are given as usual by

$$M_{\lambda,\lambda'} = \sum_{i=1}^{N} \psi_\lambda(x_i)\psi_{\lambda'}(x_i), \qquad b_\lambda = \sum_{i=1}^{N} z_i\, \psi_\lambda(x_i). \tag{2.5}$$

As for the invertibility of M, using arguments from the Theorem of Schoenberg and Whitney for B–splines, see, e.g., [20], one would assume M to be nonsingular if the data is distributed such that at least one data point is contained in each section of the support of the ψ_λ characterized by the piecewise polynomial pieces. However, since the wavelets employed here are *fixed* linear combinations of B–splines, this does not follow immediately. The above criterion which requires at least q points in the support of each newly considered wavelet would not suffice for assuring the invertibility of M. However, in all numerical examples tested with the above criterion and the set of parameters, M indeed had full rank [5]. Also in [5] extensive experiments on the performance of conjugate gradient iterations in combination with a *nested iteration* and in comparison with the tensor product hierarchical B–spline basis (see [21, 22, 29]) with respect to iteration numbers and error reduction rates can be found.

In particular, it has turned out that the $L_2(\Omega)$–orthogonality between levels (O) for the prewavelets has a very favourable effect on the amount of conjugate gradient iterations in the nested iteration procedure, although M only contains approximate $L_2(\Omega)$–inner products, see [6] for some comparisons.

To illustrate the performance of ALGORITHM, consider the data in Figure 1 which exhibits areas of different densities including one with three thousand points in the right third of the image. The Z–values of the data stem from a function which is very smooth except for an area around $x = 0.75$. Constructing only an X–adapted approximation produces the result shown in Figure 2 on the left with distribution and size of their wavelet coefficients shown in the plot on the right. It succeeds to avoid overfitting, but needs 493 wavelets and

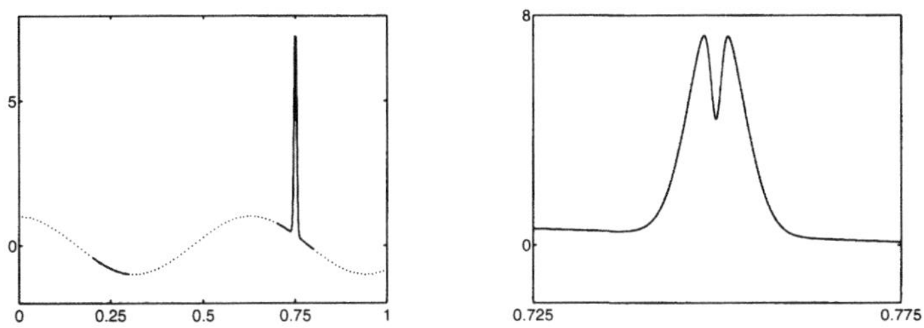

Figure 1: Left: Test data for the Algorithm. Right: zoom around $x = 0.75$.

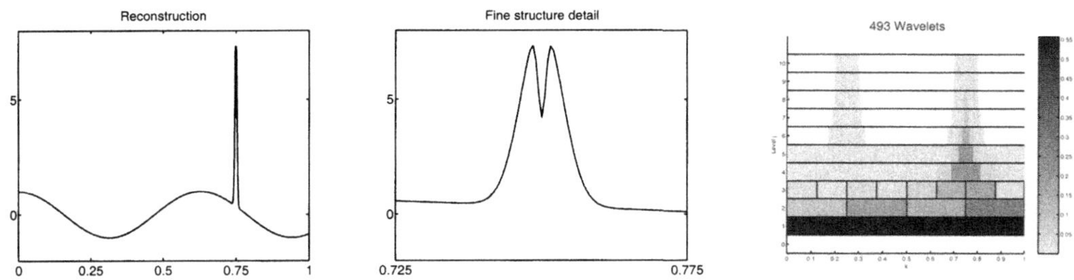

Figure 2: X–adapted fitting of data in Figure 1 (without thresholding).

26507 nonzero matrix elements in the least squares system (2.4). In contrast, the Z–adapted fitting using ALGORITHM displayed in Figure 3 produces an approximation of similar quality in the 'eye' norm and selects only 60 wavelets (with 1826 nonzero matrix elements). Inspection of the size of the wavelet coefficients in the right plots of Figure 2 and 3 according to their greyscale reveals that the X–adapted approximation produces a substantial amount of small wavelet coefficients, in particular at higher levels, which do not contribute much to the quality of the solution. For these results, the threshold 10^{-3} has been used. Furthermore, in Figure 4 the pointwise (with respect to X) errors between the exact solution and its approximations without and with thresholding are displayed.

For the bivariate case, ALGORITHM has been tested on several of the GTOPO30 data stemming from a digital elevation model [23]. The original data set showing some coastal line in Africa contains 160K points. Displayed here in Figure 5 are the approximations with respect to level $j = 2, 4$ and 6, together with the distributions and sizes of the wavelet coefficients. Further extensive comparisons reveal that the adaptive fitting procedure is very efficient when it comes to quickly approximating large data sets with a minimal amount of degrees of freedom [5].

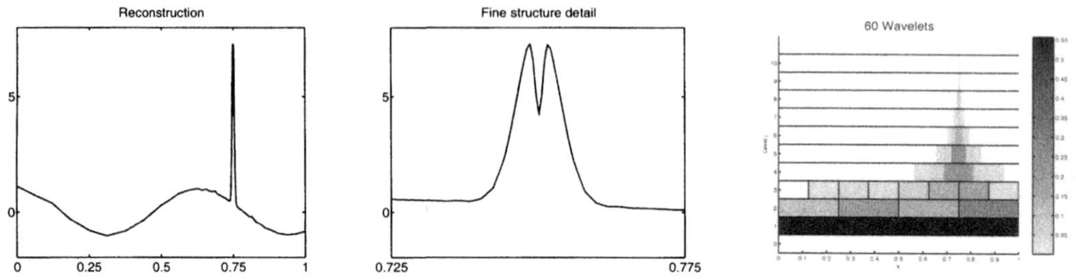

Figure 3: Z–adapted fitting of data in Figure 1 (with thresholding).

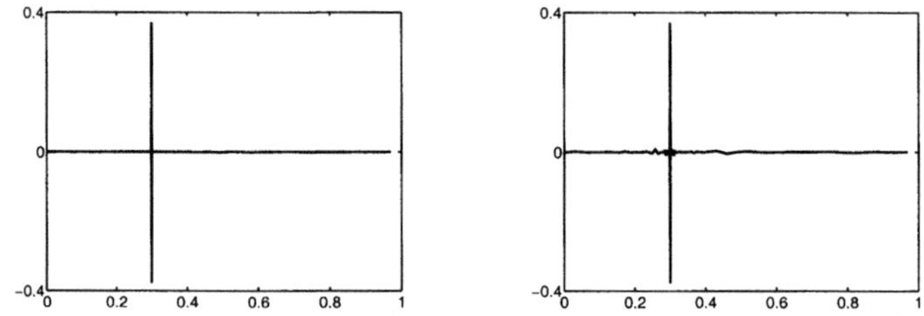

Figure 4: Pointwise error between exact data and approximation without thresholding (left) and with thresholding (right).

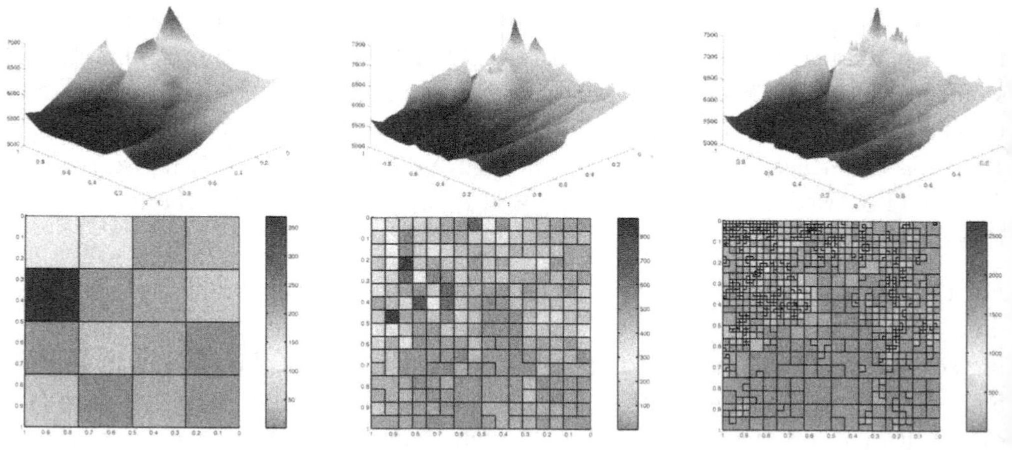

Figure 5: Z–adapted fitting of GTOPO30 data; from left to right on level $j = 2, 4, 6$ with 81, 909 and 4856 wavelets, respectively.

3 Optimal Control Problems with PDE Constraints

3.1 A Linear–Quadratic Problem

In contrast to the previous application, the class of control problems considered next allows for a solid theoretical underpinning because of its functional analytic background. While the ideas to locally refine and to threshold the produced wavelet coefficients now and then is similar in spirit, the analytic machinery allows here to prove that the resulting algorithm is *optimal* in the sense that it matches the optimal work/accuracy rate of the best N–term approximation of the solution [18]. As an additional benefit, the wavelet framework allows for a more flexible formulation of the control functional as we will see below.

Based on the theory developed in [18], the main ideas will be illustrated here in terms of a *Neumann problem with distributed control*, the simplest example of a linear–quadratic optimal control problem with an elliptic partial differential equation as constraint.

To this end, we denote by Y and Q the *state* and the *control space*, respectively, which are assumed to be (closed subspaces of) Hilbert spaces, with topological duals Y', Q' and associated dual forms $\langle\cdot,\cdot\rangle$. For a *control* u, given data $\tilde{f}$ and g, the *state* of the system under consideration is determined by the second order non–homogeneous Neumann problem

$$\begin{array}{rcll} -\Delta y + y &=& \tilde{f} + u & \text{in } \Omega, \\ \frac{\partial y}{\partial n} &=& g & \text{on } \partial\Omega, \end{array} \tag{3.1}$$

where $\frac{\partial}{\partial n}$ is the normal derivative in the direction of the outward normal at the boundary $\partial\Omega$ of Ω. In weak form, the system (3.1) may be abbreviated as a linear operator equation

$$Ay = f + u, \tag{3.2}$$

with $\langle f, v\rangle := \langle \tilde{f}, v\rangle + \int_{\partial\Omega} gv\,ds$ and $\langle Av, w\rangle := \int_\Omega (\nabla v \cdot \nabla w + vw)dx$. Thus, here we have $Y := H^1(\Omega)$ and $Q = (H^1(\Omega))' = Y'$, and A is continuous on Y and boundedly invertible from Y to Y'. Note that the state y is uniquely determined from (3.2) once u is known.

We will consider the linear system (3.2) as *constraints* for a quadratic cost functional which is described next. It usually involves a weaker norm than the one for Y corresponding to a Hilbert space Z which hosts given *observation data* y_*. In contrast, one imposes more regularity than required in a natural variational formulation on the *control* u, which is represented here by a Hilbert space U. We assume that $Y \hookrightarrow Z$ and $U \hookrightarrow Q$ are continuous embeddings. (For the present specific situation, it suffices to assume $Y \subseteq Z$ and $U \subseteq Q$. As discussed in [18], the more general case also allows for different versions of problems with *Neumann boundary control.*)

The *control problem* may now be formulated as follows.

(CoPr) *For given observations* $y_* \in Z$, *a right hand side* $f \in Y'$ *and a weight parameter* $\omega > 0$, *minimize the quadratic functional*

$$J(y,u) = \tfrac{1}{2}\|y - y_*\|_Z^2 + \tfrac{\omega}{2}\|u\|_U^2 \tag{3.3}$$

subject to the linear constraints (3.2), $Ay = f + u$.

Thus, we may choose $Z = H^r(\Omega)$ and $U = (H^t(\Omega))'$ for $0 \le r, t \le 1$. The typical case discussed in the literature is $r = t = 0$, i.e., $Z = U = L_2(\Omega)$ [26]; natural norms occur for $r = t = 1$, that is, $Z = Y = H^1(\Omega)$ and $U = Q = (H^1(\Omega))' = Y'$.

3.2 The Control Problem in Wavelet Coordinates

Control problems of the type (CoPr) are typically solved numerically by deriving first the necessary and sufficient (Karush–Kuhn–Tucker or KKT) conditions for the minimization of the functional (3.3) and then choosing finite–dimensional subspaces of Y, Z, Q, U in which (CoPr) is approximated, thereby ending up with a weakly coupled system of linear equations, see e.g. [3] for an adaptive approach based on finite elements. Alternatively, one can firstly pick finite–dimensional subspaces in which (CoPr) is discretized and then derive the KKT conditions.

Here a different strategy is proposed. Firstly, problem (CoPr) will be represented in terms of appropriately scaled wavelet bases for the whole spaces Y, Z, Q, U, yielding a representer of (CoPr) in terms of ℓ_2–norms and a linear operator which is an ℓ_2–isomorphism. Thus, the problem is discretized, that is, continuous objects are represented by sequences and operators acting on sequences, but the problem is not yet restricted to finite–dimensional subspaces. After deriving the KKT conditions, an approximation of an iterative scheme is proposed that approximates the weakly coupled infinite systems.

To this end, consider for $H \in \{Y, Q\}$ wavelet bases Ψ_H having properties (R), (L) and (CP). In view of the possible choices $Z = H^r(\Omega)$ and $U = (H^t(\Omega))'$ for

$0 \le r, t \le 1$, we can construct wavelet bases Ψ_H also for $H \in \{Z, U\}$ by simply scaling Ψ_Y, Ψ_Q accordingly, i.e., there exist diagonal matrices $\mathbf{D}^{1-r}$ and $\mathbf{D}^{t-1}$, $0 \le r, t \le 1$ with entries $(\mathbf{D}^{\beta})_{\mu,\nu} = 2^{\beta|\lambda|}\,\delta_{\mu,\nu}$ such that

$$\Psi_Z = \mathbf{D}^{1-r}\,\Psi_Y, \quad \tilde{\Psi}_Z = \mathbf{D}^{r-1}\,\tilde{\Psi}_Y, \qquad \Psi_U = \mathbf{D}^{t-1}\,\Psi_Q, \quad \tilde{\Psi}_U = \mathbf{D}^{1-t}\,\tilde{\Psi}_Q \tag{3.4}$$

form dual pairs for Z and U, respectively [18].

Now we can transform (CoPr) into wavelet coordinates. As for the constraints (3.2), expanding y in Ψ_Y (with expansion coefficients $\mathbf{y}$) and u in Ψ_Q (in terms of expansion coefficients $\check{\mathbf{u}}$), and testing with Ψ_Y, (3.2) takes the form

$$\mathbf{A}\mathbf{y} = \boldsymbol{f} + \check{\mathbf{u}}, \tag{3.5}$$

where $\mathbf{A} := \langle \Psi_Y, A\Psi_Y \rangle$ and $\boldsymbol{f} := \langle \Psi_Y, f \rangle$. Since A is boundedly invertible, it follows from the Riesz basis property (R) that the infinite matrix $\mathbf{A}$ is a boundedly invertible mapping of $\ell_2(I\!I_Y)$ onto itself, i.e., there exist finite positive constants $c_{\mathbf{A}}, C_{\mathbf{A}}$ such that

$$c_{\mathbf{A}}\,\|\mathbf{v}\| \le \|\mathbf{A}\mathbf{v}\| \le C_{\mathbf{A}}\,\|\mathbf{v}\|, \qquad \mathbf{v} \in \ell_2(I\!I_Y). \tag{3.6}$$

To simplify the notation we shall write briefly $\|\cdot\| := \|\cdot\|_{\ell_2(I\!I)}$. Since Ψ_Z is a Riesz basis for Z and Ψ_U for U, we have here $\|\mathbf{D}^{r-1}\,(\mathbf{y} - \mathbf{y}_*)\| \sim \|y - y_*\|_Z$ and $\|u\|_U \sim \|\mathbf{D}^{1-t}\,\check{\mathbf{u}}\|$. Accordingly, a *representer* of the control functional (3.3) that is equivalent to (3.3) in the sense that the involved norms are equivalent suffices to capture its main features. Thus, we arrive at the following *discrete control problem* in wavelet coordinates.

(DCP) *Minimize the quadratic functional*

$$\tilde{\mathbf{J}}(\mathbf{y}, \check{\mathbf{u}}) := \tfrac{1}{2}\|\mathbf{D}^{r-1}(\mathbf{y} - \mathbf{y}_*)\|^2 + \tfrac{\omega}{2}\|\mathbf{D}^{1-t}\,\check{\mathbf{u}}\|^2, \tag{3.7}$$

subject to the constraint (3.5), $\mathbf{A}\mathbf{y} = \boldsymbol{f} + \check{\mathbf{u}}$.

The subsequent numerical computations will be based on an equivalent formulation which is derived from (DCP) by formally inverting (3.5) and thereby eliminating $\mathbf{y}$ from (3.7), yielding a functional which only depends on $\check{\mathbf{u}}$,

$$\check{\mathbf{J}}(\check{\mathbf{u}}) \;:=\; \tfrac{1}{2}\|\mathbf{D}^{r-1}(\mathbf{A}^{-1}\check{\mathbf{u}} - (\mathbf{y}_* - \mathbf{A}^{-1}\boldsymbol{f}))\|^2 + \tfrac{\omega}{2}\|\mathbf{D}^{1-t}\,\check{\mathbf{u}}\|^2. \tag{3.8}$$

With the abbreviations $\check{\mathbf{Z}} := \mathbf{D}^{r-1}\mathbf{A}^{-1}$ and $\check{\mathbf{G}} := \mathbf{D}^{r-1}\,(\mathbf{y}_* - \mathbf{A}^{-1}\boldsymbol{f})$, $\check{\mathbf{J}}$ is of the form of a standard least squares functional

$$\check{\mathbf{J}}(\check{\mathbf{u}}) \;=\; \tfrac{1}{2}\|\check{\mathbf{Z}}\check{\mathbf{u}} - \check{\mathbf{G}}\|^2 + \tfrac{\omega}{2}\|\mathbf{D}^{1-t}\,\check{\mathbf{u}}\|^2 \tag{3.9}$$

whose minimizer is characterized by the normal equations which were in the present format derived in [24]. Indeed, $\check{\mathbf{J}}$ is convex, that is, its second variation

$$\check{\mathbf{Q}} := \delta^2 \check{\mathbf{J}}(\check{\mathbf{u}}) = \check{\mathbf{Z}}^T \check{\mathbf{Z}} + \omega \mathbf{D}^{2(1-t)} \tag{3.10}$$

is positive definite since $\check{\mathbf{Z}}^T\check{\mathbf{Z}}$ is at least positive semi–definite and $\mathbf{D}^{1-t}$ is a diagonal matrix with strictly positive diagonal entries. Consequently, the solution of (DCP) is uniquely determined by solving $\delta\check{\mathbf{J}}(\check{\mathbf{u}}) = 0$ which amounts here to the system $\check{\mathbf{Q}}\check{\mathbf{u}} = \check{\mathbf{g}}$ where $\check{\mathbf{g}} := \check{\mathbf{Z}}^T\check{\mathbf{G}}$. At this stage, one could in principle apply adaptively the approximate gradient iterations in [10] since $\check{\mathbf{Q}}$ is symmetric positive definite. However, we are still facing two obstructions, namely, (a) the spectral condition of $\check{\mathbf{Q}}$ and (b) the difficulty in applying $\check{\mathbf{Q}}$ and evaluating the right hand side due to the inverses involved in the definition of $\check{\mathbf{Z}}$ and $\check{\mathbf{G}}$. Problem (a) can be remedied by an appropriate scaling yielding an asymptotic optimal preconditioning of $\check{\mathbf{Q}}$ as follows. Defining

$$\mathbf{u} := \mathbf{D}^{1-t}\,\check{\mathbf{u}}, \qquad \mathbf{Z} := \check{\mathbf{Z}}\mathbf{D}^{t-1}, \qquad \mathbf{Q} := \mathbf{D}^{t-1}\check{\mathbf{Q}}\mathbf{D}^{t-1} = \mathbf{Z}^T\mathbf{Z} + \omega\mathbf{I}, \tag{3.11}$$

one can show that $\check{\mathbf{Q}}\check{\mathbf{u}} = \check{\mathbf{g}}$ is equivalent to

$$\mathbf{Q}\mathbf{u} = \mathbf{g} := \mathbf{Z}^T\mathbf{G} \tag{3.12}$$

where now $\mathbf{Z}$ is bounded on ℓ_2, and $\mathbf{Q}$ has uniformly bounded condition numbers. Thus, there exist finite positive constants $c_{\mathbf{Q}}, C_{\mathbf{Q}}$ such that

$$c_{\mathbf{Q}}\|\mathbf{v}\| \leq \|\mathbf{Q}\mathbf{v}\| \leq C_{\mathbf{Q}}\|\mathbf{v}\|, \qquad \mathbf{v} \in \ell_2, \tag{3.13}$$

where we can actually take $c_{\mathbf{Q}} = \omega$. Note that (3.12) yields the unique minimizer of the functional

$$\mathbf{J}(\mathbf{u}) \;:=\; \tfrac{1}{2}\|\mathbf{Z}\mathbf{u} - \mathbf{G}\|^2 + \tfrac{\omega}{2}\|\mathbf{u}\|^2, \tag{3.14}$$

which corresponds to normalizing the controls from the start in the basis Ψ_U.

In view of (3.13) and the fact that $\mathbf{Q}$ is positive definite, there exists a fixed positive parameter α such that in the *Richardson iteration* (which is a special case of a gradient method)

$$\mathbf{u}^{k+1} = \mathbf{u}^k + \alpha(\mathbf{g} - \mathbf{Q}\mathbf{u}^k) \tag{3.15}$$

the error is reduced in each step by at least a factor

$$\rho := \|\mathbf{I} - \alpha\mathbf{Q}\| < 1, \tag{3.16}$$

$$\|\mathbf{u} - \mathbf{u}^{k+1}\| \leq \rho\,\|\mathbf{u} - \mathbf{u}^k\|, \quad k = 0, 1, 2, \ldots, \tag{3.17}$$

where $\mathbf{u}$ is the exact solution of (3.12). As the involved system is still infinite, we aim at carrying out this iteration approximately with dynamically updated accuracy tolerances which will be discussed in the next subsection.

3.3 Approximate Iterations

Recalling the main conceptual tools from [10, 11], we will see how to solve (3.12) and to tackle obstruction (b), that is, the application of $\mathbf{Q}$ and the evaluation of the right hand side $\mathbf{g}$. At this point, we assume that we have a routine at our disposal with the following property. Later it will be shown how to realize this routine in the concrete case.

RES $[\eta, \mathbf{Q}, \mathbf{g}, \mathbf{v}] \to \mathbf{r}_\eta$ DETERMINES FOR A GIVEN TOLERANCE $\eta > 0$ A FINITELY SUPPORTED SEQUENCE $\mathbf{r}_\eta$ SATISFYING $$\|\mathbf{g} - \mathbf{Q}\mathbf{v} - \mathbf{r}_\eta\| \leq \eta. \quad (3.18)$$

Another ingredient will eventually play a crucial role in controlling the complexity of the scheme whose role is not yet apparent at this stage.

COARSE $[\eta, \mathbf{w}] \to \mathbf{w}_\eta$ DETERMINES FOR ANY FINITELY SUPPORTED INPUT VECTOR $\mathbf{w}$ A VECTOR $\mathbf{w}_\eta$ WITH SMALLEST POSSIBLE SUPPORT SUCH THAT $$\|\mathbf{w} - \mathbf{w}_\eta\| \leq \eta. \quad (3.19)$$

A precise description of COARSE can be found in [9]. The basic idea is to sort the entries of $\mathbf{w}$ by size and then to subtract squares of their moduli until the sum reaches η^2, starting from the smallest entry. A quasi–sorting based on binary binning can be shown to avoid the logarithmic term in the sorting procedure at the expense of the resulting support size being at most a fixed constant of the minimal size, see [1].

Next a perturbed iteration is designed that converges in the following sense. For every target accuracy ε it produces after finitely many steps a finitely supported approximate solution with accuracy ε. Following [10], to obtain a correctly balanced interplay between the routines RES and COARSE, we need the following control parameter. Given (an estimate of) the reduction rate ρ and the step size parameter α from (3.16), let $K := \min\{\ell \in \mathbb{N} : \rho^{\ell-1}(\alpha\ell+\rho) \leq \frac{1}{10}\}$.

Denoting in the following always by $\mathbf{u}$ the exact solution of (3.12), a perturbed version of (3.15) for a fixed target accuracy $\varepsilon > 0$ can be formulated as follows. In the case that no particular initial guess is known, we initialize $\overline{\mathbf{q}}^0 = \mathbf{0}$ and set $\varepsilon_0 := c_{\mathbf{Q}}^{-1}\|\mathbf{g}\|$. In this case, we briefly write SOLVE $[\varepsilon, \mathbf{Q}, \mathbf{g}] \to \overline{\mathbf{q}}_\varepsilon$.

SOLVE $[\varepsilon, \mathbf{Q}, \mathbf{g}, \overline{\mathbf{q}}^0, \varepsilon_0] \to \overline{\mathbf{q}}_\varepsilon$

(I) GIVEN AN INITIAL GUESS $\overline{\mathbf{q}}^0$ AND AN ERROR BOUND $\|\mathbf{q} - \overline{\mathbf{q}}^0\| \leq \varepsilon_0$; SET $j = 0$.

(II) IF $\varepsilon_j \leq \varepsilon$, STOP AND SET $\overline{\mathbf{q}}_\varepsilon := \overline{\mathbf{q}}^j$. OTHERWISE SET $\mathbf{v}^0 := \overline{\mathbf{q}}^j$.

(II.1) FOR $k = 0, \ldots, K-1$ COMPUTE RES $[\rho^k \varepsilon_j, \mathbf{Q}, \mathbf{g}, \mathbf{v}^k] \to \mathbf{r}^k$ AND

$$\mathbf{v}^{k+1} := \mathbf{v}^k + \alpha \mathbf{r}^k. \tag{3.20}$$

(II.2) APPLY COARSE $[\frac{2}{5}\varepsilon_j, \mathbf{v}^K] \to \overline{\mathbf{q}}^{j+1}$; SET $\varepsilon_{j+1} := \frac{1}{2}\varepsilon_j$, $j+1 \to j$ AND GO TO (II).

Straightforward perturbation arguments [10, 11] yield the convergence of the algorithm.

Proposition 3.1. *The iterates $\overline{\mathbf{q}}^j$ generated by* SOLVE $[\varepsilon, \mathbf{Q}, \mathbf{g}]$ *satisfy*

$$\|\mathbf{q} - \overline{\mathbf{q}}^j\| \leq \varepsilon_j, \qquad j \geq 0, \tag{3.21}$$

where $\varepsilon_j = 2^{-j}\varepsilon_0$.

Note that the estimates for α rely on the constants in the norm equivalences (N) and (3.13). In case there is only a poor estimate for ρ this gives rise to an overly pessimistic choice for the number K controlling the number of perturbed iterations in each block (II) of SOLVE prior to a coarsening step. As described in [11], this block may be terminated earlier, based on monitoring the approximate residuals in an appropriate way.

In order to better support numerical realizations of SOLVE, it will be helpful to derive equivalent formulations of (3.12). In fact, using a Lagrange multiplier formalism, one can derive the following [24, 25, 30].

Remark 3.2. *The solution* $\mathbf{u}$ *of the system (3.12),* $\mathbf{Q}\mathbf{u} = \mathbf{g}$, *is a component of the solution* $(\mathbf{y}, \mathbf{p}, \mathbf{u})$ *of the weakly coupled system of Euler equations in wavelet coordinates*

(EE)

$$\begin{aligned} \mathbf{A}\mathbf{y} &= \boldsymbol{f} + \mathbf{D}^{t-1}\mathbf{u} \\ \mathbf{A}^T\mathbf{p} &= -\mathbf{D}^{2(r-1)}(\mathbf{y} - \mathbf{y}_*) \qquad (3.22) \\ \omega\mathbf{u} &= \mathbf{D}^{t-1}\mathbf{p}. \qquad (3.23) \end{aligned}$$

Note that we have $0 \le r, t \le 1$ so that in these ranges the corresponding diagonal matrices appearing on the right hand sides are always bounded on ℓ_2. The first equation of (EE) is often called the *state* or *primal equation* while the second equation is the *costate* or *adjoint equation.* Although the matrix $\mathbf{A}$ is symmetric in the particular case considered here, I continue to write $\mathbf{A}^T$ for the system matrix in the costate system.

The strategy for approximating in each step the *residual* $\mathbf{g} - \mathbf{Q}\mathbf{u}^k$ will be based upon the following observation, namely, that the residual of (3.12) is just the residual of the third equation in (EE) [18].

Lemma 3.3. *For any* $\mathbf{v} \in \ell_2(I\!I_Q)$, *one has the representation*

$$\mathbf{Q}\mathbf{v} - \mathbf{g} = \omega \mathbf{v} - \mathbf{D}^{t-1}\mathbf{p}, \tag{3.24}$$

where $\mathbf{p}$ *is the solution of the first two equations in (EE). That is, for any given* $\mathbf{v}$, *the sequence* $\mathbf{p}$ *is determined by solving*

$$\mathbf{A}^T\mathbf{p} = -\mathbf{D}^{r-1}(\mathbf{D}^{r-1}\mathbf{y} - \mathbf{y}_Z), \quad \textit{where} \quad \mathbf{A}\mathbf{y} = \boldsymbol{f} + \mathbf{D}^{t-1}\mathbf{v} \quad \textit{and} \quad \mathbf{y}_Z = \mathbf{D}^{r-1}\mathbf{y}_*. \tag{3.25}$$

The realization of the routine RES for the problem (3.12) will be based on the residual representation (3.24). This in turn requires solving the two auxiliary systems in (3.25). Since the residual only has to be *approximated*, these systems will have to be solved only approximately. These approximate solutions, in turn, will be provided again by employing SOLVE but this time with respect to suitable residual schemes tailored to the systems in (3.25). In our special case, the matrix $\mathbf{A}$ is symmetric positive definite, and the choice of wavelet bases ensures the validity of (3.6). Thus, (3.16) holds for $\mathbf{A}$ and $\mathbf{A}^T$ so that the scheme SOLVE can indeed be invoked. As pointed out in [18], the theory is neither confined to symmetric $\mathbf{A}$ nor to the positive definite case. Although we conceptually use the fact that a gradient iteration for the reduced problem (3.12) reduces the error for $\mathbf{u}$ in each step by a fixed amount, employing (EE) for the evaluation of the residuals will generate as byproducts approximate solutions to the exact solution triple $(\mathbf{y}, \mathbf{p}, \mathbf{u})$ of (EE).

Under this hypothesis, we have to formulate next the ingredients for suitable versions $\text{SOLVE}_{\text{PRM}}$ and $\text{SOLVE}_{\text{ADJ}}$ of SOLVE for the systems in (EE). Specifically, this requires identifying residual routines RES_{PRM} and RES_{ADJ} for the systems $\text{SOLVE}_{\text{PRM}}$ and $\text{SOLVE}_{\text{ADJ}}$. The main task in both cases is to apply the operators $\mathbf{A}, \mathbf{A}^T, \mathbf{D}^{t-1}$ and $\mathbf{D}^{r-1}$. Again we assume for the moment that routines for the application of these operators are available, i.e., that for any

$\mathbf{L} \in \{\mathbf{A}, \mathbf{A}^T, \mathbf{D}^{t-1}, \mathbf{D}^{r-1}\}$ we have a scheme at our disposal with the following property. We shall later discuss their realization.

> APPLY$[\eta, \mathbf{L}, \mathbf{v}] \to \mathbf{w}_\eta$ DETERMINES FOR ANY FINITELY SUPPORTED INPUT VECTOR $\mathbf{v}$ AND ANY TOLERANCE $\eta > 0$ A FINITELY SUPPORTED OUTPUT $\mathbf{w}_\eta$ WHICH SATISFIES
> $$\|\mathbf{L}\mathbf{v} - \mathbf{w}_\eta\| \leq \eta. \tag{3.26}$$

The scheme SOLVE$_{\text{PRM}}$ for the first system in (EE) is then defined by

$$\text{SOLVE}_{\text{PRM}}\,[\eta, \mathbf{A}, \mathbf{D}^{t-1}, \boldsymbol{f}, \mathbf{v}, \overline{\mathbf{y}}^0, \varepsilon_0] := \text{SOLVE}\,[\eta, \mathbf{A}, \boldsymbol{f} + \mathbf{D}^{t-1}\mathbf{v}, \overline{\mathbf{y}}^0, \varepsilon_0],$$

where $\overline{\mathbf{y}}^0$ is an initial guess for the solution $\mathbf{y}$ of $\mathbf{A}\mathbf{y} = \boldsymbol{f} + \mathbf{D}^{t-1}\mathbf{v}$ with accuracy ε_0. The scheme RES for Step (II) in SOLVE is in this case realized by a new routine RES$_{\text{PRM}}$ which is defined as follows.

> RES$_{\text{PRM}}\,[\eta, \mathbf{A}, \mathbf{D}^{t-1}, \boldsymbol{f}, \mathbf{v}, \overline{\mathbf{y}}] \to \mathbf{r}_\eta$ DETERMINES FOR ANY POSITIVE TOLERANCE η, A GIVEN FINITELY SUPPORTED $\mathbf{v}$ AND ANY FINITELY SUPPORTED INPUT $\overline{\mathbf{y}}$ A FINITELY SUPPORTED APPROXIMATE RESIDUAL $\mathbf{r}_\eta$ SATISFYING (3.18), THAT IS,
> $$\|\boldsymbol{f} + \mathbf{D}^{t-1}\mathbf{v} - \mathbf{A}\overline{\mathbf{y}} - \mathbf{r}_\eta\| \leq \eta, \tag{3.27}$$
> AS FOLLOWS:
> (I) APPLY$[\frac{1}{3}\eta, \mathbf{A}, \overline{\mathbf{y}}] \to \mathbf{w}_\eta$;
> (II) COARSE$[\frac{1}{3}\eta, \boldsymbol{f}] \to \boldsymbol{f}_\eta$; APPLY$[\frac{1}{3}\eta, \mathbf{D}^{t-1}, \mathbf{v}] \to \mathbf{z}_\eta$;
> (III) SET $\mathbf{r}_\eta := \boldsymbol{f}_\eta + \mathbf{z}_\eta - \mathbf{w}_\eta$.

By triangle inequality, one can for RES$_{\text{PRM}}$and the subsequent variants of RES show that indeed (3.27) or (3.18) hold.

Similarly, we need a version of SOLVE for the approximate solution of the second system in (EE), $\mathbf{A}^T\mathbf{p} = -\mathbf{D}^{r-1}(\mathbf{D}^{r-1}\mathbf{T}_Z\mathbf{y} - \mathbf{y}_Z)$, which therefore depends on $\mathbf{y}_Z = \mathbf{D}^{r-1}\mathbf{y}_*$, an approximate solution $\overline{\mathbf{y}}$ of the primal system and possibly on some initial guess $\overline{\mathbf{p}}^0$ with accuracy ε_0. Here we set

$$\text{SOLVE}_{\text{ADJ}}\,[\eta, \mathbf{A}, \mathbf{D}^{r-1}, \mathbf{y}_Z, \overline{\mathbf{y}}, \overline{\mathbf{p}}^0, \varepsilon_0] := \text{SOLVE}\,[\eta, \mathbf{A}^T, \mathbf{D}^{r-1}(\mathbf{y}_Z - \mathbf{D}^{r-1}\overline{\mathbf{y}}), \overline{\mathbf{p}}^0, \varepsilon_0].$$

As usual we assume that the data $\boldsymbol{f}, \mathbf{y}_Z$ are approximated in a preprocessing step with sufficient accuracy. Again we have to identify a suitable residual approximation scheme RES$_{\text{ADJ}}$ for Step (II) of this version of SOLVE where the main issue is the approximate evaluation of the right hand side. Thus, we define the routine RES$_{\text{ADJ}}$ as follows.

$\text{RES}_{\text{ADJ}}\,[\eta, \mathbf{A}, \mathbf{D}^{r-1}, \mathbf{y}_Z, \overline{\mathbf{y}}, \overline{\mathbf{p}}] \to \mathbf{r}_\eta$ DETERMINES FOR ANY POSITIVE TOLERANCE η, GIVEN FINITELY SUPPORTED DATA $\overline{\mathbf{y}}, \mathbf{y}_Z$ AND ANY FINITELY SUPPORTED INPUT $\overline{\mathbf{p}}$ AN APPROXIMATE RESIDUAL $\mathbf{r}_\eta$ SATISFYING (3.18), I.E.,

$$\| -\mathbf{D}^{r-1}(\mathbf{D}^{r-1}\overline{\mathbf{y}} - \mathbf{y}_Z) - \mathbf{A}^T\overline{\mathbf{p}} - \mathbf{r}_\eta\| \leq \eta, \tag{3.28}$$

AS FOLLOWS:

(i) $\text{APPLY}[\frac{1}{3}\eta, \mathbf{A}^T, \overline{\mathbf{p}}] \to \mathbf{w}_\eta$;

(ii) $\text{APPLY}[\frac{1}{6}\eta, \mathbf{D}^{r-1}, \overline{\mathbf{y}}] \to \mathbf{z}_\eta$; $\text{COARSE}[\frac{1}{6}\eta, \mathbf{y}_Z] \to (\mathbf{y}_Z)_\eta$; set $\mathbf{d}_\eta := (\mathbf{y}_Z)_\eta - \mathbf{z}_\eta$; $\text{APPLY}[\frac{1}{3}\eta, \mathbf{D}^{r-1}, \mathbf{d}_\eta] \to \mathbf{v}_\eta$;

(iii) set $\mathbf{r}_\eta := \mathbf{v}_\eta - \mathbf{w}_\eta$.

Now we are in a position to define the residual scheme for the version of SOLVE applied to (3.12). We shall refer to this specification as $\text{SOLVE}_{\text{DCP}}$. Likewise the corresponding residual scheme is denoted by RES_{DCP}. Since the scheme is based on Lemma 3.3, it will therefore involve several parameters stemming from the auxiliary systems (EE).

$\text{RES}_{\text{DCP}}[\eta, \mathbf{Q}, \mathbf{g}, \tilde{\mathbf{y}}, \delta_y, \tilde{\mathbf{p}}, \delta_p, \mathbf{v}, \delta_v] \to (\mathbf{r}_\eta, \tilde{\mathbf{y}}, \delta_y, \tilde{\mathbf{p}}, \delta_p)$ DETERMINES FOR ANY APPROXIMATE SOLUTION TRIPLE $(\tilde{\mathbf{y}}, \tilde{\mathbf{p}}, \mathbf{v})$ OF THE SYSTEM (EE) SATISFYING

$$\|\mathbf{y} - \tilde{\mathbf{y}}\| \leq \delta_y, \quad \|\mathbf{p} - \tilde{\mathbf{p}}\| \leq \delta_p, \quad \|\mathbf{u} - \mathbf{v}\| \leq \delta_v, \tag{3.29}$$

AN APPROXIMATE RESIDUAL $\mathbf{r}_\eta$ SUCH THAT

$$\|\mathbf{g} - \mathbf{Q}\mathbf{v} - \mathbf{r}_\eta\| \leq \eta. \tag{3.30}$$

MOREOVER, THE INITIAL APPROXIMATIONS $\tilde{\mathbf{y}}, \tilde{\mathbf{p}}$ ARE OVERWRITTEN BY NEW APPROXIMATIONS $\tilde{\mathbf{y}}, \tilde{\mathbf{p}}$ SATISFYING (3.29) WITH NEW BOUNDS δ_y AND δ_p DEFINED IN (3.31) BELOW, AS FOLLOWS:

(I) $\text{SOLVE}_{\text{PRM}}[\frac{1}{3}c_{\mathbf{A}}\,\eta, \mathbf{A}, \boldsymbol{f}, \mathbf{v}, \tilde{\mathbf{y}}, \delta_y] \to \mathbf{y}_\eta$;

(II) $\text{SOLVE}_{\text{ADJ}}[\frac{1}{3}\eta, \mathbf{A}, \mathbf{D}^{r-1}, \mathbf{y}_Z, \mathbf{y}_\eta, \tilde{\mathbf{p}}, \delta_p] \to \mathbf{p}_\eta$;

(III) $\text{APPLY}[\frac{1}{3}\eta, \mathbf{D}^{t-1}, \mathbf{p}_\eta] \to \mathbf{q}_\eta$; SET $\mathbf{r}_\eta := \mathbf{q}_\eta - \omega\mathbf{v}$;

(IV) SET $\xi_y := c_{\mathbf{A}}^{-1}\,\delta_v + \frac{1}{3}c_{\mathbf{A}}\eta$, $\xi_p := c_{\mathbf{A}}^{-2}\,\delta_v + \frac{2}{3}\eta$; REPLACE $\tilde{\mathbf{y}}, \delta_y$ AND $\tilde{\mathbf{p}}, \delta_p$ BY

$$\begin{aligned} \tilde{\mathbf{y}} &:= \text{COARSE}[4\xi_y, \mathbf{y}_\eta], & \delta_y &:= 5\,\xi_y, \\ \tilde{\mathbf{p}} &:= \text{COARSE}[4\xi_p, \mathbf{p}_\eta], & \delta_p &:= 5\,\xi_p. \end{aligned} \tag{3.31}$$

Step (IV) already indicates the conditions on the tolerance η and the accuracy bound δ_v under which the new error bounds in (3.31) are actually tighter. The precise relation between η and δ_v in the context of $\text{SOLVE}_{\text{DCP}}$ is not apparent yet and emerges from the complexity analysis in [18] as well as the claimed estimates (3.30) and (3.31).

Finally, the scheme $\text{SOLVE}_{\text{DCP}}$ attains the following form with the error reduction factor ρ from (3.16) and α from (3.15), see [18] for details.

$\text{SOLVE}_{\text{DCP}}\,[\varepsilon, \mathbf{Q}, \mathbf{g}] \to \overline{\mathbf{u}}_\varepsilon$

(I) LET $\overline{\mathbf{q}}^0 := \mathbf{0}$ AND $\varepsilon_0 := c_{\mathbf{A}}^{-1}(\|\mathbf{y}_Z\| + c_{\mathbf{A}}^{-1}\|\boldsymbol{f}\|)$. MOREOVER, LET $\tilde{\mathbf{y}} := \mathbf{0}$, $\tilde{\mathbf{p}} := \mathbf{0}$ AND SET $j = 0$. DEFINE $\delta_y := \delta_{y,0} := c_{\mathbf{A}}^{-1}(\|\boldsymbol{f}\| + \varepsilon_0)$ AND $\delta_p := \delta_{p,0} := c_{\mathbf{A}}^{-1}(\delta_{y,0} + \|\mathbf{y}_Z\|)$.

(II) IF $\varepsilon_j \leq \varepsilon$, STOP AND SET $\overline{\mathbf{u}}_\varepsilon := \overline{\mathbf{u}}^j$, $\overline{\mathbf{y}}_\varepsilon = \tilde{\mathbf{y}}$, $\overline{\mathbf{p}}_\varepsilon = \tilde{\mathbf{p}}$.
OTHERWISE SET $\mathbf{v}^0 := \overline{\mathbf{u}}^j$.

(II.1) FOR $k = 0, \ldots, K-1$, COMPUTE
$\text{RES}_{\text{DCP}}\,[\rho^k \varepsilon_j, \mathbf{Q}, \mathbf{g}, \tilde{\mathbf{y}}, \delta_y, \tilde{\mathbf{p}}, \delta_p, \mathbf{v}^k, \delta_k] \to (\mathbf{r}^k, \tilde{\mathbf{y}}, \delta_y, \tilde{\mathbf{p}}, \delta_p)$,
WHERE $\delta_0 := \varepsilon_j$ AND $\delta_k := \rho^{k-1}(\alpha k + \rho)\varepsilon_j$;
SET

$$\mathbf{v}^{k+1} := \mathbf{v}^k + \alpha \mathbf{r}^k. \tag{3.32}$$

(II.2) APPLY $\text{COARSE}\,[\tfrac{2}{5}\varepsilon_j, \mathbf{v}^K] \to \overline{\mathbf{u}}^{j+1}$; SET $\varepsilon_{j+1} := \tfrac{1}{2}\varepsilon_j$, $j+1 \to j$ AND GO TO (II).

By overwriting $\tilde{\mathbf{y}}$, $\tilde{\mathbf{p}}$ at the last stage prior to the termination of $\text{SOLVE}_{\text{DCP}}$ one has $\delta_v \leq \varepsilon$, $\eta \leq \varepsilon$, so that the following fact is an immediate consequence of (3.31).

Remark 3.4. *The outputs $\overline{\mathbf{y}}_\varepsilon$ and $\overline{\mathbf{p}}_\varepsilon$ produced by $\text{SOLVE}_{\text{DCP}}$ in addition to $\mathbf{u}_\varepsilon$ are approximations to the exact solutions $\mathbf{y}, \mathbf{p}$ of (EE) satisfying*

$$\|\mathbf{y} - \overline{\mathbf{y}}_\varepsilon\| \leq 5\varepsilon\,(c_{\mathbf{A}}^{-1} + \tfrac{1}{3}c_{\mathbf{A}}), \qquad \|\mathbf{p} - \overline{\mathbf{p}}_\varepsilon\| \leq 5\varepsilon\,(c_{\mathbf{A}}^{-2} + \tfrac{2}{3}).$$

3.4 Complexity Analysis

Proposition 3.1 states that the routine SOLVE converges for an arbitrary given accuracy provided that there is a routine RES with the property (3.18). Then we have broken down step by step the necessary ingredients to derive computable versions which satisfy these requirements. What we finally want to show is that the routines are *optimal* in the sense that they provide the optimal work/accuracy rate in terms of best N–term approximation. The complexity analysis given next also reveals the role of the routines COARSE within the algorithms and the particular choices of thresholds in Step (IV) of RES_{DCP}.

In order to be able to assess the quality of the adaptive algorithm, the notion of *optimality* has to be clarified. The scheme SOLVE is said to have an *optimal work/accuracy rate* s if the following holds: Whenever the error of *best N–term approximation* satisfies

$$\sigma_N(\mathbf{q}) := \|\mathbf{q} - \mathbf{q}_N\| := \min_{\#\,\mathrm{supp}\,\mathbf{v} \le N} \|\mathbf{q} - \mathbf{v}\| \lesssim N^{-s},$$

then the solution $\overline{\mathbf{q}}_\varepsilon$ is generated by SOLVE at an expense that also stays proportional to $\varepsilon^{-1/s}$ and in that sense matches the best N–term approximation rate. Note that this implies that $\#\mathrm{supp}\,\overline{\mathbf{q}}_\varepsilon$ also stays proportional to $\varepsilon^{-1/s}$. Thus, our benchmark is that whenever the solution of (3.12) can be approximated by N terms at rate s, SOLVE recovers that rate asymptotically. Of course, the best N–term approximation $\mathbf{q}_N$ of $\mathbf{q}$ is given by picking the N largest terms in modulus from $\mathbf{q}$. However, when $\mathbf{q}$ is the (unknown) solution of (3.12) this information is certainly not available.

Since we are here in the framework of sequence spaces ℓ_2, the formulation of appropriate criteria for complexity will be based on a characterization of sequences which are in a certain sense *sparse*. Following [9], we consider sequences $\mathbf{v}$ for which the best N–term approximation error decays at a particular rate (*Lorentz spaces*). That is, for any given threshold $0 < \eta \le 1$, say, the number of terms exceeding that threshold is controlled by some function of this threshold. In particular, set for some $0 < \tau < 2$

$$\ell_\tau^w := \{\mathbf{v} \in \ell_2 : \ \#\{\lambda \in I\!I : |v_\lambda| > \eta\} \le C_{\mathbf{v}}\, \eta^{-\tau}, \text{ for all } 0 < \eta \le 1\}. \qquad (3.33)$$

This determines a strict subspace of ℓ_2 only when $\tau < 2$. Smaller τ's indicate sparser sequences. Let $C_{\mathbf{v}}$ for a given $\mathbf{v} \in \ell_\tau^w$ be the smallest constant for which (3.33) holds. Then one has $|\mathbf{v}|_{\ell_\tau^w} := \sup_{n\in\mathbb{N}} n^{1/\tau} v_n^* = C_{\mathbf{v}}^{1/\tau}$,where $\mathbf{v}^* = (v_n^*)_{n\in\mathbb{N}}$ is a non–decreasing rearrangement of $\mathbf{v}$. Furthermore, $\|\mathbf{v}\|_{\ell_\tau^w} := \|\mathbf{v}\| + |\mathbf{v}|_{\ell_\tau^w}$ is a quasi–norm for ℓ_τ^w. Since the continuous embeddings $\ell_\tau \hookrightarrow \ell_\tau^w \hookrightarrow \ell_{\tau+\varepsilon} \hookrightarrow \ell_2$ hold for $\tau < \tau + \varepsilon < 2$, ℓ_τ^w is 'close' to ℓ_τ and is therefore called *weak* ℓ_τ. The following crucial result connects sequences in ℓ_τ^w to best N–term approximation [9].

Proposition 3.5. *Let positive real numbers s and τ be related by*

$$\frac{1}{\tau} = s + \frac{1}{2}. \qquad (3.34)$$

Then $\mathbf{v} \in \ell_\tau^w$ if and only if $\sigma_N(\mathbf{v}) = \|\mathbf{v} - \mathbf{v}_N\| \lesssim N^{-s}\, \|\mathbf{v}\|_{\ell_\tau^w}$.

The property that an array of wavelet coefficients $\mathbf{v}$ belongs to ℓ_τ is equivalent to the fact that the expansion $\mathbf{v}^T\Psi_H$ in terms of a wavelet basis Ψ_H for a

Hilbert space H belongs to a certain *Besov space* which describes a much weaker regularity measure than a Sobolev space of corresponding order [8]. Thus, Proposition 3.5 expresses how much loss of regularity can be compensated by judiciously placing the degrees of freedom in a nonlinear way so as to retain a certain optimal order of error decay.

A key criterion for a scheme SOLVE to exhibit an optimal work/accuracy rate can be formulated through the following property of the respective residual approximation. The routine RES is called τ^**–sparse* for some $0 < \tau^* < 2$ if the following holds: Whenever the solution $\mathbf{q}$ of (3.12) belongs to ℓ^w_τ for some $\tau^* < \tau < 2$, then for any $\mathbf{v}$ with finite support the output $\mathbf{r}_\eta$ of RES $[\eta, \mathbf{Q}, \mathbf{g}, \mathbf{v}]$ satisfies $\|\mathbf{r}_\eta\|_{\ell^w_\tau} \lesssim \max\{\|\mathbf{v}\|_{\ell^w_\tau}, \|\mathbf{q}\|_{\ell^w_\tau}\}$, $\#\,\mathrm{supp}\,\mathbf{r}_\eta \lesssim \eta^{-1/s} \max\{\|\mathbf{v}\|^{1/s}_{\ell^w_\tau}, \|\mathbf{q}\|^{1/s}_{\ell^w_\tau}\}$, where s and τ are related by (3.34), and the number of floating point operations needed to compute $\mathbf{r}_\eta$ stays proportional to $\#\,\mathrm{supp}\,\mathbf{r}_\eta$.

The analysis in [10] then yields the following result.

Theorem 3.6. *If* RES *is* τ^**–sparse and if the exact solution* $\mathbf{q}$ *of (3.12) belongs to* ℓ^w_τ *for some* $\tau > \tau^*$*, then for every* $\varepsilon > 0$ *algorithm* SOLVE $[\varepsilon, \mathbf{Q}, \mathbf{g}]$ *produces after finitely many steps an output* $\overline{\mathbf{q}}_\varepsilon$ *(which, according to Proposition 3.1, always satisfies* $\|\mathbf{q} - \overline{\mathbf{q}}_\varepsilon\| < \varepsilon$*) with the following properties: For* s *and* τ *related by (3.34), one has*

$$\#\,\mathrm{supp}\,\overline{\mathbf{q}}_\varepsilon \lesssim \varepsilon^{-1/s}\|\mathbf{q}\|^{1/s}_{\ell^w_\tau}, \qquad \|\overline{\mathbf{q}}_\varepsilon\|_{\ell^w_\tau} \lesssim \|\mathbf{q}\|_{\ell^w_\tau}, \tag{3.35}$$

and the number of floating point operations needed to compute $\overline{\mathbf{q}}_\varepsilon$ *remains proportional to* $\#\,\mathrm{supp}\,\overline{\mathbf{q}}_\varepsilon$.

Thus, τ^*–sparsity of the routine RES implies asymptotically optimal work/accuracy rates for SOLVE for a certain range of decay rates given by τ^*. We stress that the algorithm itself does *not* require any a–priori knowledge about the solution such as its actual best N–term approximation rate. Theorem 3.6 also states that controlling the ℓ^w_τ–norm of the quantities generated in the computations is crucial. This finally explains the role of COARSE in Step (II.2) of SOLVE in terms of the following result [9].

Lemma 3.7. *Let* $\mathbf{v} \in \ell^w_\tau$ *and let* $\mathbf{w}$ *be any finitely supported approximation such that* $\|\mathbf{v} - \mathbf{w}\| \le \frac{1}{5}\eta$*. Then the output* $\mathbf{w}_\eta$ *of* COARSE $[\frac{4}{5}\eta, \mathbf{w}]$ *satisfies*

$$\#\,\mathrm{supp}\,\mathbf{w}_\eta \lesssim \|\mathbf{v}\|^{1/\tau}_{\ell^w_\tau}\eta^{-1/s}, \quad \|\mathbf{v} - \mathbf{w}_\eta\| \lesssim \eta, \quad \text{and} \quad \|\mathbf{w}_\eta\|_{\ell^w_\tau} \lesssim \|\mathbf{v}\|_{\ell^w_\tau}. \tag{3.36}$$

This can be interpreted as follows. If an error bound for a given finitely supported approximation $\mathbf{w}$ is known, a certain coarsening using only knowledge about $\mathbf{w}$ produces a new approximation to (the possibly unknown) $\mathbf{v}$ which gives rise to a slightly larger error but realizes the optimal relation between support and accuracy up to a uniform constant. In the scheme SOLVE, this means that by the coarsening step the ℓ_τ^w–norms of the iterates $\mathbf{v}^K$ are controlled.

It remains to establish that for $\text{SOLVE}_{\text{DCP}}$ the corresponding routine RES_{DCP} is τ^*–sparse. The following results from [18] reduce this question to the efficiency of APPLY. We say that $\text{APPLY}[\cdot, \mathbf{L}, \cdot]$ is *τ^*–efficient* for some $0 < \tau^* < 2$ if for any finitely supported $\mathbf{v} \in \ell_\tau^w$, for $0 < \tau^* < \tau < 2$, the output $\mathbf{w}_\eta$ of $\text{APPLY}[\eta, \mathbf{L}, \mathbf{v}]$ satisfies $\|\mathbf{w}_\eta\|_{\ell_\tau^w} \lesssim \|\mathbf{v}\|_{\ell_\tau^w}$ and $\#\operatorname{supp} \mathbf{w}_\eta \lesssim \eta^{-1/s} \|\mathbf{v}\|_{\ell_\tau^w}^{1/s}$ for $\eta \to 0$. Here the constants depend only on τ as $\tau \to \tau^*$ and s, τ satisfy (3.34). Moreover, the number of floating point operations needed to compute $\mathbf{w}_\eta$ is to remain proportional to $\#\operatorname{supp} \mathbf{w}_\eta$.

Proposition 3.8. *If the* APPLY *schemes in* RES_{PRM} *and* RES_{ADJ} *are τ^*–efficient for some $\tau^* < 2$, then* RES_{DCP} *is τ^*–sparse whenever there exists a constant C such that $C\eta \geq \max\{\delta_v, \delta_p\}$ and*

$$\max\{\|\tilde{\mathbf{p}}\|_{\ell_\tau^w}, \|\tilde{\mathbf{y}}\|_{\ell_\tau^w}, \|\mathbf{v}\|_{\ell_\tau^w}\} \leq C \left(\|\mathbf{y}\|_{\ell_\tau^w} + \|\mathbf{p}\|_{\ell_\tau^w} + \|\mathbf{u}\|_{\ell_\tau^w}\right),$$

where $\mathbf{v}$ is the current finitely supported input and $\tilde{\mathbf{y}}, \tilde{\mathbf{p}}$ are the initial guesses for the exact solution components $(\mathbf{y}, \mathbf{p})$.

Theorem 3.9. *If the* APPLY *schemes appearing in* RES_{PRM} *and* RES_{ADJ} *are τ^*–efficient for some $\tau^* < 2$ and the components of the solution $(\mathbf{y}, \mathbf{p}, \mathbf{u})$ of (EE) all belong to the respective space ℓ_τ^w for some $\tau > \tau^*$, then the approximate solutions $\mathbf{y}_\varepsilon, \mathbf{p}_\varepsilon, \mathbf{u}_\varepsilon$, produced by* $\text{SOLVE}_{\text{DCP}}$ *for any target accuracy ε, satisfy*

$$\|\mathbf{y}_\varepsilon\|_{\ell_\tau^w} + \|\mathbf{p}_\varepsilon\|_{\ell_\tau^w} + \|\mathbf{u}_\varepsilon\|_{\ell_\tau^w} \lesssim \|\mathbf{y}\|_{\ell_\tau^w} + \|\mathbf{p}\|_{\ell_\tau^w} + \|\mathbf{u}\|_{\ell_\tau^w}, \tag{3.37}$$

and

$$(\#\operatorname{supp} \mathbf{y}_\varepsilon) + (\#\operatorname{supp} \mathbf{p}_\varepsilon) + (\#\operatorname{supp} \mathbf{u}_\varepsilon) \lesssim \left(\|\mathbf{y}\|_{\ell_\tau^w}^{1/s} + \|\mathbf{p}\|_{\ell_\tau^w}^{1/s} + \|\mathbf{u}\|_{\ell_\tau^w}^{1/s}\right) \varepsilon^{-1/s}, \tag{3.38}$$

where the constants only depend on τ when τ approaches τ^*. Moreover, the number of floating point operations required during the execution of $\text{SOLVE}_{\text{DCP}}$ *remains proportional to the right hand side of (3.38).*

Thus, the practical realization of $\text{SOLVE}_{\text{DCP}}$ providing optimal work/accuracy rates for a possibly large range of decay rates of the error of best N–term approximation hinges on the availability of τ^*–efficient schemes APPLY with possibly small τ^* for the involved operators. For the approximate application of wavelet representations of a wide class of operators, including differential operators, one can indeed devise efficient schemes which is a consequence of the cancellation properties (CP) together with the norm equivalences (1.3) for the relevant function spaces. For the example considered above, the τ^*–efficiency of $\mathbf{A}$ defined in (3.5) can be shown whenever $\mathbf{A}$ is s^*–compressible where τ^* and s^* are related by (3.34). One knowns that s^* is the larger the higher the 'regularity' of the operator and the order of cancellation properties of the wavelets are. Estimates for s^* in terms of these quantities for spline wavelets and the above differential operator A can be found in [2]. Hence, Theorem 3.9 guarantees asymptotically optimal complexity bounds for $\tau > \tau^*$. This means that the scheme $\text{SOLVE}_{\text{DCP}}$ recovers rates of the error of best N–term approximation of order N^{-s} for $s < s^*$.

When describing the control problem, it has been pointed out that the wavelet framework allows for a flexible choice of norms in the control functional which is reflected in the parameters r and t in (3.7). In [4] first numerical experiments with algorithms of the above form on uniform refinements explore the influence of different norms on the quality of the solution and also the role of the weight parameter ω. As for the influence of the diagonal matrices $\mathbf{D}^{r-1}, \mathbf{D}^{t-1}$ appearing in (EE) within the adaptive algorithm, in the range $0 \leq r, t \leq 1$ scaling with either one of these matrices makes a sequence more compressible, that is, the diagonal matrices produce a shift in weak ℓ_τ spaces [18].

Proposition 3.10. *For $\beta > 0$, $\mathbf{p} \in \ell_\tau^w$ implies $\mathbf{D}^{-\beta}\mathbf{p} \in \ell_{\tau'}^w$, where $\frac{1}{\tau'} := \frac{1}{\tau} + \frac{\beta}{d}$.*

We can conclude the following. Whatever the sparsity class of the adjoint variable $\mathbf{p}$ is, the control $\mathbf{u}$ is in view of (3.23) even sparser. This means also that although the control $\mathbf{u}$ may be accurately recovered with relatively few degrees of freedom, the overall solution complexity is in the above case bounded from below by the less sparse auxiliary variable $\mathbf{p}$.

Acknowledgments. I want to thank an anonymous referee for the valuable remarks and suggestions, and Carsten Burstedde and Daniel Castaño Diez for their assistance during the preparation of this manuscript. This work has been supported in part by the Deutsche Forschungsgemeinschaft (SFB 611 and KU 1028/7–1).

References

[1] A. Barinka: *Fast Evaluation Tools for Adaptive Wavelet Schemes*, Dissertation, RWTH Aachen, 2003, in preparation.

[2] A. Barinka, T. Barsch, Ph. Charton, A. Cohen, S. Dahlke, W. Dahmen, K. Urban: *Adaptive wavelet schemes for elliptic problems — Implementation and numerical experiments*, SIAM J. Sci. Comp. **23** (2001), 910–939.

[3] R. Becker, H. Kapp, R. Rannacher: *Adaptive finite element methods for optimal control of partial differential equations: Basic concept*, SIAM J. Contr. Optim. **39** (2000), 113–132.

[4] C. Burstedde, A. Kunoth: *Wavelet methods for linear–quadratic control problems*, Manuscript, in preparation.

[5] D. Castaño Diez: *Adaptive data fitting with wavelets*, Manuscript, October 2002.

[6] D. Castaño Diez, A. Kunoth: *Adaptive fitting of scattered data by spline–wavelets*, Manuscript, October 2002, to appear in: *Curves and Surfaces*, L.L. Schumaker et. al. (eds.), Vanderbilt University Press, 2003.

[7] C. K. Chui, E. G. Quak: *Wavelets on a bounded interval*, in: *Numerical Methods of Approximation Theory*, Vol. 9, D. Braess, L. L. Schumaker (eds.), Int. Ser. Numer. Math. **105**, Birkhäuser, Basel, 1992, 53–75.

[8] A. Cohen: *Numerical Analysis of Wavelet Methods*, Elsevier, 2003.

[9] A. Cohen, W. Dahmen, R. DeVore: *Adaptive wavelet methods for elliptic operator equations — Convergence rates*, Math. Comp. **70** (2001), 27–75.

[10] A. Cohen, W. Dahmen, R. DeVore: *Adaptive wavelet methods II — Beyond the elliptic case*, Found. Computat. Math. **2** (2002), 203–245.

[11] A. Cohen, W. Dahmen, R. DeVore: *Adaptive wavelet scheme for nonlinear variational problems*, Preprint, July 2002.

[12] A. Cohen, R. Masson: *Wavelet adaptive methods for second order elliptic problems, boundary conditions and domain decomposition*, Numer. Math. **86** (2000), 193–238.

[13] W. Dahmen: *Wavelet and multiscale methods for operator equations*, Acta Numerica (1997), 55–228.

[14] W. Dahmen: *Wavelet methods for PDEs — Some recent developments*, J. Comput. Appl. Math. **128** (2001), 133–185.

[15] W. Dahmen: *Multiscale and Wavelet Methods for Operator Equations*, C.I.M.E. Lecture Notes, Springer, 2003.

[16] W. Dahmen, A. Kunoth: *Multilevel preconditioning*, Numer. Math. **63** (1992), 315–344.

[17] W. Dahmen, A. Kunoth, R. Schneider: *Wavelet least squares methods for boundary value problems*, SIAM J. Numer. Anal. **39** (2002), 1985–2013.

[18] W. Dahmen, A. Kunoth: *Adaptive wavelet methods for linear–quadratic elliptic control problems: Convergence rates*, Preprint No. 42, SFB 611, Universität Bonn, December 2002, submitted for publication.

[19] W. Dahmen, A. Kunoth, K. Urban: *Biorthogonal spline–wavelets on the interval – Stability and moment conditions*, Appl. Comput. Harm. Anal. **6** (1999), 132–196.

[20] P. Dierckx: *Curve and Surface Fitting with Splines*, Oxford University Press, 1993.

[21] D. R. Forsey, R. H. Bartels: *Hierarchical B–spline refinement*, Computer Graphics **22** (1988), 205–212.

[22] S. J. Gortler, M. Cohen: *Hierarchical and variational geometric modeling with wavelets*, in: *Proc.* 1995 *Symposium on Interactive* 3*D Graphics*, 35 ff.

[23] GTOPO30 data at U.S. Geological Survey's EROS Data Center, http://edcdaac.usgs.gov/gtopo30/gtopo30.html.

[24] A. Kunoth: *Fast iterative solution of saddle point problems in optimal control based on wavelets*, Comput. Optim. Appl. **22** (2002), 225–259.

[25] A. Kunoth: *Wavelet Methods — Boundary Value Problems and Control Problems*, Advances in Numerical Mathematics, Teubner, 2001.

[26] J. L. Lions: *Optimal Control of Systems Governed by Partial Differential Equations*, Springer, Berlin, 1971.

[27] E. J. Stollnitz, T.D. DeRose, D.H. Salesin: *Wavelets for Computer Graphics*, Morgan Kaufmann Publishers, San Francisco, 2000.

[28] G. Wahba: *Spline Models for Observational Data*, Series in Applied Mathematics **59**, SIAM, Philadelphia, 1990.

[29] H. Yserentant: *On the multilevel splitting of finite element spaces*, Numer. Math. **49** (1986), 379–412.

[30] E. Zeidler: *Nonlinear Functional Analysis and its Applications* III: *Variational Methods and Optimization*, Springer, 1985.

Address:

Angela Kunoth
Institut für Angewandte Mathematik
Universität Bonn
Wegelerstr. 6
D–53115 Bonn
Germany

International Series of Numerical Mathematics
Vol. 145,

Almost Interpolation and Radial Basis Functions

Alain Le Méhauté

Pour Manfred Reimer, à l'occasion de son 70e anniversaire

Abstract

After a review of some basic facts for Radial Basis Interpolation, we introduce the idea of almost interpolation for RBF, and show that in this setting it is possible to enlarge quite a lot the set of basic radial functions that can be used as basis. Our main result provides a Schoenberg-Whitney type condition for almost interpolation sets.

Introduction

We are interested here in multivariate interpolation by sums of radial basis functions. In the literature the focus has been on the special case where one uses the translates of one single basis function φ, and where the centers of the different translates of φ are usually the given data points. The main result in this case is that the interpolation problem, whatever the dimension is, is always solvable provided that the basis function is conditionally positive definite. This is a very restrictive hypothesis on both, the location of the interpolation points and the nature of φ. In this present paper, we want to show that it is possible to deal with interpolation points in a more general position and with a very broad class of radial functions, if we replace interpolation by what we call almost interpolation.

Let us recall the basic setup for interpolation with radial basis functions.

Let $\mathbf{T} := \{t_1, t_2, \ldots, t_N\}$ be a set of *distinct* points in $\mathbb{R}^d$, and let us consider a set $\mathbf{\Phi} := \{\phi_1, \phi_2, \ldots, \phi_N\}$ of continuous, d–variate functions. Given the data $\mathbf{Z} := \{z_1, z_2, \ldots, z_N\} \subset \mathbb{R}$, there exists a unique $f \in \text{span}\{\phi_1, \phi_2, \ldots, \phi_N\}$ which interpolates the value z_i at the knot t_i, for $i = 1, 2, \ldots, N$, if and only if the matrix

$$\mathbf{M} := \mathbf{M}(\mathbf{\Phi}, \mathbf{T}) := \begin{pmatrix} \phi_1(t_1) & \phi_2(t_1) & \cdots & \phi_N(t_1) \\ \phi_1(t_2) & \phi_2(t_2) & \cdots & \phi_N(t_2) \\ \vdots & \vdots & & \vdots \\ \phi_1(t_N) & \phi_2(t_N) & \cdots & \phi_N(t_N) \end{pmatrix}$$

is non–singular. In the radial basis setting, we are interested in the special case where the ϕ_i's are radially symmetric, *i.e.*,

$$\phi_i(t) = \varphi_i(\|t - \xi_i\|), \quad i = 1, 2, \ldots, N.$$

Here, $\xi_1, \xi_2, \ldots, \xi_N$ are given points in $\mathbb{R}^d$ (called the centers), and $\varphi_i : [0, \infty) \longmapsto \mathbb{R}$, $i = 1, 2, \ldots, N$, are continuous univariate functions.

1 Interpolation by a Sum of Translates of a Single Given Radial Function

This subject has been comprehensively reviewed in several recent papers (see [6] and references therein), and it is sufficient here to mention a few basic facts only.

In most of the cases, people are dealing with a single basis function $\varphi_i = \varphi$, $i = 1, 2, \ldots, N$, and almost anytime the centers are identical with the given interpolation points, *i.e.*, $\xi_i = t_i$, $i = 1, 2, \ldots, N$. Then the interpolating function is

$$s(t) = \sum_{i=1}^{N} \lambda_i \, \varphi(\|t - t_i\|) \; + p_m(t)$$

where $p_m \in \mathbb{P}_m[\mathbb{R}^d]$, *i.e.*, p_m is a polynomial of d variables, of degree $\leq m$.

In this case, the relevant interpolation matrix is of the form

$$\mathbf{A_m}(\mathbf{\Phi}, \mathbf{T}) = \begin{pmatrix} \mathbf{M}(\mathbf{\Phi}, \mathbf{T}) & \mathbf{B} \\ \mathbf{B}^\top & 0 \end{pmatrix}.$$

It is well known (see [21], for example) that

Theorem 1.1. *A sufficient condition for this matrix to be non–singular for any data set* $\mathbf{T}$ *is* φ *to be* $CP(m, d)$, *i.e.*, φ *is a conditionally stictly positive definite function of order* m, *defined on* $\mathbb{R}^d$.

Let us recall that a continuous function $F : \mathbb{R}^d \to \mathbb{R}$ is said to be $CP(m,d)$ if for any integer $p > 0$, any set of points $x_1, x_2, \ldots, x_p \in \mathbb{R}^d$ and any non–zero $c = (c_1, c_2, \ldots, c_p) \in \mathbb{R}^p$ such that $\sum_{i=1}^p c_i x_i^\alpha = 0$ for all $|\alpha| < m$, we have $\sum_{i=1}^p \sum_{j=1}^p c_i c_j F(x_i - x_j) > 0$.

Now, from [19] and [12] we get a nice characterization, which is a generalization of a former result of Schoenberg, for a radial function to be $CP(m,d)$ *for any* $d \geq 1$:

Theorem 1.2. *Given $g : [0,\infty) \longmapsto \mathbb{R}$, the associated radial basis function $G(x) = g(\|x\|)$ is $CP(m,d)$ for all $d \geq 1$, if and only if the function $\psi(t) := (-1)^m g^{(m)}(t)$ is strictly completely monotone on $(0,\infty)$, i.e., ψ is C^∞ and the successive derivatives of ψ alternate in sign.*

Properties 1.3. *It is very easy to check that*

1) *For the dimension d being fixed:*

$$CP(0,d) \subset CP(1,d) \subset \cdots \subset CP(m,d) \subset \cdots$$

2) *For the order m being fixed:*

$$CP(m,1) \supset CP(m,2) \supset \cdots \supset CP(m,d) \supset \cdots$$

3) *For $f, g \in CP(m,d)$ and $\lambda > 0$:*

$$f+g,\ fg,\ \lambda f \in CP(m,d)\ .$$

Various well–known radial basis functions are widely used. Among them there are the Gaussians $\phi(r) = e^{-r^2}$, the multiquadrics [13] $\phi(r) = (h^2 + r^2)^{\pm\frac{1}{2}}$ for $h > 0$, and the thin plate splines [9]

$$\phi(r) = \begin{cases} r^{2m+2s-d}\log r, & \text{whenever } 2m+2s-d \text{ is an even integer,} \\ r^{2m+2s-d}, & \text{otherwise,} \end{cases}$$

for $2m + 2s - d \in \mathbb{R}^+$. Also shifted versions of these functions are considered, [10]. Compactly supported radial basis functions where introduced more recently, but we lack a satisfactory theory for them [5, 16, 22].

2 Hilbertian Spaces of Functions and Reproducing Kernels

Let $X \subset \mathbb{R}^d$ and $E = \mathbb{R}^X$ be the set of real valued functions defined on X. We denote by $\mathcal{H}(E)$ the set of all the hilbertian subspaces of E, by $G^+(X)$

the set of the reproducing kernels on X, and by $\mathcal{L}^+(E)$ the set of the Schwartz kernels on E ([2]). More precisely:

• $H \in \mathcal{H}(E)$ is a subspace of E which is a Hilbert space where the pointwise evaluation is a continuous functional, *i.e.*,

$$\forall x \in X, \exists M > 0 \ s.t. \ \forall f \in H, \quad |f(x)| \leq M\|f\|_H.$$

• $g \in G^+(X)$ is said to be a reproducing kernel (one should better say a Bergman–Aronszajn kernel [1, 3]) if

$$g : X \times X \longmapsto \mathbb{R} \quad \text{and} \quad \begin{cases} \forall M \in X, g(\bullet, M) \in H \quad \text{and} \\ \forall M \in X, \forall \psi \in H, (\psi | g(\bullet, M)) = \psi(M). \end{cases}$$

• $L \in \mathcal{L}^+(E)$ is said to be a Schwartz kernel if $L : E' \longmapsto E$ is linear and if the bilinear form

$$E' \times E' \ni (e', f') \longmapsto \langle e', L[f'] \rangle \in \mathbb{R}$$

is symmetric and positive (*i.e.*, $\langle e', L[f'] \rangle = \langle f', L[e'] \rangle$ and $\langle e', L[e'] \rangle \geq 0$).

It is easy to show that a radial basis function $\phi \in CP(0, d)$, associated with the univariate φ, corresponds to a reproducing kernel in a hilbert space H (see [2]):

Theorem 2.1. *Let $h \in CP(0, d)$, and let $\omega = \widehat{h}$ denote its Fourier transform. Then $g(x, y) := h(x - y)$ is the reproducing kernel of the Hilbert space $H_\omega := \{f, \widehat{f} \in L^2_{(\frac{1}{\omega})}\}$ endowed with the scalar product*

$$(f, g)_\omega := \int_{\mathbb{R}^d} \frac{1}{\omega} \ \widehat{f}\widehat{g} \ dx.$$

When ϕ is $CP(m, d)$ for $m > 0$, one has to deal with a seminorm instead of a norm, and with a semi–Hilbertian space. In [8], the null space of the semi–norm is $\mathbb{P}_m$, and the semireproducing kernel is linked to the Green's functions of an elliptic operator (see [4] for a generalization of this). In another setting, Madych and Nelson [18] introduced the idea of generalized Fourier transform: Let $\mathcal{S}_m$ be the set of fast decreasing functions which are orthogonal to $\mathbb{P}_m$. For $f \in \mathcal{C}(\mathbb{R}^d)$, let $\widehat{(f)}_m$ be such that

$$\forall \varphi \in \mathcal{S}_m, \quad \int_{\mathbb{R}^d} f \ \varphi \ dx = \int_{\mathbb{R}^d} \widehat{(f)}_m \ \widehat{\varphi} \ dx \ .$$

Obviously, for any $p \in \mathbb{P}_m$, we have $\widehat{(p)}_m = 0$ and thus $\widehat{(f+p)}_m = \widehat{(f)}_m$. From this follows

Theorem 2.2. *Let $h \in \mathcal{C}(\mathbb{R}^d)$ be such that $\omega = \widehat{(h)_m}$ is positive, and assume that*

$$\int_{\|\xi\|>1} \omega(\xi)\, d\xi < +\infty \quad \text{and} \quad \int_{\|\xi\|<1} \|\xi\|^{2m}\, \omega(\xi)\, d\xi < +\infty\,.$$

Then the space $H_{\omega,m} = \{f, \widehat{(f)_m} \in L^2_{(\frac{1}{\omega})}\}$ is a semi–Hilbertian space endowed with the semi–scalar product

$$(f,g)_{\omega,m} = \int_{\mathbb{R}^d} \frac{1}{\omega}\, \widehat{(f)_m}\, \widehat{(g)_m}\, dx\,,$$

and h is its semi–reproducing kernel.

This situation covers most of the usual radial basis functions, except for the inverse multiquadrics. Now, the next theorems give us the opportunity to play a lot with the different approaches [17]:

Theorem 2.3. *In the above setting, interpolation by radial basis functions of type $CP(m,d)$, for $m \geq 0$, is nothing but interpolation by a spline in a Hilbert space (or a semi–Hilbertian space), i.e., interpolation minimizing the norm (the semi–norm, respectively).*

Theorem 2.4. *The set $\mathcal{H}(E)$ of the hilbertian subspaces of E, the set $G^+(X)$ of the reproducing kernels on X and the set $\mathcal{L}^+(E)$ of the Schwartz kernels on E are isomorphic to each other.*

Shortly, we have: from the Riesz representation theorem, any $H \in \mathcal{H}(E)$ has a unique reproducing kernel g. Then, any $g \in \mathcal{G}^+(X)$ is a symetric positive function which is the reproducing kernel of a $H \in \mathcal{H}(E)$ $H \in \mathcal{H}(E)$ defined as the set of functions ψ on X limits of Cauchy sequences of functions from the prehilbertian space $H_0 = \{\sum_{finite} \lambda_i g(.,P_i), P_i \in X\}$, endowed with the following inner product: for $f_1 = \sum_{i=1}^n a_i g(.,P_i)$ and $f_2 = \sum_{j=1}^m b_j g(.,Q_j)$, write $[f_1,f_2] = \sum_{i=1}^n \sum_{j=1}^m a_i b_j g(Q_j,P_i)$. Next $g \in \mathcal{G}^+(X)$ and $L \in \mathcal{L}^+(E)$ are related together through $L\delta_P = g(.,P), \forall P \in X$, and for $L \in \mathcal{L}^+(E)$ and $H \in \mathcal{H}(E)$, we have $(h, Le\prime) = \langle e', h\rangle$, for all $e' \in E'$ and for all $h \in H$.

So we can use properties from one of the settings to find properties for the other settings, for example:

- Let $H \in \mathcal{H}(E)$, and $\lambda > 0$. The space $\lambda H \in \mathcal{H}(E)$ is $\lambda H = \{\lambda x, x \in H\}$ with the scalar product $(h,k)_{\lambda H} = \frac{1}{\lambda}(h,k)_H$. So, if g is the Bergman–Aronszajn kernel for H, the kernel of λH is λg, i.e., $g_{\lambda H} = \lambda g_H$.

- For $H_1, H_2 \in \mathcal{H}(E)$ we have $H_1 + H_2 = \{x = x_1 + x_2, x_1 \in H_1, x_2 \in H_2\} \in \mathcal{H}(E)$, with the norm $\|h\|^2_{H_1+H_2} = \inf_{h_1+h_2=h} \{\|h_1\|^2_{H_1} + \|h_2\|^2_{H_2}\}$, called the inf–

convolution norm. The reproducing kernel for $H_1 + H_2$ is the sum of the two reproducing kernels, *i.e.*, $g_{H_1+H_2} = g_{H_1} + g_{H_2}$ (see [16]).

• Let $E = \mathbb{R}^X$ and $F = \mathbb{R}^Y$. For a given $u : E \longmapsto F$, linear and continuous, and a given $H \in \mathcal{H}(E)$, we put $u(H) = \{u(x), x \in H\}$. The norm $\|k\|_{u(H)} = \inf\{\|h\|_H, h \in H \text{ and } k = u(h)\}$, and the associated scalar product $(.,.)_H$, realizes that $u(H) \in \mathcal{H}(F)$. The Schwartz kernels transform to $L_{u(H)} = uL_H u^\top$, and the Aronszajn kernels transform to $g_{u(H)}(R,S) = (Lu^\top \delta_R, Lu^\top \delta_S)_H$.

This can provide a framework for the Hilbertian aspect (spaces, norms,...) of some experiments with radial basis functions such as:

⋆ *restriction to a subspace*, (truncation, for the construction of compactly supported radial basis function, e.g.): Let $E = \mathbb{R}^X$ and $F = \mathbb{R}^Y$ with $Y \subset X$. If

$$u = r_{|Y} : f \in E \longmapsto r_{|Y}[f] = f_{|Y} \in F$$

is the restriction from X to Y, then for any $H \in \mathcal{H}(E)$, let $H_{|Y} = u(H) = r_{|Y}(H)$. Then $g_{H_{|Y}} = g_{H|_Y}$, *i.e.*, the kernel of the space of the restrictions is the restriction of the original kernel.

⋆ *product by a function*: Let $E = F = \mathbb{R}^X$, pick $a \in E$ and define

$$u : f \in E \longmapsto af \in E\ .$$

Then, for any $H \in \mathcal{H}(E)$, we have $aH = u(H)$ and

$$g_{aH}(R,S) = a(R)a(S)g_H(R,S)\ .$$

⋆ *mapping in X*: Let $E = \mathbb{R}^X$, $F = \mathbb{R}^Y$, consider a mapping $v : Y \longmapsto X$, and let us choose $u : f \in E \longmapsto f \circ v \in F$ (as the composition of functions). Then for $H \circ v = u(H)$, we obtain

$$g_{H\circ v}(R,S) = g_H(v(R), v(S))\ .$$

For example, if v is an affine transformation in $\mathbb{R}^d$, we use ellipses instead of circles in a new kind of 'radial' basis functions, and the euclidean distance is replaced by an ellipses–based distance. Also, this approach enables us to use a Gaussian radial basis function defined by $\phi(x) = \exp^{-\langle Ax,x\rangle}$ which is strictly $CP(0,d)$ whenever A is a positive definite matrix. These two examples are common to people dealing with Neural Networks and with Learning Theory, see [20] and the references there.

3 Almost Interpolation with Radial Functions

In the above sections, we have met most of the usual radial basis functions, but there are only sufficient conditions for the interpolation matrix to be non–singular. The obvious question is: Is it possible to enlarge the set of basis functions, for instance by replacing the requirement of 'interpolation' by the the weaker concept of 'almost interpolation' as introduced in [7]?

Definition 3.1. $\mathbf{T} = \{t_1, t_2, \ldots, t_N\}$ *is said to be an almost interpolation set with respect to* $\mathbf{\Phi} = \{\phi_1, \phi_2, \ldots, \phi_N\}$ *provided that for any set of neighborhoods* $B(t_1), B(t_2), \ldots, B(t_N)$ *of the points in* $\mathbf{T}$*, there exist points* $t'_i \in B(t_i)$*,* $i = 1, \ldots, N$*, such that the matrix* $\mathbf{M}(\mathbf{\Phi}, \mathbf{T}')$ *is nonsingular for* $\mathbf{T}' = (t'_1, t'_2, \ldots, t'_N)$.

Let

$$\mathcal{F} \subset C[0,\infty) \cap C^1(0,\infty) \tag{3.1}$$

denote the class of real–valued functions φ with the property that φ' does not vanish identically on any subinterval of supp φ. The following is the main result of this paper. Here, $(\text{supp}\ \phi)^\circ$ denotes the interior of the support of ϕ.

Theorem 3.2. *Given* $\xi_1, \ldots, \xi_N \in I\!R^d$*, suppose that* $\varphi_1, \ldots, \varphi_N$ *belong to the class* $\mathcal{F}$*, and let* $\phi_1, \ldots, \phi_N$ *be the associated radial basis functions defined on* $I\!R^d$*. Assume that for some permutation* $j_1, \ldots, j_N$ *of* $\{1, 2, \ldots, N\}$*,*

$$t_{j_i} \in (\text{supp}\ \phi_i)^\circ \quad \textit{for} \quad i = 1, \ldots, N. \tag{3.2}$$

Then the set $\mathbf{T} = \{t_1, t_2, \ldots, t_N\}$ *is an almost interpolation set with respect to* $\mathbf{\Phi} = \{\phi_1, \phi_2, \ldots, \phi_N\}$.

The proof is based on the following elementary observation from linear algebra:

Lemma 3.3. *Given an* $(n \times n)$*–matrix* M*, with row vectors* $r_1, \ldots, r_n$*, let* H *be the hyperplane spanned by* $r_1, \ldots, r_{n-1}$*. Suppose* H *is non degenerate,* i.e.*, the vectors* $r_1, \ldots, r_{n-1}$ *are linearly independent. Then* $\det M = 0$ *if and only if* r_n *is parallel to* H.

Assuming that each t_i lies in the interior of the support of ϕ_i, for $i = 1, \ldots, N$, we now sketch the *proof of Theorem* 3.2, which is by induction on N. The result is trivial for $N = 1$. Consider the case $N = 2$. If

$$\det\ \mathbf{M} = \det\ \mathbf{M}(t_1, t_2) = \det \begin{pmatrix} \phi_1(t_1) & \phi_2(t_1) \\ \phi_1(t_2) & \phi_2(t_2) \end{pmatrix} \neq 0$$

there is nothing to show. So, let us assume that $\det \mathbf{M}(t_1, t_2) = 0$.

Given any neighborhoods $B(t_1)$ and $B(t_2)$ of the points t_1 and t_2, in the interior of the support of ϕ_1 and ϕ_2 respectively, we can choose points $t_1' \in B(t_1)$ and $t_2' \in B(t_2)$ such that the determinant det $\mathbf{M}(t_1', t_2')$ is nonzero, as follows. If $\phi_1(t_1) = 0$, then we can choose $t_1' \in B(t_1)$ with $\phi_1(t_1') \neq 0$ since $\varphi \in \mathcal{F}$. Otherwise, we put $t_1' = t_1$. In case det $\mathbf{M}(t_1', t_2) \neq 0$ we are done. Assuming the contrary, by Lemma 3.3, the second row of $\mathbf{M}$ is parallel to the first one. In this case we choose t_2' to be any point in $B(t_2)$ satisfying

$$\|t_2' - \xi_1\| = \|t_2 - \xi_1\| \quad \text{and} \quad \phi_2(t_2') \neq \phi_2(t_2).$$

This is possible because ϕ_2 is a radial basis function associated with a function $\varphi_2 \in \mathcal{F}$. Since

$$\frac{\phi_1(t_2)}{\phi_1(t_2')} = 1 \quad \text{and} \quad \frac{\phi_2(t_2)}{\phi_2(t_2')} \neq 1,$$

we have det $\mathbf{M}(t_1', t_2') \neq 0$.

Now let $N \geq 3$, and assume that $t_1, t_2, \ldots, t_N$ be such that the determinant of $\mathbf{M}(t_1, t_2, \ldots, t_N)$ vanishes (otherwise we are done). We can assume that, for all permutations of the points, $\det \mathbf{M}(t_1, t_2, \ldots, t_{N-1}) \neq 0$ (since otherwise, due to the induction hypothesis, we solve first the problem with $N-1$ points). Thus the rows $r_1, \ldots, r_{N-1}$ of $\mathbf{M}(t_1, t_2, \ldots, t_N)$ are linearly independent. As vectors in $\mathbb{R}^N$, they span a hyperplane P, and r_N belongs to this P.

Let Q denote the affine plane $\{x_N = \phi_N(t_N)\}$. The two planes P and Q intersect along a line Δ. Indeed, assuming they are parallel planes implies that $\phi_N(t_N) = 0$; because $\phi_N \in \mathcal{F}$, there exists t_N' in a neighborhood of t_N s.t. $\phi_N(t_N') \neq 0$. Then wether $\phi_k(t_N') = 0, k = 0, \ldots, N-1$, and we are done , or $detM(t_1, t_2, /ldots, t_{N-1}, t_N')$ is $\neq 0$, done, or it is 0. In this case the corresponding plane P' is no longuer parallel to Q. Let δ and $\rho_1, \ldots, \rho_N$ be the projections of Δ and of the vectors $r_1, \ldots, r_N$, respectively, on the plane $Z = \{x_N = 0\}$. Each vector r_i ends at a point $Q_i \in Q$ and each ρ_i at a point $q_i \in Z$, and the q_i's are all on the line δ. Because the q_i's are distinct (if not, the problem will not be solved in $\mathbb{R}^{N-1}$), there exists $\xi \geq 0$ such that the ball $B(q_N, \xi)$ does not intersect with any of the lines supporting the vectors $\rho_1, \ldots, \rho_{N-1}$. Now the ball $B(Q_N, \xi)$ intersects Δ along one of its diameter D. Take any Q_N' in $B(Q_N, \xi) \cap Q$ but not in D. Thus Q_N' corresponds to a vector r_N', which is linearly independent of the vectors $r_1, \ldots, r_{N-1}$. Because the ϕ_i's in $\mathcal{F}$ are continuous functions, there exists a point t_N' in the neighborhood of t_N such that r_N' is the last row of $\mathbf{M}(t_1, t_2, \ldots, t_{N-1}, t_N')$. Whence, $\mathbf{M}(t_1, t_2, \ldots, t_{N-1}, t_N')$ is non–singular.

Remark 3.4. Condition (3.2) can be regarded as an analog of the classical Schoenberg–Whitney condition for (univariate) spline interpolation. This was generalized in [8] to almost interpolation with bivariate splines.

Remark 3.5. Condition (3.2) is automatically fulfilled whenever all the φ_i's have full support

$[0, \infty)$.

Remark 3.6. We claim that condition (3.2) is necessary. Indeed,

$$\det M = \sum (-1)^{(j_1+\cdots+j_N)} \quad \phi_1(t_{j_1})\phi_2(t_{j_2})\cdots\phi_N(t_{j_N})$$

where the sum is taken over all permutations $j_1, \ldots, j_N$ of $\{1, 2, \ldots, N\}$. Thus, if (3.2) fails to hold, then each term of the sum vanishes, and this remains true for all points $t'_1, \ldots, t'_N$ which are sufficiently close to

$t_1, \ldots, t_N$.

Example 3.7. In [14, 15], Ed Kansa suggested to use multiquadrics (MQ) for the solution of different PDE problems.

For instance, he was using a domain decomposition technic as follows: Let $\Omega \subset \mathbb{R}^2$ be the union of a finite number of subdomains Ω_j, $j = 1, \ldots, J$, and let us consider some interpolation problem over the set of data points

$$T = \bigcup_{j=1}^{J} T_j = \{t_i^j, i = 1, \ldots, N_j, j = 1, \ldots, J\}.$$

Then, defining $\phi_j(t - t_i^j) = \sqrt{\|t - t_i^j\|^2 + r_j^2}$, the interpolating function is choosen to be $s(t) = \sum_{j=1}^{J} \sum_{i=1}^{N_j} a_i^j \; \phi_j(t - t_i^j)$. Kansa claims that "numerical experiments show that we can greatly improve the condition number of the MQ–matrix by choosing carefully the parameters r_j". Obviously, this example does not fit with the usual MQ interpolation, since different basis functions are used in different arrays.

Another example is by the same author [15], where multiquadrics are used as a basis for a meshless approximation "as the spatial approximation scheme for parabolic, hyperbolic PDE's and the elliptic Poisson's equations". He wrote that "MQ is not only exceptionnally accurate, but is more efficient than other methods". The MQ expansion is given as

$$s(t) = \sum_{j=1}^{N} a_j \sqrt{\|t - t_i^j\|^2 + r_j^2} \,,$$

and " by numerical experimentation, (he) found that the best results occured with the following"

$$r_j^2 = r_{min}^2 \left(\frac{r_{max}^2}{r_{min}^2}\right)^{\left(\frac{j-i}{N-1}\right)}, \quad j = 1, \dots, N,$$

where r_{max}^2 and r_{min}^2 are input parameters.

Example 3.8. Another example is the interpolating function of Floater and Iske [11]. In order to have some multistep interpolation process, they consider the set of data points T as a nested sequence of data sets $T_1 \subset T_2 \subset \cdots \subset T_M = T$; they affix to each $T_k = \{x_1^k, \dots, x_{N_k}^k\}$ a different radial basis function ϕ_k and the k–th interpolant

$$s_k[f](x) = \sum_{j=1}^{N_k} c_j^k \; \phi_k(\,\|x - x_j^k\|\,)$$

is assumed to interpolate $f - \sum_{i=1}^{k-1}[f]$ on T_k, where f is the underlying function to be interpolated. This guarantees that $S = \sum_{k=1}^{M} s_k[f]$ matches f at each of the centers.

One can put all the interpolation constraints in the same linear system. Let $I_j = T_j \setminus T_1 \cup \cdots T_{j-1}$, and ordering the data points as $I_1, I_2, \dots, I_M$, we get a matrix A, such that $A_{nm} = \phi_q(x_n - x_m)$ whenever $x_m \in I_q$. Using this for interpolating the original data, we are dealing with a block matrix and the Floater–Iske algorithm is nothing but the first iteration of a block Gauss Seidel method, the starting values being taken to be zero. Another point of view for this multistep process is also to consider the final interpolant operator as a boolean sum of the different interpolants.

Aknowledgement: I want to thank Larry Schumaker for introducing me to the idea of almost interpolation, during a short visit to Vanderbilt University some time ago.

References

[1] N. Aronszajn: *Theory of reproducing kernels,* Trans. Amer. Math. Soc. **68** (1950), 337–404.

[2] M. Atteia: *Hilbertian Kernels and Spline Functions,* North–Holland, Amsterdam, 1992.

[3] S. Bergman: *The Kernel Function and Conformal Mapping,* Math. Surveys AMS **V** (1950).

[4] A. Bouhamidi, A. Le Méhauté: *Multivariate interpolating (m,l,s) splines,* Adv. Comput. Math. **11** (1999), 287–314.

[5] M. D. Buhmann: *Radial functions on compact support,* Proc. Edinburgh Math. Soc. **41** (1998), 33–46.

[6] M. D. Buhmann: *Radial basis functions,* Acta Numerica (2000), 1–38.

[7] O. Davydov: *On almost interpolation,* J. Approx. Theory **91** (1997), 398–418.

[8] O. Davydov, M. Sommer, H. Strauss: *Locally linearly independent systems and almost interpolation,* In: *Multivariate Approximation and Splines,* G. Nürnberger, J. W. Schmidt, G. Walz (eds.), Birkhäuser, Basel, 1997.

[9] J. Duchon: *Interpolation des fonctions de deux variables suivant le principe de la flexion des plaques minces,* RAIRO Anal. Numer. **10** (1976), 5–12.

[10] N. Dyn, D. Levin, S. Rippa: *Numerical procedures for global surface fitting of scattered data by radial functions,* SIAM J. Sci. Statist. Comput. **7** (1986), 639–659.

[11] M. Floater, A. Iske: *Multistep scattered data interpolation using compactly supported radial basis functions*, J. Comput. Appl. Math. **73** (1996), 65–78.

[12] K. Guo, S. Hu, X. Sun: *Conditionally positive definite functions and Laplace–Stieltjes integral,* J. Approx. Theory **74** (1993), 249–265.

[13] R. L. Hardy: *Multiquadric equations of topographic and other irregular surfaces,* J. Geophys. Res. **76** (1971), 1905–1915.

[14] E. Kansa: *A strictly conservative spatial approximation scheme for the governing engineering and physics equations over irregular regions and inhomogeneous scattered nodes,* UCRI 101634, Lawrence Livermore National Laboratory (1989).

[15] E. Kansa: *Multiquadrics. A scattered data approximation scheme with applications to computational fluid dynamics. I. Surface approximation and partial derivatives estimates, II. Solutions to parabolic, hyperbolic and elliptic PDE's.*, Comput. Math. Appl. **19** (1990), 127–145 and 147–161.

[16] A. Le Méhauté: *Inf–convolution and radial basis functions,* in: *New Developments in Approximation Theory,* M. W. Müller, M. Buhmann, D. Mache, M. Felten (eds.), ISNM, vol. **132**, Birkhäuser, Basel, (1999).

[17] A. Le Méhauté: *Approximation par splines Hilbertiennes,* Cours de DEA, Postgraduate lectures, Université de Nantes, 2002.

[18] W. R. Madych, S. A. Nelson: *Multivariate interpolation and conditionally positive definite functions,* J. Approx. Theory and its Applications **4** (1988), 77–89.

[19] C. Micchelli: *Interpolation of scattered data: distance matrices and conditionally positive definite functions,* Constr. Approx. **2** (1986), 11–22.

[20] T. Poggio, F. Girosi: *Networks for approximation and learning,* in: Proceed. IEEE **78** (9) (1990), 1481–1497.

[21] M. Powell: *The theory of radial basis functions approximation in 1990,* in: *Advances in Numerical Analysis* **2** (1992), 105–208.

[22] R. Schaback: *Reconstruction of multivariate functions from scattered data,* manuscript.

Address:

Alain Le Méhauté
Laboratoire Jean Leray
UMR 6629 CNRS & Université de Nantes
2 rue de la Houssinière
46–44322 Nantes
France

International Series of Numerical Mathematics
Vol. 145, ©2003 Birkhäuser Verlag Basel/Switzerland

Fundamental Splines on Triangulations

Günther Nürnberger and Frank Zeilfelder

Dedicated to Professor Dr. Manfred Reimer on the occasion of his 70th birthday

Abstract

We analyse properties of the fundamental splines associated with local Lagrange interpolation (cf. Nürnberger and Zeilfelder [14–16]) for C^1 splines on triangulations. These splines are zero except at one point of an interpolation set and form a basis of the spline space. It is proved that the support of the fundamental splines is small. Moreover, we show that the interpolation methods can be further simplified for cubic C^1 splines on separable triangulations. In this case, the supports of the fundamental spline basis are even smaller. Finally, we describe an algorithm for modifying an arbitrary triangulation such that the resulting triangulation is separable.

Introduction

A basic problem in multivariate approximation is to construct interpolation sets for finite–dimensional spaces which are useful for applications (cf. Chui [2]). In this paper, we consider interpolation by bivariate polynomial splines, which are a natural generalization of splines in one variable. Let Δ be a triangulation of a planar domain Ω. Then, the space of *bivariate C^r splines of degree q* is defined by

$$\mathcal{S}_q^r(\Delta) = \{s \in C^r(\Omega) : \; s|_T \in \mathcal{P}_q, \; T \in \Delta\},$$

where $\mathcal{P}_q$ is the *space of bivariate polynomials*, i. e.,

$$\mathcal{P}_q = \operatorname{span}\{x^i y^j : i \geq 0, \; j \geq 0, \; i + j \leq q\}.$$

The (*Lagrange*) *interpolation problem* for $\mathcal{S}_q^r(\Delta)$ is to determine point sets $\{z_1, \dots, z_d\}$ in Ω, where d is the dimension of $\mathcal{S}_q^r(\Delta)$, such that for any $f \in C(\Omega)$, a unique spline $s_f \in \mathcal{S}_q^r(\Delta)$ exists such that

$$s_f(z_i) = f(z_i), \; i = 1, \dots, d. \qquad (0.1)$$

In this case, the point set $\{z_1, \ldots, z_d\} \subseteq \Omega$ is called a *Lagrange interpolation set for* $\mathcal{S}_q^r(\Delta)$. Interpolation methods for bivariate splines are said to be local, if the change of a single value $f(z_i)$ only influences the interpolating spline s_f in a neighborhood of the interpolation point z_i. In this case, we call them *local interpolation sets.* Analogously, we define Hermite interpolation by bivariate splines. In this case, certain partial derivatives at points $z_1, \ldots, z_m \in \Omega$ with $m < d$ are involved, and the total number of conditions is equal to d.

There exists a vast literature on local Hermite interpolation by bivariate splines (cf. Nürnberger and Zeilfelder [13], Zeilfelder [18], and the references therein). A main difficulty in constructing local Lagrange interpolation sets for smooth splines is that the known local Hermite interpolation schemes (cf. Clough and Tocher [3], Fraeijs de Veubeke [7], Sander [17]) cannot be transformed directly into local Lagrange interpolation schemes on the whole triangulation. On the other hand, local Lagrange interpolation methods are important for many applications, since only data is needed and no derivatives. If in practice, a surface is given by a linear spline on a fine triangulation (with many triangles), then the Lagrange interpolating spline on a coarse subtriangulation can be constructed by taking the Lagrange data directly from the linear spline. This stands in contrast to Hermite interpolation methods, where (cross–boundary) derivatives have to be computed approximately.

The problem of constructing local Lagrange interpolation sets for C^1 splines was solved for arbitrary triangulations (cf. Nürnberger and Zeilfelder [14–16]), and for C^1 cubic splines on triangulations obtained from arbitrary quadrangulations (cf. Nürnberger, Schumaker and Zeilfelder [10–12]) by inserting diagonals in the convex quadrilaterals. The numerical examples given in Nürnberger, Schumaker and Zeilfelder [10, 11] and Nürnberger and Zeilfelder [14–16] show that the interpolating splines can be computed efficiently in real world applications involving large data sets consisting of millions of points.

In this paper, we analyse properties of the fundamental spline basis associated with the local Lagrange interpolation methods of Nürnberger and Zeilfelder [14, 16] in the case of cubic C^1 splines. In these methods the interpolating splines are defined w.r.t. a triangulation $\tilde{\Delta}$ obtained from an arbitrary triangulation, where approximately half of the triangles contain a Clough–Tocher split. Given $f \in C(\Omega)$ and a (local) Lagrange interpolation set $\{z_1, \ldots, z_d\}$ for $\mathcal{S}_3^1(\tilde{\Delta})$, the interpolating spline $s_f \in \mathcal{S}_3^1(\tilde{\Delta})$ with (0.1) can be written as

$$s = \sum_{i=1}^{d} f(z_i)\; s_i$$

where the (*Lagrange*) *fundamental splines* $s_i \in \mathcal{S}_3^1(\tilde{\Delta})$, $i = 1, \ldots, d$, are determined by the conditions

$$s_i(z_j) = \delta_{i,j}, \ j = 1, \ldots, d.$$

Here, $\delta_{i,j}$ denotes Kronecker's symbol. It is obvious that the splines s_i form a basis $\mathcal{S} = \{s_i : i = 1, \ldots, d\}$ of the spline space $\mathcal{S}_3^1(\tilde{\Delta})$. We call such a basis *fundamental spline basis.* We will prove that the support of the fundamental splines (i.e., roughly speaking the subset of Ω where s_i is non–zero) is small. Moreover, we show that the algorithm of constructing local interpolation sets for C^1 splines can be further simplified for separable triangulations. In this case, the supports of the fundamental spline basis associated with the interpolation scheme become even smaller. Finally, we show that any triangulation can be modified such that the resulting triangulation is separable. For doing this, in general a few vertices and edges have to be added.

1 Interpolation on Arbitrary Triangulations

We first recall the algorithm for constructing local Lagrange interpolation sets (cf. Nürnberger and Zeilfelder [16]). This method has been developed for C^1 splines of degree q, $q \geq 3$. In this paper, we only consider the case when $q = 3$.

Let an arbitrary triangulation Δ be given. In the following, we construct an interpolation set $\mathcal{L}$ for cubic C^1 splines. This is done by Algorithm 1. Here, we successively add points to some set $\mathcal{L}$ and denote the resulting set again by $\mathcal{L}$. Moreover, in Step 1 of the algorithm we successively mark the vertices of Δ whenever interpolation points are chosen near the vertices.

Algorithm 1: Construction of local interpolation sets $\mathcal{L}$ for C^1 cubic splines for given triangulations Δ.

Step 1: We start with the set $\mathcal{L}$ consisting of all vertices of Δ, and set all vertices of Δ to be unmarked. For $\ell = 3, 2, 1$, repeat as often as possible: If there exists $T \in \Delta$ with ℓ unmarked vertices $v_1, \ldots, v_\ell$, then for each $i = 1, \ldots, \ell$, add the points $v_i + \frac{1}{3}(v - v_i)$, for all vertices $v \neq v_i$ of T, and mark $v_1, \ldots, v_\ell$.

Step 1 terminates when all the vertices are marked. Then we color the triangles of Δ with two colors, say black and white.

Coloring of triangles: First, choose a large subset of Δ which does not contain two *neighboring triangles*, i.e. triangles with a common edge. The triangles of this subset are colored black, and the remaining triangles are colored white. Successively apply the following algorithmic rule: If a triangle T in Δ exists with more than one neighboring triangle of the same color, then we flip the color of T. Finally, we flip the color of all white triangles which have two boundary edges and a neighboring white triangle.

The coloring method terminates, since the number of edges shared by two triangles with the same color decreases at each step. Then we apply a Clough–Tocher split to every white triangle of Δ (i.e., for these triangles we connect the barycenter with the three vertices). The resulting triangulation is denoted by $\tilde{\Delta}$. By using the coloring additional interpolation points are chosen inside some of the triangles of Δ.

Step 2: Choose the barycenter of every black triangle T with no neighboring black triangle. For every pair of neighboring black triangles and for every pair of neighboring white triangles, choose the barycenter for exactly one of its triangles. For every white triangle T which contains a boundary edge of Δ, choose the barycenter of T, if this point has not already been chosen. Otherwise, choose an interpolation point on an edge of $\tilde{\Delta}$ which connects the barycenter of T with a vertex of a black triangle. In addition, choose such an interpolation point if a white triangle contains two boundary edges. Add all these points to $\mathcal{L}$.

The final set of all points chosen in Step 1 and Step 2 of Algorithm 1 is denoted by $\mathcal{L}$.

Figure 1 shows an example of an interpolation set $\mathcal{L}$ obtained from Algorithm 1. To the left, it is shown how the interpolation points are chosen such that the interpolating spline is uniquely and locally determined on the edges of Δ. This is an example of a point set resulting from the first step of Algorithm 1. Note that the order of choosing interpolation points in Step 1 is important for the locality of the method. To the right of Figure 1 an example for $\mathcal{L}$ is shown including the coloring of the triangles used as a preparation of the second step of Algorithm 1. In this step, some additional points are chosen in the interior of the triangles of Δ, such that the method is local. The algorithmic complexity of the coloring is linear in the number of edges, and therefore Algorithm 1 is fast.

We note that the number d of interpolation points in $\mathcal{L}$, and hence the dimen-

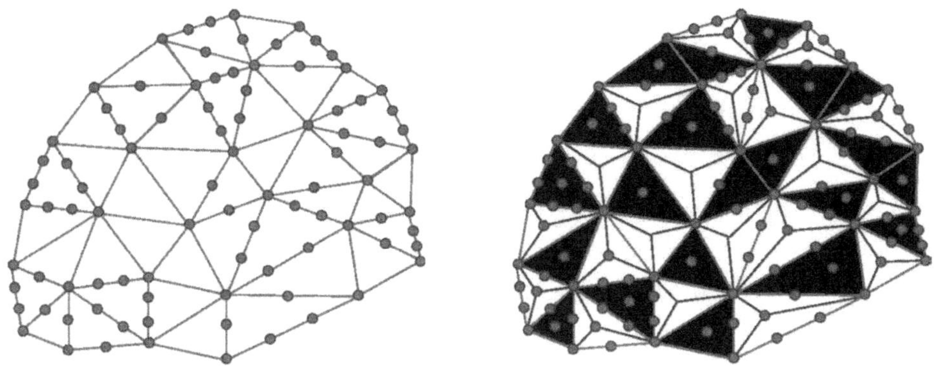

Figure 1: Construction of interpolation points in the two steps of the algorithm. The interpolation points are marked by dots.

sion of the resulting spline space $\mathcal{S}_3^1(\tilde{\Delta})$ is

$$d = 3V_\Delta + E_\Delta - 2\ B,$$

where V_Δ is the number of vertices of Δ, E_Δ is the number of edges of Δ, and B is the number of black triangles in Δ. (In the example shown in Figure 1, right, we have $V_\Delta = 30,\ E_\Delta = 70,\ B = 21$, and $d = 118$.)

The following result was given in Nürnberger and Zeilfelder [16].

Theorem 1.1. *The set $\mathcal{L} = \{z_1, \ldots, z_d\}$ constructed in Algorithm 1 is a local Lagrange interpolation set for $\mathcal{S}_3^1(\tilde{\Delta})$. Given $f \in C^4(\Omega)$, the interpolating spline $s_f \in \mathcal{S}_3^1(\tilde{\Delta})$ with*

$$s_f(z_i) = f(z_i),\ i = 1, \ldots, d, \tag{1.1}$$

yields optimal approximation order, i.e., there exists a constant $K > 0$, depending only on f and the smallest angle of the triangles in $\tilde{\Delta}$, such that for all $\omega \in \{0, \ldots, 3\}$,

$$\|D^\omega(f - s_f)\| \leq Kh^{4-\omega},$$

where h denotes the maximal diameter of the triangles in Δ. (Here, $\|.\|$ denotes the uniform norm, and D^ω denotes the ωth derivative.)

2 Support of Fundamental Splines

Given $f \in C(\Omega)$, the interpolating spline $s_f \in \mathcal{S}_3^1(\tilde{\Delta})$ from Theorem 1.1 with (1.1) can be written as

$$s = \sum_{i=1}^{d} f(z_i)\, s_i,$$

where s_i, $i = 1, \ldots, d$, are the (Lagrange) fundamental splines with $s_i(z_j) = \delta_{i,j}$, which form a basis $\mathcal{S} = \{s_i : \; i = 1, \ldots, d\}$ of $\mathcal{S}_3^1(\tilde{\Delta})$. Such bases are sometimes also called cardinal spline bases.

The proof of Theorem 1.1 given in Nürnberger and Zeilfelder [16] uses weak interpolation techniques to show the optimal approximation order. These arguments also show that the interpolation method described in the previous section is local, and it follows that the fundamental splines s_i, $i = 1, \ldots, d$, have local support. These supports are quite different depending on the associated interpolation points chosen in the two steps of the algorithm.

In the following, we will show that the supports are small. For doing this, we denote by star(T) the union of all triangles which intersect with a triangle $T \in \Delta$ in at least one point, and set $\text{star}^m(T){:=}\text{star}^{m-1}(\text{star}(T))$. We say that a spline s is $\text{star}^m(T)$ supported, if its support, i.e. the set

$$\text{supp}(s) = \overline{\{s(z) \neq 0 : z \in \Omega\}}$$

is contained in $\text{star}^m(T)$ for a triangle $T \in \Delta$.

The next result is on the supports of the fundamental splines s_i, $i = 1, \ldots, d$, associated with the interpolation set $\mathcal{L}$ from Theorem 1.1.

Theorem 2.1. *The supports of the fundamental splines $s_i \in \mathcal{S}$, $i = 1, \ldots, d$, satisfy $\text{supp}(s_i) \subseteq \text{star}^2(T)$, if the corresponding interpolation point $z_i \in \mathcal{L}$ lies strictly inside a triangle $T \in \Delta$, and $\text{supp}(s_i) \subseteq \text{star}^4(T)$ if $z_i \in \mathcal{L}$ lies on an edge of a triangle $T \in \Delta$.*

Proof. We will analyse the case when the support of s_i is of maximal size. For doing this, we consider two cases.

Case 1. Suppose z_i lies strictly inside the triangle $T \in \Delta$, i.e. z_i is constructed in the second step of the Algorithm 1. In this case, we have $s_i|_e \equiv 0$ for all edges e of Δ. This implies that the first derivatives $\frac{\partial s_i}{\partial x}$ and $\frac{\partial s_i}{\partial y}$ of s_i vanish at all the vertices of Δ. Let us assume that z_i lies inside a black triangle T with a neighboring black triangle T^*. The C^1 continuity across the common edge of

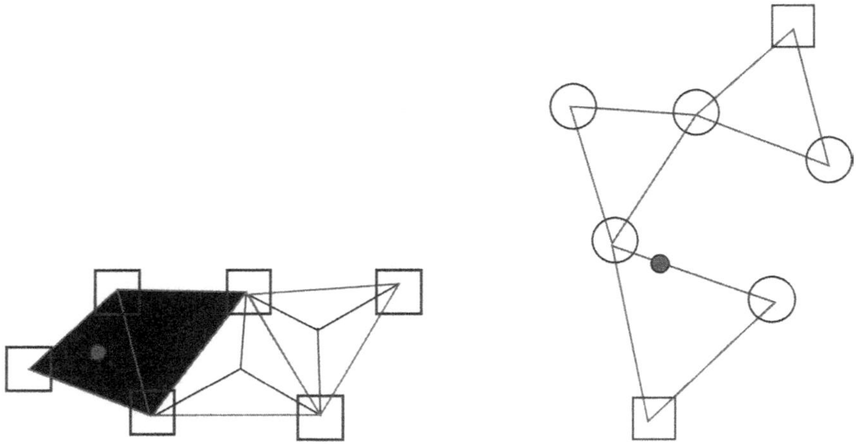

Figure 2: Propagation of fundamental basis splines – in the interior (left), and along the edges (right). The dot indicates the interpolation point associated with the fundamental spline s_i. The circles indicate derivatives of s_i which are possibly non–zero, while the boxes indicate zero derivatives of s_i at the vertices.

T and T^* implies that $s_i|_{T^*} \not\equiv 0$. A similar argument now implies that s_i can be non–zero in a neighboring white triangle of T^*. If this white triangle has a white neighboring triangle $\tilde{T}$, then the smoothness across the involved edges of $\tilde{\Delta}$ implies that the non–zero property can propagate also into $\tilde{T}$. By the choice of interpolation points and the coloring this propagation can not go further. It follows that in this case the support of s_i can be as much as $\text{star}^2(T)$ (an illustration of this case is given in Figure 2, left). Since it is easy to see that this is the case of a support with maximal size under the assumption that z_i lies strictly inside a triangle $T \in \Delta$, this completes the proof of Case 1.

Case 2. Suppose z_i lies on an edge $e = [v_1, v_2]$ of a triangle $T \in \Delta$, i.e. z_i is constructed in the first step of Algorithm 1. Again we analyse supports of maximal size. Therefore, let us further assume that z_i lies in the interior of e and T is a triangle which had three unmarked vertices when carrying out Step 1 of Algorithm 1. Since $s_i(z_i) = 1$, it follows that the partial derivatives $\frac{\partial s_i}{\partial x}$ and $\frac{\partial s_i}{\partial y}$ can be non–zero at the vertices v_1 and v_2. If there exists a triangle $T^* \in \Delta$ with vertex v_1 or v_2 which had exactly two unmarked vertices when carrying out Step 1 of Algorithm 1, then it follows that the partial derivatives $\frac{\partial s_i}{\partial x}$ and $\frac{\partial s_i}{\partial y}$ can be non–zero at the two vertices of T^* different from v_1 and v_2, respectively. The same argument applies again, if there exists a triangle

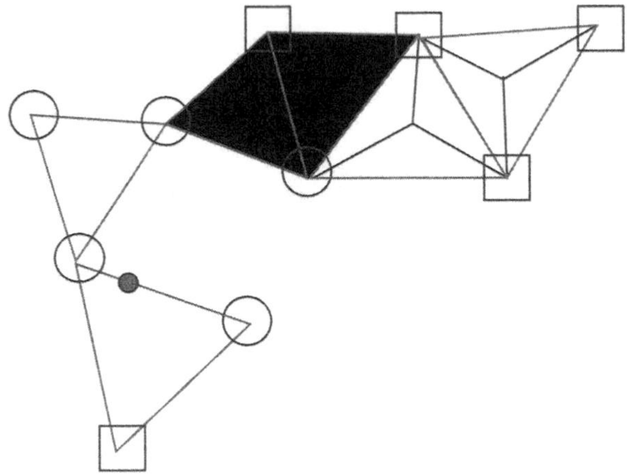

Figure 3: Support of a fundamental spline: the maximal support may appear when propagation along the edges and in the interior are combined. The dot indicates the interpolation point associated with s_i.The circles indicate derivatives of s_i which are possibly non–zero, while the boxes indicate zero derivatives of s_i at the vertices.

$\tilde{T}$ such that T^* has a common vertex different from v_1 and v_2, respectively, such that $\tilde{T}$ had exactly one unmarked vertex when carrying out Step 1 of Algorithm 1. In particular, it can be that $\frac{\partial s_i}{\partial x}$ and $\frac{\partial s_i}{\partial y}$ are non–zero at a vertex v_3 of $\tilde{T}$. The order of choosing points in Step 1 of the algorithm guarantess that for all the remaining vertices v of Δ, we have $\frac{\partial s_i(v)}{\partial x} = 0$ and $\frac{\partial s_i(v)}{\partial y} = 0$ (an illustration of this case is given in Figure 2, right). Therefore propagation along the edges of Δ can not go further along the edges of Δ. It remains to consider propagation through the interior of the triangles of Δ. Using the arguments from Case 1, it can be seen that non–zero derivatives at v_3 possibly initiate further chains of propagation through at most two black triangles, and two white ones. Combining these observations, we see that the maximal support s_i is $\text{star}^4(T)$ as asserted (an illustration is given in Figure 3). Since it is easy to see that this is the case of a support with maximal size under the assumption that z_i lies on an edge of Δ, this completes the proof of Case 2. □

Remark 2.2. Theorem 2.1 shows that the supports of the basis splines associated with the interpolation method are small. Note that in applications we often consider spline spaces, where the corresponding triangulations consist of up to 10^5 triangles, and more. According to our practical experience, fundamental splines of largest possible support are actually quite rare, and most of

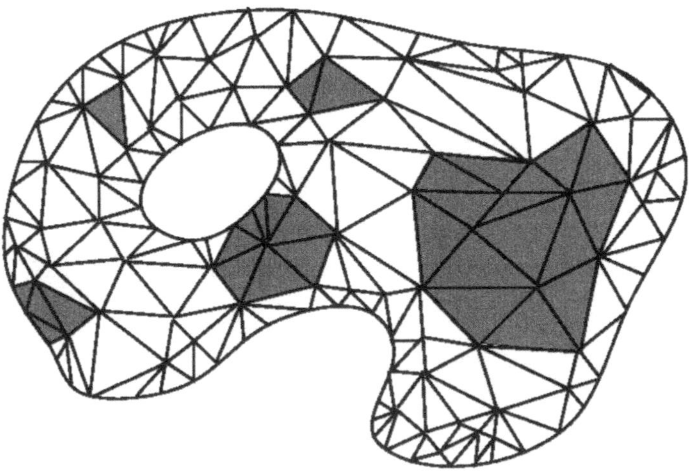

Figure 4: Supports of some fundamental splines s_i.

the splines s_i have much smaller supports than those analysed in the proof of Theorem 2.1. In particular, there exist cases where the support is only one triangle or a pair of neighboring triangles from Δ. Some typical supports are shown in Figure 4. In addition, we note that for fundamental C^1 splines of higher degree supports are essentially the same. Moreover, for C^1 quadratic splines (cf. Nürnberger and Zeilfelder [15]), it can be shown by arguments similar as above that the largest possible support is in $\text{star}^3(T)$.

Remark 2.3. In practical applications, it is not necessary to compute the fundamental basis splines. In our implementations (cf. Nürnberger, Schumaker and Zeilfelder [10, 11], Nürnberger and Zeilfelder [14–16]), we use the Bernstein–Bézier representation (cf. de Boor [1], Farin [6], Nürnberger [9]) of the polynomial pieces of the interpolating splines, and determine the Bernstein–Bézier coefficients by solving locally small linear systems resulting from the interpolation conditions. For cubic C^1 splines many of these coefficients are given by simple equations.

3 Interpolation on Separable Triangulations

We discuss the interpolation method from Section 1 for the general class of separable triangulations. We show that by using such triangulations the method becomes simpler and the supports of the fundamental splines are even smaller.

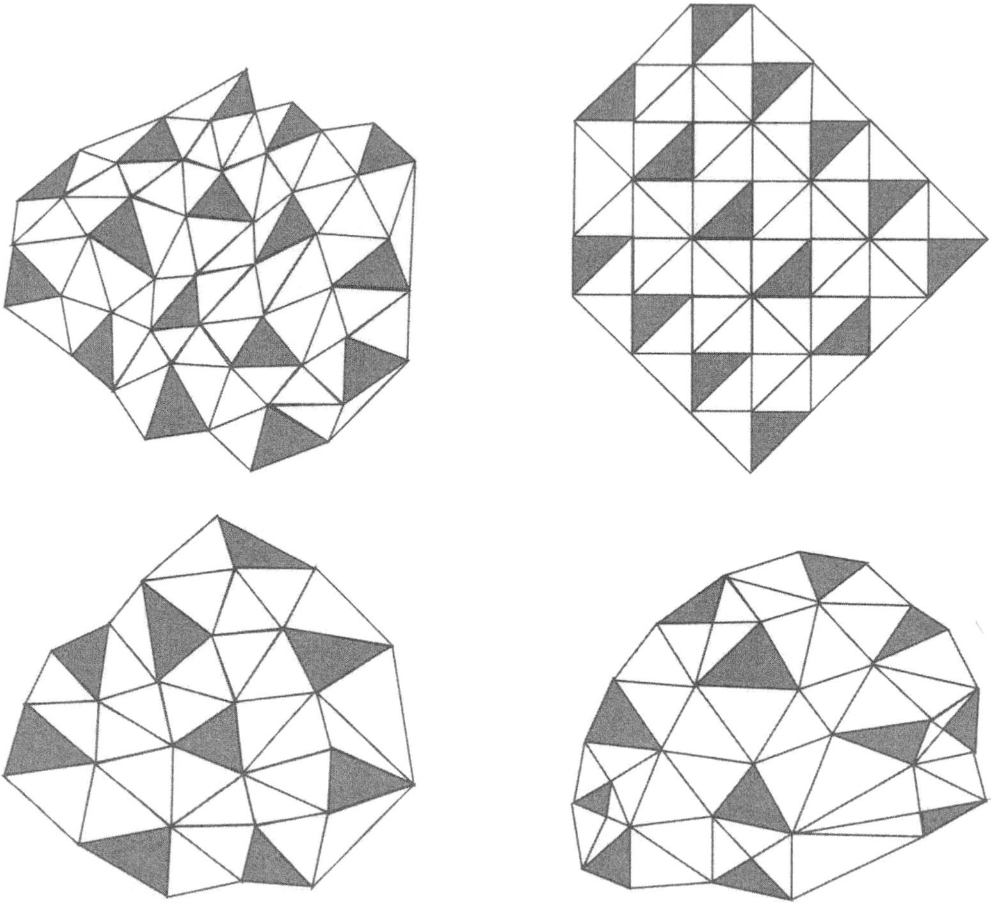

Figure 5: Examples of separable triangulations. The triangles from Δ_0 are colored grey.

Definition 3.1. A triangulation Δ is said to be *separable* provided a subset $\Delta_0 \subseteq \Delta$ of triangles exists such that for every vertex v of Δ, there is a unique triangle $T \in \Delta_0$ with vertex at v.

We note that many natural triangulations are separable. In many cases, it suffices to add a few triangles at the boundary to obtain a separable triangulation from a given triangulation. Some examples of separable triangulations are shown in Figure 5. Note that the triangulations considered in the first paper on local Lagrange interpolation by C^1 splines (cf. Nürnberger and Zeilfelder [14]) are included in this class.

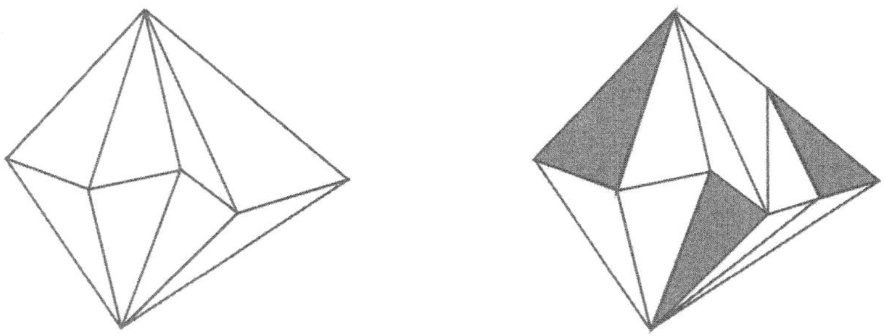

Figure 6: Example of a triangulation which is not separable (left), and a modification of this triangulation which is separable (right). In the latter case, the triangles from Δ_0 are colored grey.

However, not every given triangulation is separable as it can be seen from Figure 6, left. In Section 4, we describe a method for constructing a separable triangulation from a given triangulation. A simple example where we applied this algorithm is shown in Figure 6, right.

For separable triangulations, Algorithm 1 can be simplied by the following modification. As above, we successively add points to some set $\mathcal{L}$ and denote the resulting set again by $\mathcal{L}$.

Algorithm 2: Construction of local interpolation sets $\mathcal{L}$ for cubic C^1 splines for separable triangulations Δ.

Let $\Delta_0 \subseteq \Delta$ be a subtriangulation as in Definition 3.1 and initialize $\mathcal{L}$ by including all the points $\frac{i}{3}u + \frac{j}{3}v + \frac{k}{3}w$, $i + j + k = 3$, where $T \in \Delta_0$ is an arbitrary triangle with vertices u, v, and w.

Choose a subset $\Delta_1 \subseteq \Delta \setminus \Delta_0$ of triangles with the following properties:

(i): the intersection of every $T \in \Delta_1$ with each triangle from Δ_0 is empty or a common vertex,

(ii): Δ_1 contains no neighboring triangles, and

(iii): there does not exist a set of triangles $\Delta_1^* \neq \Delta_1$ with $\Delta_1 \subset \Delta_1^* \subseteq \Delta \setminus \Delta_0$ which satisfies (i) and (ii).

Choose the barycenter of the triangles from Δ_1, and add these points to $\mathcal{L}$.

Choose a subset $\Delta_2 \subseteq \Delta \setminus (\Delta_0 \cup \Delta_1)$ of triangles with the following properties:

(i): for every $T \in \Delta_2$, there exists exactly one neighboring triangle of T from $\Delta_0 \cup \Delta_1$,

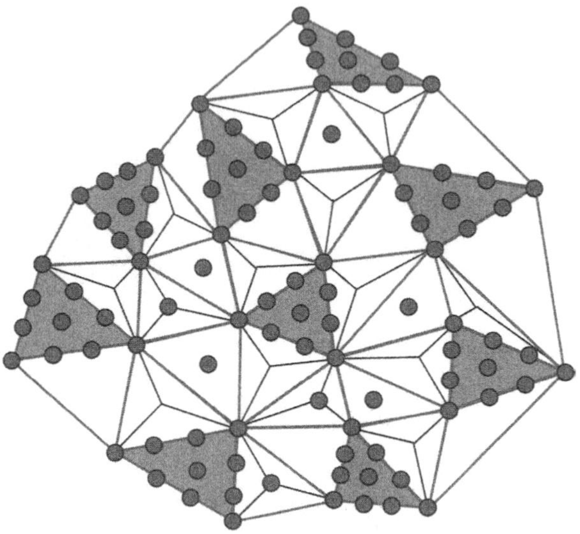

Figure 7: Example of interpolation points and the triangulation $\tilde{\Delta}$ resulting from a given, separable triangulation. The interpolation points are marked by dots.

(ii): Δ_2 contains no neighboring triangles, and

(iii): there does not exist a set of triangles $\Delta_2^* \neq \Delta_2$ with $\Delta_2 \subset \Delta_2^* \subseteq \Delta \setminus (\Delta_0 \cup \Delta_1)$ which satisfies (i) and (ii).

With the aid of these sets, we modify the given triangulation Δ, and denote the resulting triangulation by $\tilde{\Delta}$. The triangulation $\tilde{\Delta}$ is obtained from Δ by applying the Clough–Tocher split to every triangle of $\Delta \setminus (\Delta_0 \cup \Delta_1 \cup \Delta_2)$.

For every pair of neighboring triangles in $\Delta \setminus (\Delta_0 \cup \Delta_1 \cup \Delta_2)$, choose the barycenter for exactly one of its triangles. For every triangle $T \in \Delta \setminus (\Delta_0 \cup \Delta_1 \cup \Delta_2)$ which contains a boundary edge of Δ, choose the barycenter of T. Add all these points to $\mathcal{L}$.

The final set of all points chosen in Algorithm 2 is denoted by $\mathcal{L}$.

Figure 7 shows an example of an interpolation set $\mathcal{L}$ resulting from Algorithm 2 for separable triangulations. The number d of interpolation points in $\mathcal{L}$, and hence the dimension of the spline space $\mathcal{S}_3^1(\tilde{\Delta})$ is

$$d = 3V_\Delta + E_\Delta - 2\ \#(\Delta_0 \cup \Delta_1 \cup \Delta_2),$$

where V_Δ is the number of vertices of Δ, E_Δ is the number of edges of Δ, and $\#$ denotes the cardinality of a set of triangles. (In the example shown in Figure 7, we have $V_\Delta = 27$, $E_\Delta = 65$, $\#(\Delta_0) = 9$, $\#(\Delta_1) = 5$, $\#(\Delta_2) = 10$, and $d = 98$.)

Given $f \in C(\Omega)$, it is easy to see that the polynomial pieces $s_f|_T$ of a spline $s_f \in \mathcal{S}_3^1(\tilde{\Delta})$ are uniquely determined by the ten interpolation conditions on every triangle T of Δ_0. Since Δ is separable, this implies that the function and gradient value of s_f at the vertices of Δ are uniquely determined. Therefore, it follows from the choice of interpolation points that the spline is uniquely determined on the triangles of Δ_1 and Δ_2. Note that there is no undesired propagation between the triangles of Δ_1 and Δ_2, respectively, since property (ii) is satisfied. For the remaining triangles of Δ the Clough–Tocher split is applied. No triangle of this type has more than one neighboring triangle of the same type. This follows from the definition of Δ_0 and the choice of the sets Δ_1 and Δ_2. Therefore, the arguments developed in Nürnberger and Zeilfelder [16] can be applied, and the method for separable triangulations is completely local.

Hence, we obtain the following result by using similar arguments as in the proof of Theorem 1.1.

Theorem 3.2. *Let Δ be a separable triangulation. The set $\mathcal{L} = \{z_1, \ldots, z_d\}$ constructed in Algorithm 2 is a local Lagrange interpolation set for $\mathcal{S}_3^1(\tilde{\Delta})$, and for $f \in C^4(\Omega)$, the interpolating spline $s_f \in \mathcal{S}_3^1(\tilde{\Delta})$ with*

$$s_f(z_i) = f(z_i), \ i = 1, \ldots, d,$$

yields optimal approximation order.

As in Section 2, let s_i, $i = 1, \ldots, d$, the Lagrange fundamental splines from $\mathcal{S}_3^1(\tilde{\Delta})$ determined by the conditions

$$s_i(z_j) = \delta_{i,j}, \ j = 1, \ldots, d,$$

and $\mathcal{S} = \{s_i : \ i = 1, \ldots, d\}$ be the fundamental spline basis of $\mathcal{S}_3^1(\tilde{\Delta})$, where $\tilde{\Delta}$ results from applying Algorithm 2 to a separable triangulation.

Theorem 3.3. *Let Δ be a separable triangulation. The supports of the basis splines $s_i \in \mathcal{S}$, $i = 1, \ldots, d$, satisfy $supp(s_i) \subseteq star^3(T)$, where $T \in \Delta$ is an appropriate triangle.*

Proof. The proof follows from a close inspection of the proof of Theorem 2.1, and the fact that for separable triangulations the function and gradient value of every s_i is non–zero for at most three vertices of Δ. □

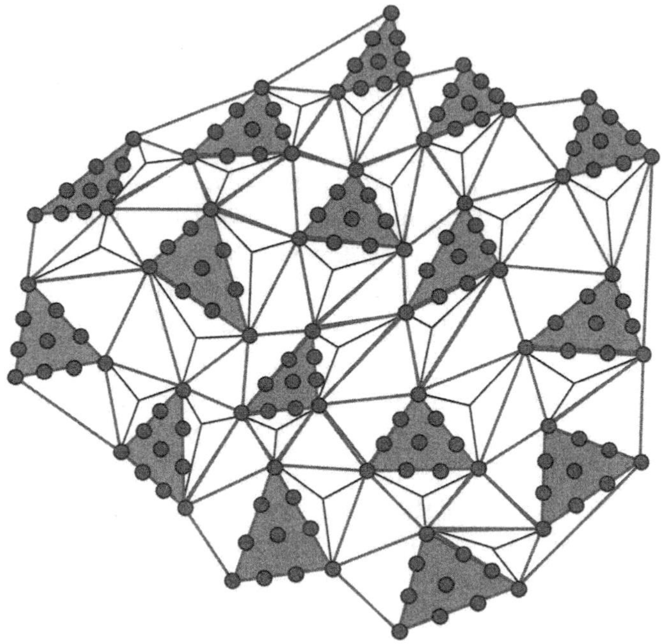

Figure 8: An interpolation set for $\mathcal{S}_3^1(\tilde{\Delta})$ which consists of points from the triangles of Δ_0, only. The interpolation points are marked by dots.

Remark 3.4. For some triangulations, it is possible to slightly modify the method from this section such that the Lagrange interpolating spline is uniquely and locally determined by choosing the ten points in the triangles of Δ_0, only, and no additional interpolation points have to be chosen. Figure 8 shows such an example. In a different context, a similar observation has been used for an approximation method (cf. Haber, Zeilfelder, Davydov and Seidel [5], Davydov and Zeilfelder [8], see also Zeilfelder [18]) using extension of local approximating polynomials to C^1 and C^2 splines on a uniform type triangulation, where the polynomials of varying degrees are adapted to given (local) data.

4 Construction of Separable Triangulations

We describe an algorithm for constructing a separable triangulation Δ_{sep} from a given triangulation Δ. The algorithm shows that this is done iteratively by adding a few vertices and edges to Δ. During the algorithm, the vertices of Δ are successively marked.

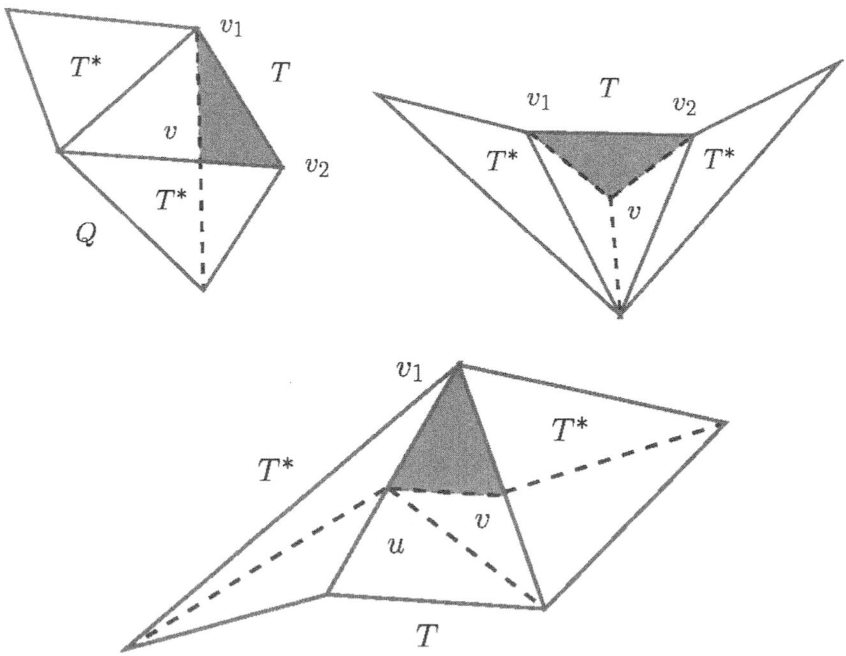

Figure 9: Modification of Δ in Case 1 (top) and Case 2 (bottom) of Algorithm 3. The dotted edges are added to the current triangulation Δ_{sep}, and the grey triangle is added to the current subset Δ_0 of Δ_{sep}.

Algorithm 3: Construction of separable triangulations Δ_{sep} from given triangulations Δ.

Let $\Delta_0 \subseteq \Delta$ be a subset of triangles such that for every vertex v of Δ_0, there is a unique triangle $T \in \Delta_0$ with vertex at v, and there does not exist a set of triangles $\Delta_0^* \neq \Delta_0$ with $\Delta_0 \subset \Delta_0^* \subseteq \Delta$ with the same property. Set all vertices of $\Delta \setminus \Delta_0$ to be unmarked (the remaining vertices are marked), and let $\Delta_{sep} = \Delta$.

For $\ell = 2, 1$, repeat as often as possible: If there exists $T \in \Delta_{sep}$ with ℓ unmarked vertices $v_1, \ldots, v_\ell$, then proceed as in the following two cases.

Case 1. $\ell = 2$. If T has a neighboring triangle $T^* \in \Delta_{sep}$ such that T and T^* form a strictly convex quadrilateral Q, then add the second diagonal to Q, add the four resulting triangles to Δ_{sep}, and remove T and T^* from Δ_{sep}. Moreover, in this case, let v be the intersection of the two diagonals in Q, and add the triangle with vertices v_1, v_2, and v to Δ_0. Otherwise, apply the Clough–Tocher split to T, add the three resulting triangles to Δ_{sep}, and remove T from Δ_{sep}. Moreover, in this case, let v be the barycenter of T, and add the triangle with vertices v_1, v_2, and v to Δ_0. Set the vertices v_1, v_2, and v to be marked. (Figure 9 (top) illustrates these two subcases.)

Case 2. $\ell = 1$. Let u, v be the midpoints of the edges of T with endpoint v_1. Connect u with v and with the vertex of T opposite to u, and add the triangle with vertices v_1, u, and v to Δ_0. For any neighboring triangle T^* of T with vertex v_1 connect u and v with the vertex of T^* opposite to u and v, respectively. Add all the new triangles to Δ_{sep}, and remove T as well as its neighboring triangles T^* from Δ_{sep}. Set the vertices v_1, u, and v to be marked. (Figure 9 (bottom) illustrates this case.)

Algorithm 3 terminates when all the vertices of Δ are marked.

The order of marking vertices is important in Algorithm 3 since it guarantees that only few edges and vertices are added to Δ in order to obtain a separable triangulation Δ_{sep}. Note that the structure of this algorithm is similar to the first step of Algorithm 1 where the order of the algorithmic steps is also important.

On the other hand, Algorithm 1 is an algorithm for constructing interpolation points for splines on a given triangulation, while Algorithm 3 finds a new triangulation from a given triangulation which is particularly useful for interpolation by splines.

Theorem 4.1. *Let Δ be an arbitrary triangulation. Then, the triangulation Δ_{sep} resulting from Algorithm 3 is separable.*

Proof. We consider the subset $\Delta_0 \subseteq \Delta_{sep}$ of triangles resulting from Algorithm 3 and show that for every vertex w of Δ_{sep}, there is a unique triangle $T \in \Delta_0$ with vertex at w.

Let w be an arbitrary vertex of Δ_{sep}.

If w has been marked when Algorithm 3 starts, then it is evident that w has the desired property, since it follows from the construction that no triangle from Δ_0 generated by a later step of Algorithm 3 has w as a vertex. Otherwise, w has been marked once in a later step when carrying out Algorithm 3. In this case, it follows from the constructions in Case 1 and 2, respectively, that exactly one triangle has been added to Δ_0 which has w as a vertex. This property is also satisfied in the case when w is not a vertex of Δ. This shows the assertation. □

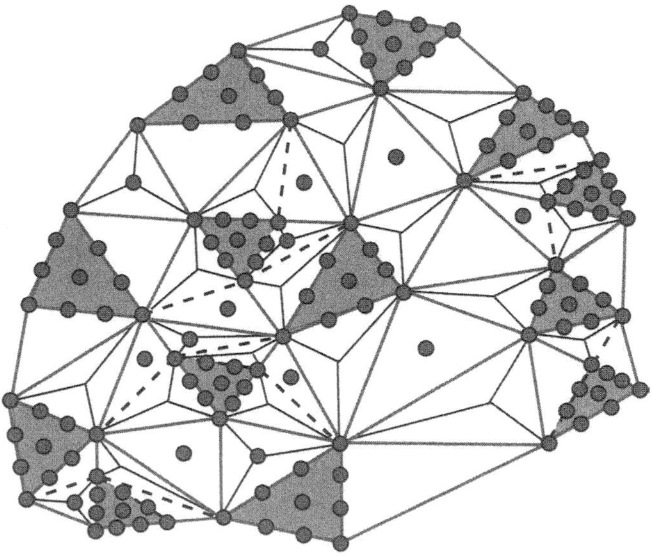

Figure 10: An interpolation set $\mathcal{L}$ for $\mathcal{S}_3^1(\tilde{\Delta})$, where a separable triangulation is used which is obtained from Algorithm 3 applied to the triangulation from Figure 1, left.

Figure 5 (bottom,right) shows a separable triangulation Δ_{sep} which results from applying Algorithm 3 to the triangulation from Figure 1, left. Figure 10 shows a different and more complex example of a separable triangulation Δ_{sep} obtained from this triangulation, including the triangulation $\tilde{\Delta}$ and a set of interpolation points $\mathcal{L}$ for $\mathcal{S}_3^1(\tilde{\Delta})$ resulting from Algorithm 2 (applied to Δ_{sep}). In this example, Algorithm 3 inserts 9 additional vertices, and generates 25 additional edges. Here, we have $V_{\Delta_{sep}} = 39$, $E_{\Delta_{sep}} = 95$, $\#(\Delta_0) = 13$, $\#(\Delta_1) = 8$, $\#(\Delta_2) = 13$, and $d = 144$.

References

[1] C. de Boor: *B–form basics,* In: *Geometric Modeling*, G. Farin, (ed.), 131–148, SIAM, Philadelphia, 1987.

[2] C. K. Chui: *Multivariate Splines*, CBMS 54, SIAM, Philadelphia, 1988.

[3] R. W. Clough, J. L. Tocher: *Finite element stiffness matrices for analysis of plates in bending,* In: *Proc. Conf. on Matrix Methods in Structural Mechanics*, Wright Patterson A. F. B., Ohio, 1965.

[4] O. Davydov, G. Nürnberger, F. Zeilfelder: *Bivariate spline interpolation with optimal approximation order,* Constr. Approx. **17** (2001), 181–208.

[5] O. Davydov, F. Zeilfelder: *Scattered data fitting by direct extension of local polynomials with bivariate splines,* to appear in: Advances in Comp. Math., 2003.

[6] G. Farin: *Triangular Bernstein–Bézier patches,* Comp. Aided Geom. Design **3** (1986), 83–127.

[7] G. Fraeijs de Veubeke: *Bending and stretching of plates,* In: *Proc. Conf. on Matrix Methods in Structural Mechanics*, Wright Patterson A. F. B., Ohio, 1965.

[8] J. Haber, F. Zeilfelder, O. Davydov, H.–P. Seidel: *Smooth approximation and rendering of large scattered data sets,* In: *Proceedings of IEEE Visualization* 2001, T. Ertl, K. Joy, A. Varshney (eds.), 341–347, 571, IEEE, 2001.

[9] G. Nürnberger: *Approximation by Spline Functions,* Springer, Berlin, 1989.

[10] G. Nürnberger, L. L. Schumaker, F. Zeilfelder: *Local Lagrange interpolation by bivariate C^1 cubic splines,* In: *Mathematical Methods in CAGD: Oslo* 2000, T. Lyche, L.L. Schumaker, (eds.), 393–404, Vanderbilt University Press, Nashville, 2001.

[11] G. Nürnberger, L. L. Schumaker, F. Zeilfelder: *Lagrange Interpolation by C^1 cubic splines on Triangulations of Separable Quadrangulations,* In: *Approximation Theory X: Splines, Wavelets, and Applications*, C. K. Chui, L.L. Schumaker, J. Stöckler, (eds.), 405–424, Vanderbilt University Press, Nashville, 2002.

[12] G. Nürnberger, L. L. Schumaker, F. Zeilfelder: *Lagrange interpolation by C^1 cubic splines on triangulated quadrangulations,* to appear in: Advances in Comp. Math., 2003.

[13] G. Nürnberger, F. Zeilfelder: *Developments in bivariate spline interpolation,* J. Comput. Appl. Math. **121** (2000), 125–152.

[14] G. Nürnberger, F. Zeilfelder: *Local Lagrange interpolation by cubic splines on a class of triangulations*, In: *Proc. Conf. Trends in Approximation Theory*, K. Kopotun, T. Lyche, M. Neamtu, (eds.), 341–350, Vanderbilt University Press, Nashville, 2001.

[15] G. Nürnberger, F. Zeilfelder: *Local Lagrange interpolation on Powell–Sabin triangulations and terrain modelling*, In: *Recent Progress in Multivariate Approximation*, W. Haußmann, K. Jetter, M. Reimer (eds.), 227–244, ISNM **137**, Birkhäuser, Basel 2001.

[16] G. Nürnberger, F. Zeilfelder: *Lagrange interpolation by bivariate C^1-splines with optimal approximation order*, to appear in: Advances in Comp. Math., 2003.

[17] G. Sander: *Bornes supérieures et inférieures dans l'analyse matricielle des plaques en flexion–torsion*, Bull. Soc. Royale Science Liège **33** (1964), 456–494.

[18] F. Zeilfelder: *Scattered data fitting with bivariate splines*, In: *Principles of Multiresolution in Geometric Modelling*, M. Floater, A. Iske, E. Quak (eds.), 243–286, Springer, Berlin, 2002.

Address:

Günther Nürnberger, Frank Zeilfelder
University of Mannheim
Institute for Mathematics
D–68131 Mannheim
Germany

International Series of Numerical Mathematics
Vol. 145,

Integer DCT–II by Lifting Steps

Gerlind Plonka and Manfred Tasche

Dedicated to Manfred Reimer on the occasion of his 70th birthday

Abstract

In image compression, the discrete cosine transform of type II (DCT–II) is of special interest. In this paper we use a new approach to construct an integer DCT–II first considered in [15]. Our method is based on a factorization of the cosine matrix of type II into a product of sparse, orthogonal matrices. The construction of the integer DCT–II of length 8 works with lifting steps and rounding–off. We are especially interested in the normwise error and the componentwise error when the integer DCT–II is compared with the exact DCT–II.

1 Introduction

The discrete cosine transform of type II (DCT–II) has found a wide range of applications in signal and image processing (see [17, 18]), especially in image compression. It has become the heart of international standards in image compression such as JPEG and MPEG. In some applications, the input data consist of integer vectors or integer matrices. But the output of DCT–II algorithm consists no longer of integers. For lossless coding it would be of great interest to be able to characterize the output completely again with integers. Lossless coding schemes are hardly based on integer DCT–II which have been studied in recent years (see [2, 3, 4, 5, 7, 8, 9, 12, 13, 15, 16, 19, 20]). Especially, integer DCT–II of length 8 and 16 and integer wavelets (see e.g. [1]) have been proposed.

An integer DCT–II of length n is a nonlinear, left–invertible mapping which acts on $\mathbb{Z}^n$ and approximates the classical DCT–II of length n. Integer DCT–II possesses some features of the classical DCT–II, whereas its computational cost is not higher than in the classical case. As known the lossy JPEG compression algorithm divides the source image into 8×8 blocks. Then the two–dimensional DCT–II of each block is performed and the result is compressed by a quantization step. Hence $n = 8$ is the most interesting case which we consider in detail.

Note that in the JPEG–2000 proposal [11], the use of the integer DCT–II for lossless image coding is recommended.

Observe that the procedure of rounding an output vector of DCT–II to the next integer vector used in lossy signal compression is not invertible. Take for example a unit vector $\mathbf{e}_j := (\delta(k-j))_{k=0}^{n-1}$, then the transformed vector $C_n^{II}\,\mathbf{e}_j$ will for all $j = 0, \ldots, n-1$ be rounded to the zero vector for $n > 4$. An injective mapping can be obtained by admitting an appropriate scaling factor (see e.g. [13]). But this procedure enlarges the entries of the transformed vector and the goal that it may be decoded more efficiently than the original vector will be quickly defeated.

One method for developing an integer DCT–II algorithm is based on the idea to approximate the components of the cosine matrix C_n^{II} of order n by dyadic rationals (see e.g. [19]), paying attention that the symmetry relations are kept. This method destroys the orthonormality of the matrix and the challenge is to find an invertible approximation $\tilde{C}_n$ of C_n^{II} such that its inverse $\tilde{C}_n^{-1}$ again only consists of dyadic rationals. For this reason, suitable approximations of the cosine matrix C_n^{II} of order $n = 8$ or $n = 16$ have only been given.

In [3, 5, 7, 9, 20], different factorizations of the transform matrix into products of so–called lifting matrices and simple matrices are applied. Here a lifting matrix is a matrix whose diagonal elements are 1, and only one nondiagonal element is nonzero. Simple matrices are permutation matrices or sparse matrices whose nonzero entries are only integers or half integers. Then the noninteger entries of the lifting matrices are rounded to dyadic rationals, and the inverse matrix factors are easy to determine. This method has the advantage that it works for arbitrary radix–2 lengths.

Due to the rounding of matrix entries, these two methods can lead to high errors, if one compares the integer DCT–II output with the classical DCT–II result, especially if the range for the components of the input vector is large. Explicit error estimates for these algorithms have not been considered.

In this paper, we use a new approach to integer DCT–II introduced in [15]. We are not building integer DCT–II in integer arithmetic. Thus the computations are still done with floating point numbers, but the result is guaranteed to be an integer and the left–invertibility is preserved. Note that in some papers the notion *integer* DCT just means that floating point operations are avoided while the resulting vector consists of dyadic rationals (see [5, 9, 12, 19, 20]).

Our algorithms are based on a factorization of C_8^{II} (see [14]) into sparse orthogonal matrices of simple structure. By suitable permutations, each matrix factor can be transferred to a block–diagonal matrix, where every block is an orthogonal matrix of order 2. Now the idea for construction of integer DCT–II of length 8 is very simple. For each block R_2 of order 2 and for arbitrary $\mathbf{x} \in \mathbb{Z}^2$, find a suitable integer approximation of $R_2\mathbf{x}$ such that this process is left–invertible.

The paper is organized as follows. In Section 2 we introduce the cosine matrix of type II and present a factorization of C_8^{II} into products of sparse, orthogonal matrices. In Section 3, we present integer transforms of length 2. Applying the lifting technique (see e.g. [6]) and rounding–off, we construct an integer approximation of $R_2\mathbf{x}$ for arbitrary $\mathbf{x} \in \mathbb{Z}^2$ and estimate the truncation error. Further, we give some properties of the corresponding nonlinear mapping which allow the conjecture that vanishing components of the exact vector $2\,C_8^{II}\mathbf{x}$ with $\mathbf{x} \in \mathbb{Z}^8$ are preserved in the proposed integer DCT–II algorithm.

The results of Sections 2 and 3 are applied to integer DCT–II of length 8 (in Section 4) and to two–dimensional integer DCT–II of size 8×8 (in Section 5). We present an algorithm for the integer DCT–II and estimate the normwise error as well as the componentwise error when the integer DCT–II result is compared with the exact DCT–II. Our worst case estimates for the absolute error show that the proposed integer DCT–II algorithm is very close to the exact DCT–II and hence preserves the features of frequency decorrelation. In particular, a detailed consideration of the componentwise error will lead us to better error estimates than given in [15]. The numerical results illustrate the performance of our new integer DCT–II algorithm.

2 Factorization of Cosine Matrix

Let $n \geq 2$ be a given integer. In the following, we consider the *cosine matrix of type* II *with order* n which is defined by

$$C_n^{II} := \sqrt{\tfrac{2}{n}} \left(\varepsilon_n(j) \cos \tfrac{j(2k+1)\pi}{2n} \right)_{j,k=0}^{n-1}, \tag{2.1}$$

where $\varepsilon_n(0) := \sqrt{2}/2$ and $\varepsilon_n(j) := 1$ for $j \in \{1, \ldots, n-1\}$. In our notation a subscript of a matrix denotes the corresponding order. Observe that these matrices are orthogonal (see e.g. [17], pp. 13 – 14; [18]). The *discrete cosine transform of type* II (DCT–II) *with length* n is a linear mapping of $\mathbb{R}^n$ onto $\mathbb{R}^n$, which is generated by C_n^{II}. In [14], simple split–radix algorithms are proposed for these transforms of radix–2 length n. They are based on factorization of

C_n^{II} into a product of sparse, orthogonal matrices. In this paper, we want to restrict ourselves to $n = 8$ and use the orthogonal factorization of C_8^{II} in order to present an integer DCT–II of length 8, which is very close to the original DCT–II and maps integer vectors to integer vectors (see also [15]). Naturally, this integer DCT–II is not longer a linear mapping.

First, we introduce some notations. Let I_n denote the identity matrix and $J_n := (\delta(j+k-n+1))_{j,k=0}^{n-1}$ the counteridentity matrix, where δ means the Kronecker symbol. Blanks in a matrix indicate zeros. The direct sum of two matrices A, B is defined to be a block–diagonal matrix $A \oplus B := \text{diag}\,(A,\, B)$. For even $n \geq 4$, P_n denotes the *even–odd permutation matrix* (or 2*–stride permutation matrix*) defined by

$$P_n\mathbf{x} := (x_0, x_2, \dots, x_{n-2}, x_1, x_3, \dots, x_{n-1})^T, \qquad \mathbf{x} = (x_j)_{j=0}^{n-1}.$$

An orthogonal factorization of the cosine matrix C_8^{II} looks as follows (see [14]):

$$C_8^{II} = B_8\,(I_4 \oplus A_4(1))\,(C_2^{II} \oplus C_2^{IV} \oplus C_2^{II} \oplus C_2^{II})\,(T_4(0) \oplus T_4(1))\,T_8(0) \quad (2.2)$$

with the bit reversal matrix $B_8 := P_8^T(P_4 \oplus P_4)$, determined by

$$B_8\mathbf{x} = (x_0, x_4, x_2, x_6, x_1, x_5, x_3, x_7)^T, \quad \mathbf{x} = (x_j)_{j=0}^{7},$$

$$A_4(1) = \frac{1}{\sqrt{2}}\begin{pmatrix} \sqrt{2} & & & \\ & 1 & 1 & \\ & 1 & -1 & \\ & & & \sqrt{2} \end{pmatrix}, \quad T_4(1) = \begin{pmatrix} \cos\frac{\pi}{16} & & & \sin\frac{\pi}{16} \\ & \cos\frac{3\pi}{16} & \sin\frac{3\pi}{16} & \\ & -\sin\frac{3\pi}{16} & \cos\frac{3\pi}{16} & \\ \sin\frac{\pi}{16} & & & -\cos\frac{\pi}{16} \end{pmatrix},$$

$$T_4(0) = \frac{1}{\sqrt{2}}\begin{pmatrix} I_2 & J_2 \\ I_2 & -J_2 \end{pmatrix}, \qquad T_8(0) = \frac{1}{\sqrt{2}}\begin{pmatrix} I_4 & J_4 \\ I_4 & -J_4 \end{pmatrix},$$

and with

$$C_2^{II} = \frac{1}{\sqrt{2}}\begin{pmatrix} 1 & 1 \\ 1 & -1 \end{pmatrix}, \quad C_2^{IV} := \begin{pmatrix} \cos\frac{\pi}{8} & \sin\frac{\pi}{8} \\ \sin\frac{\pi}{8} & -\cos\frac{\pi}{8} \end{pmatrix}.$$

Let us denote the five *orthogonal* matrix factors of C_8^{II} in (2.2) in this order by

$$C_8^{II} = B_8\,A_8(0,1)\,T_8(0,1,0,0)\,T_8(0,1)\,T_8(0). \quad (2.3)$$

These notions are analogous to those in [14] and are motivated by observing that $T_8(0,1) = T_4(0) \oplus T_4(1)$ and that $T_8(0,1,0,0) = T_2(0) \oplus T_2(1) \oplus T_2(0) \oplus T_2(0)$ with $T_2(0) := C_2^{II}$ and $T_2(1) := C_2^{IV}$ (see [14]).

Note that the factorization (2.3) implies a fast algorithm for computing the DCT–II of length 8 with 11 multiplications and 29 additions (see [14]). This algorithm is very similar to that of C. Loeffler et al. [10].

3 Integer Transforms of Length 2

Considering the factorization (2.3) of C_8^{II}, we observe that each of the matrix factors possesses at most two nonzero entries in one row. The main idea to obtain an integer DCT–II is now as follows. First we consider rotation matrices of the form

$$R_2(\omega) := \begin{pmatrix} \cos\omega & \sin\omega \\ -\sin\omega & \cos\omega \end{pmatrix}, \qquad \omega \in (0, \tfrac{\pi}{4}].$$

For the selected case $n = 8$ we need $R_2(\omega)$ for $\omega \in \{\frac{\pi}{16}, \frac{\pi}{8}, \frac{3\pi}{16}, \frac{\pi}{4}\}$ only.

For $R_2(\omega)$ and for arbitrary $\mathbf{x} \in \mathbb{Z}^2$ we want to find a suitable integer approximation of $R_2(\omega)\mathbf{x}$ such that this process is left–invertible. Afterwards, we construct an integer DCT–II of length 8 from these partial algorithms.

We use the following notations. For $a \in \mathbb{R}$ let $\lfloor a \rfloor := \max\{x \le a;\, x \in \mathbb{Z}\}$. Further let $\mathrm{rd}\,(a) := \lfloor a + 1/2 \rfloor$ be the integer next to a.

Let $s \in \mathbb{R}$ with $s \neq 0$ be given. Then matrices of the form

$$\begin{pmatrix} 1 & s \\ 0 & 1 \end{pmatrix}, \qquad \begin{pmatrix} 1 & 0 \\ s & 1 \end{pmatrix}$$

are called *lifting matrices of order* 2 (see e.g. [3, 4, 5, 6, 9, 15, 20]). Note that the inverse of a lifting matrix is again a lifting matrix:

$$\begin{pmatrix} 1 & s \\ 0 & 1 \end{pmatrix}^{-1} = \begin{pmatrix} 1 & -s \\ 0 & 1 \end{pmatrix}, \qquad \begin{pmatrix} 1 & 0 \\ s & 1 \end{pmatrix}^{-1} = \begin{pmatrix} 1 & 0 \\ -s & 1 \end{pmatrix}.$$

Every rotation matrix $R_2(\omega)$ of order 2 can be represented as a product of three lifting matrices (see [6]):

$$R_2(\omega) = \begin{pmatrix} 1 & \tan\frac{\omega}{2} \\ 0 & 1 \end{pmatrix} \begin{pmatrix} 1 & 0 \\ -\sin\omega & 1 \end{pmatrix} \begin{pmatrix} 1 & \tan\frac{\omega}{2} \\ 0 & 1 \end{pmatrix}. \tag{3.1}$$

Note that the above factorization of $R_2(\omega)$ consists of *nonorthogonal* matrix factors. This factorization is used as follows.

A *lifting step* of the form

$$\hat{\mathbf{y}} := \begin{pmatrix} 1 & s \\ 0 & 1 \end{pmatrix} \mathbf{x}$$

with $\mathbf{x} = (x_0,\, x_1)^T \in \mathbb{Z}^2$ can be approximated by $\mathbf{y} = (y_0,\, y_1)^T \in \mathbb{Z}^2$ with

$$y_0 = x_0 + \lfloor s\,x_1 + \tfrac{1}{2} \rfloor = x_0 + \mathrm{rd}\,(s x_1), \qquad y_1 = x_1.$$

This transform is left–invertible and a left–inverse reads as follows

$$x_0 = y_0 - \lfloor s\, y_1 + \tfrac{1}{2} \rfloor = y_0 - \mathrm{rd}\,(s y_1), \qquad x_1 = y_1.$$

Indeed, we have that

$$y_0 - \lfloor s\, y_1 + \tfrac{1}{2} \rfloor = x_0 + \lfloor s\, x_1 + \tfrac{1}{2} \rfloor - \lfloor s\, x_1 + \tfrac{1}{2} \rfloor = x_0.$$

We obtain

Theorem 3.1. *Let $R_2(\omega)$ with $\omega \in (0, \frac{\pi}{4}]$ be a rotation matrix. Then for arbitrary $\mathbf{x} = (x_0, x_1)^T \in \mathbb{Z}^2$, a suitable integer approximation $\mathbf{y} = (y_0, y_1)^T \in \mathbb{Z}^2$ of $\hat{\mathbf{y}} := R_2(\omega)\mathbf{x}$ is given by $y_0 = z_2$, $y_1 = z_1$, where*

$$\begin{aligned} z_0 &:= x_0 + \mathrm{rd}\,(x_1 \tan \tfrac{\omega}{2}), \\ z_1 &:= x_1 + \mathrm{rd}\,(-z_0 \sin \omega), \\ z_2 &:= z_0 + \mathrm{rd}\,(z_1 \tan \tfrac{\omega}{2}). \end{aligned} \tag{3.2}$$

This integer transform is left–invertible and a left–inverse reads $x_0 = w_2$, $x_1 = w_1$, where

$$\begin{aligned} w_0 &:= y_0 - \mathrm{rd}\,(y_1 \tan \tfrac{\omega}{2}), \\ w_1 &:= y_1 - \mathrm{rd}\,(-w_0 \sin \omega), \\ w_2 &:= w_0 - \mathrm{rd}\,(w_1 \tan \tfrac{\omega}{2}). \end{aligned}$$

Further, the error estimates

$$(\hat{y}_0 - y_0)^2 + (\hat{y}_1 - y_1)^2 \le \tfrac{1}{4}(3 + 4\sin\omega + 2\cos\omega + (\tan\tfrac{\omega}{2})^2) \tag{3.3}$$

and

$$\begin{aligned} |\hat{y}_0 - y_0| &\le \tfrac{1}{2}(1 + \tan\tfrac{\omega}{2} + \cos\omega), \\ |\hat{y}_1 - y_1| &\le \tfrac{1}{2}(1 + \sin\omega) \end{aligned} \tag{3.4}$$

hold.

Proof. The formulas for y_0, y_1 directly follow by applying the lifting steps to the three matrices in (3.1). Using the inverse of (3.1), we obtain the formulas for x_0, x_1 analogously. For a proof of the error estimates see [15]. The estimates for the componentwise errors in (3.4) directly follow from the proof of Theorem 3.8 in [15]. □

Remark 3.2. (1) Let $\hat{\mathbf{y}} := R_2(\omega)\mathbf{x}$ with arbitrary $\mathbf{x} \in \mathbb{Z}^2$ and $\mathbf{y}$ its integer approximation constructed via (3.2). The special values for the error $\|\hat{\mathbf{y}} - \mathbf{y}\|_2$ and the componentwise absolute errors $\|\hat{\mathbf{y}} - \mathbf{y}\|_\infty$ via the lifting procedure for $\omega \in \{\frac{\pi}{4}, \frac{\pi}{8}, \frac{\pi}{16}, \frac{3\pi}{16}\}$ follow by inserting into formulas (3.3) – (3.4). In particular, we obtain

$$\|\hat{\mathbf{y}} - \mathbf{y}\|_2 \leq \begin{cases} 1.361453 & \text{for } \omega = \frac{\pi}{4}, \\ 1.266694 & \text{for } \omega = \frac{\pi}{8}, \\ 1.199128 & \text{for } \omega = \frac{\pi}{16}, \\ 1.320723 & \text{for } \omega = \frac{3\pi}{16}, \end{cases}$$

$$|\hat{y}_0 - y_0| \leq \begin{cases} 1.060660 & \text{for } \omega = \frac{\pi}{4}, \\ 1.061396 & \text{for } \omega = \frac{\pi}{8}, \\ 1.039638 & \text{for } \omega = \frac{\pi}{16}, \\ 1.067408 & \text{for } \omega = \frac{3\pi}{16}. \end{cases} \qquad |\hat{y}_1 - y_1| \leq \begin{cases} 0.853553 & \text{for } \omega = \frac{\pi}{4}, \\ 0.691342 & \text{for } \omega = \frac{\pi}{8}, \\ 0.597545 & \text{for } \omega = \frac{\pi}{16}, \\ 0.777785 & \text{for } \omega = \frac{3\pi}{16}. \end{cases}$$

(2) In particular, for $\omega = \frac{\pi}{4}$, we have

$$R_2(\tfrac{\pi}{4}) = \tfrac{1}{\sqrt{2}} \begin{pmatrix} 1 & 1 \\ -1 & 1 \end{pmatrix}$$

such that $\mathbf{y} := \sqrt{2}R_2(\frac{\pi}{4})\mathbf{x}$ remains to be an integer vector for arbitrary $\mathbf{x} \in \mathbb{Z}^2$. The multiplication of $\sqrt{2}R_2(\frac{\pi}{4})$ with $\mathbf{x} \in \mathbb{Z}^2$ is left–invertible and moreover, it does not produce any rounding error. Further, this procedure to handle rotation matrices with angle $\frac{\pi}{4}$ requires 2 additions only instead of 3 multiplications and 3 additions for 3 lifting steps. Thus we shall use the multiplication of $\sqrt{2}R_2(\frac{\pi}{4})$ with $\mathbf{x} \in \mathbb{Z}^2$ instead of the lifting method of Theorem 3.1 for the integer DCT–II as often as possible, when the rotation matrix with angle $\frac{\pi}{4}$ occurs in the factorization (2.3).

Let us now consider the nonlinear mapping $f_\omega : \mathbb{Z}^2 \to \mathbb{Z}^2$ with $f_\omega((x_0, x_1)^T) = (y_0, y_1)^T$ where $y_0 := z_2$, $y_1 := z_1$ and z_0, z_1, z_2 are given in (3.2). Note that for arbitrary $\mathbf{x} \in \mathbb{Z}^2$, $f_\omega(\mathbf{x})$ is an integer approximation of $R_2(\omega)\mathbf{x}$. Then we observe the following properties of f_ω.

Lemma 3.3. *For all* $\omega \in (0, \frac{\pi}{4}]$ *we have*

$$f_\omega((0,0)^T) = (0,0)^T.$$

Further, for $\omega = \frac{\pi}{4}$ *and* $x_0 \in \mathbb{Z}$ *we have*

$$\begin{aligned} f_{\pi/4}((x_0, x_0)^T) &= (\mathrm{rd}(\sqrt{2}\,x_0),\, 0)^T, \\ f_{\pi/4}((x_0, -x_0)^T) &= (0,\, -\mathrm{rd}(\sqrt{2}\,x_0) + \delta)^T, \end{aligned}$$

where

$$\delta := \begin{cases} 1 & if \quad \sqrt{2}x_0 - \mathrm{rd}\,(\sqrt{2}x_0) \in (-\frac{1}{2}, -1 + \frac{\sqrt{2}}{2}], \\ 0 & if \quad \sqrt{2}x_0 - \mathrm{rd}(\sqrt{2}x_0) \in (-1 + \frac{\sqrt{2}}{2}, 1 - \frac{\sqrt{2}}{2}], \\ -1 & if \quad \sqrt{2}x_0 - \mathrm{rd}\,(\sqrt{2}x_0) \in (1 - \frac{\sqrt{2}}{2}, \frac{1}{2}). \end{cases}$$

Proof. The first assertion follows immediately from the definition of z_0, z_1, z_2, since for $x_0 = x_1 = 0$ we obviously have $z_0 = z_1 = z_2 = 0$.

We consider the second assertion. Let $\varepsilon \in (-\frac{1}{2}, \frac{1}{2})$ be defined by

$$\sqrt{2}\,x_0 = \mathrm{rd}(\sqrt{2}\,x_0) + \varepsilon.$$

Then we obtain for $\omega = \frac{\pi}{4}$ and $x_0 = x_1$ by $\tan\frac{\pi}{8} = \sqrt{2} - 1$ that

$$z_0 = x_0 + \mathrm{rd}(x_0(\sqrt{2} - 1)) = \mathrm{rd}(\sqrt{2}x_0) = \sqrt{2}x_0 - \varepsilon$$

and

$$z_1 = x_0 + \mathrm{rd}(-z_0\tfrac{\sqrt{2}}{2}) = x_0 + \mathrm{rd}((-\sqrt{2}x_0 + \varepsilon)\tfrac{\sqrt{2}}{2}) = x_0 - x_0 + \mathrm{rd}(\tfrac{\sqrt{2}}{2}\varepsilon) = 0$$

as well as $z_2 = z_0 = \mathrm{rd}(\sqrt{2}x_0)$.

For $x_1 = -x_0$ we observe that

$$z_0 = x_0 + \mathrm{rd}(-x_0(\sqrt{2} - 1)) = 2x_0 + \mathrm{rd}(-\sqrt{2}x_0) = 2x_0 - \mathrm{rd}(\sqrt{2}x_0),$$

since for $s \in \mathbb{R} \setminus (\frac{1}{2} + \mathbb{Z})$ we simply find that $\mathrm{rd}(-s) = -\mathrm{rd}(s)$. Further,

$$\begin{aligned} z_1 &= -x_0 + \mathrm{rd}((-2x_0 + \mathrm{rd}(\sqrt{2}x_0))\tfrac{\sqrt{2}}{2}) \\ &= -x_0 + \mathrm{rd}(-\sqrt{2}x_0 + (\sqrt{2}x_0 - \varepsilon)\tfrac{\sqrt{2}}{2}) \\ &= \mathrm{rd}(-\sqrt{2}x_0 - \tfrac{\sqrt{2}}{2}\varepsilon) = \mathrm{rd}(-\mathrm{rd}(\sqrt{2}x_0) - \varepsilon - \tfrac{\sqrt{2}}{2}\varepsilon) \\ &= -\mathrm{rd}(\sqrt{2}x_0) + \mathrm{rd}(-(1 + \tfrac{\sqrt{2}}{2})\varepsilon). \end{aligned}$$

Finally it follows that

$$\begin{aligned} z_2 &= z_0 + \mathrm{rd}((\sqrt{2} - 1)\,z_1) \\ &= 2x_0 - \mathrm{rd}(\sqrt{2}x_0) + \mathrm{rd}\left((\sqrt{2} - 1)\left(-\mathrm{rd}(\sqrt{2}x_0) + \mathrm{rd}(-(1 + \tfrac{\sqrt{2}}{2})\varepsilon)\right)\right) \\ &= 2x_0 + \mathrm{rd}\left(-\sqrt{2}\,\mathrm{rd}(\sqrt{2}x_0) + \sqrt{2}\,\mathrm{rd}(-(1 + \tfrac{\sqrt{2}}{2})\varepsilon) - \mathrm{rd}(-(1 + \tfrac{\sqrt{2}}{2})\varepsilon)\right) \\ &= 2x_0 + \mathrm{rd}\left(-2x_0 + \sqrt{2}\varepsilon + \sqrt{2}\,\mathrm{rd}(-(1 + \tfrac{\sqrt{2}}{2})\varepsilon)\right) - \mathrm{rd}(-(1 + \tfrac{\sqrt{2}}{2})\varepsilon) \\ &= \mathrm{rd}\left(\sqrt{2}\varepsilon + \sqrt{2}\,\mathrm{rd}(-(1 + \tfrac{\sqrt{2}}{2})\varepsilon)\right) - \mathrm{rd}\left(-(1 + \tfrac{\sqrt{2}}{2})\varepsilon\right). \end{aligned}$$

Now for $\varepsilon \in (-1+\frac{\sqrt{2}}{2}, 1-\frac{\sqrt{2}}{2}]$ we have $-(1+\frac{\sqrt{2}}{2})\varepsilon \in [-\frac{1}{2}, \frac{1}{2})$, i.e., $\mathrm{rd}(-(1+\frac{\sqrt{2}}{2})\varepsilon) = 0$, and we find $z_2 = \mathrm{rd}(\sqrt{2}\varepsilon) = 0$, since

$$|\sqrt{2}\,\varepsilon| \leq \sqrt{2}\,(1-\tfrac{\sqrt{2}}{2}) = \sqrt{2}-1 < \tfrac{1}{2}.$$

For $\varepsilon \in (1-\frac{\sqrt{2}}{2}, \frac{1}{2})$ we have $-\frac{1}{2}-\frac{\sqrt{2}}{4} < -(1+\frac{\sqrt{2}}{2})\varepsilon < -\frac{1}{2}$, i.e., $\mathrm{rd}(-(1+\frac{\sqrt{2}}{2})\varepsilon) = -1$. Hence we obtain

$$z_2 = \mathrm{rd}(\sqrt{2}\varepsilon - \sqrt{2}) + 1 = 0$$

by $\sqrt{2}(\varepsilon - 1) \in (-1, -\frac{\sqrt{2}}{2})$.
For $\varepsilon \in (-\frac{1}{2}, -1+\frac{\sqrt{2}}{2}]$ we have $\mathrm{rd}(-(1+\frac{\sqrt{2}}{2})\varepsilon) = 1$ and $z_2 = \mathrm{rd}(\sqrt{2}\varepsilon+\sqrt{2})-1 = 0$, since $\sqrt{2}(\varepsilon+1) \in (\frac{\sqrt{2}}{2}, 1]$. □

Remark 3.4. Lemma 3.3 shows that for the special cases $\mathbf{x} = (x_0, x_0)^T$ and $\mathbf{x} = (x_0, -x_0)^T$, respectively, the error $R_2(\frac{\pi}{4})\mathbf{x} - f_{\pi/4}(\mathbf{x})$ vanishes in the second (resp. first) component and is smaller than $\frac{1}{2}$ in the other component. This property of f_ω implies the conjecture that the integer DCT–II Algorithm 4.1, presented in the next section, preserves vanishing components, i.e., if a component $\hat{y}_j$ ($j \in \{0, \ldots, 7\}$) of $\hat{\mathbf{y}} = 2C_8^{II}\mathbf{x}$ vanishes, then the j–th component of the integer DCT–II equals zero too. Recently it has been shown in [16] that this conjecture is true indeed.

4 Integer DCT–II of Length 8

Now we present an integer DCT–II algorithm, where the lifting method of Theorem 3.1 is used. Further, we estimate the truncation errors in the worst case.

Based on the factorization

$$2C_8^{II} = B_8 A_8(0,1)(I_4 \oplus \sqrt{2}I_4)\,T_8(0,1,0,0)(\sqrt{2}I_4 \oplus I_4)T_8(0,1)\sqrt{2}T_8(0) \quad (4.1)$$

we apply lifting to the submatrix $T_4(1)$ of $T_8(0,1)$, to the submatrix $C_2^{II} \oplus C_2^{IV}$ of $T_8(0,1,0,0)$ and to the submatrix $A_4(1)$ of $A_8(0,1)$. The other matrix–vector products are computed directly. The above factorization (4.1) ensures that the most rotation matrices with angle $\frac{\pi}{4}$ can be computed without lifting using the fact that $\sqrt{2}R_2(\frac{\pi}{4})$ has only ± 1 as entries (see Remark 3.2, (2)).

Algorithm 4.1.

Input: $\mathbf{x} \in \mathbb{Z}^8$.

1. Compute $\mathbf{x}^{(1)} := \sqrt{2}\, T_8(0)\, \mathbf{x}$.
2. Put

$$\mathbf{w}^{(0)} := (x_0^{(1)}, x_1^{(1)}, x_2^{(1)}, x_3^{(1)})^T,\ \mathbf{w}^{(1)} := (x_4^{(1)}, x_5^{(1)})^T,\ \mathbf{w}^{(2)} := (x_7^{(1)}, x_6^{(1)})^T.$$

Compute $\mathbf{z} := \sqrt{2}\, T_4(0)\mathbf{w}^{(0)}$ and

$$\begin{aligned} \mathbf{z}^{(0)} &:= \mathrm{rd}\,((\tan\tfrac{\pi}{32} \oplus \tan\tfrac{3\pi}{32})\mathbf{w}^{(2)}) + \mathbf{w}^{(1)}, \\ \mathbf{z}^{(1)} &:= \mathrm{rd}\,(((-\sin\tfrac{\pi}{16}) \oplus (-\sin\tfrac{3\pi}{16}))\mathbf{z}^{(0)}) + \mathbf{w}^{(2)}, \\ \mathbf{z}^{(2)} &:= \mathrm{rd}((\tan\tfrac{\pi}{32} \oplus \tan\tfrac{3\pi}{32})\mathbf{z}^{(1)}) + \mathbf{z}^{(0)}. \end{aligned}$$

Put $\mathbf{x}^{(2)} := (\mathbf{z}^T,\ z_0^{(2)}, z_1^{(2)}, z_1^{(1)}, -z_0^{(1)})^T$.

3. Put

$$\mathbf{w}^{(0)} := (x_0^{(2)}, x_2^{(2)})^T,\ \mathbf{w}^{(1)} := (x_1^{(2)}, x_3^{(2)})^T,\ \mathbf{w}^{(2)} := (x_4^{(2)}, x_5^{(2)}, x_6^{(2)}, x_7^{(2)})^T.$$

Compute $\mathbf{z} := (\sqrt{2}\, C_2^{II} \oplus \sqrt{2}\, C_2^{II})\mathbf{w}^{(2)}$ and

$$\begin{aligned} \mathbf{z}^{(0)} &:= \mathrm{rd}\,((\tan\tfrac{\pi}{8} \oplus \tan\tfrac{\pi}{16})\mathbf{w}^{(1)}) + \mathbf{w}^{(0)}, \\ \mathbf{z}^{(1)} &:= \mathrm{rd}\,(((-\sin\tfrac{\pi}{4}) \oplus (-\sin\tfrac{\pi}{8}))\mathbf{z}^{(0)}) + \mathbf{w}^{(1)}, \\ \mathbf{z}^{(2)} &:= \mathrm{rd}\,((\tan\tfrac{\pi}{8} \oplus \tan\tfrac{\pi}{16})\mathbf{z}^{(1)}) + \mathbf{z}^{(0)}. \end{aligned}$$

Put $\mathbf{x}^{(3)} := (z_0^{(2)}, -z_0^{(1)}, z_1^{(2)}, -z_1^{(1)},\ \mathbf{z}^T)^T$.

4. For $j = 0, \ldots, 4$ put $x_j^{(4)} := x_j^{(3)}$ and $x_7^{(4)} := x_6^{(3)}$. Compute

$$\begin{aligned} z_0 &:= \mathrm{rd}\,(x_7^{(3)} \tan\tfrac{\pi}{8}) + x_5^{(3)}, \quad x_6^{(4)} := -\mathrm{rd}\,(-z_0 \sin\tfrac{\pi}{4}) - x_7^{(3)}, \\ x_5^{(4)} &:= \mathrm{rd}\,(-x_6^{(4)} \tan\tfrac{\pi}{8}) + z_0 \end{aligned}$$

and put $\mathbf{x}^{(4)} := (x_j^{(4)})_{j=0}^7$.

5. Compute $\mathbf{y} := B_8\, \mathbf{x}^{(4)}$.

Output: $\mathbf{y} \in \mathbb{Z}^8$ integer approximation of $\hat{\mathbf{y}} = 2\, C_8^{II}\mathbf{x}$.

Algorithm 4.1 needs only 15 multiplications, 31 additions and 15 rounding operations. Hence, its arithmetical complexity is nearly optimal, keeping in mind that best algorithms for DCT–II of length 8 need 11 multiplications and 29 additions without counting the final scaling by $2\sqrt{2}$ (see [10, 14, 17]).

The left–inverse integer DCT–II algorithm for Algorithm 4.1 simply follows by going backward and taking the left–inverse lifting procedure of Theorem 3.1. Now we analyze the error caused by Algorithm 4.1 comparing the resulting integer vector $\mathbf{y}$ with the exact result $\hat{\mathbf{y}} = 2\,C_8^{II}\mathbf{x}$ of the scaled DCT–II of length 8. A detailed consideration of the componentwise error will lead us to better error estimates than given in [15].

Theorem 4.2. *Let $\mathbf{x} \in \mathbb{Z}^8$ be an arbitrary vector of integers. Using Algorithm 4.1, the resulting integer approximation $\mathbf{y}$ of $\hat{\mathbf{y}} = 2C_8^{II}\,\mathbf{x}$ satisfies the error estimate*

$$\|\hat{\mathbf{y}} - \mathbf{y}\|_2 \le 5.743824. \tag{4.2}$$

Further an analysis of the componentwise error gives

$$\begin{aligned} |\hat{y}_0 - y_0| &\le 1.060660, & |\hat{y}_1 - y_1| &\le 2.107046,\\ |\hat{y}_2 - y_2| &\le 1.061396, & |\hat{y}_3 - y_3| &\le 3.523072,\\ |\hat{y}_4 - y_4| &\le 0.853553, & |\hat{y}_5 - y_5| &\le 3.315965,\\ |\hat{y}_6 - y_6| &\le 0.691342, & |\hat{y}_7 - y_7| &\le 1.375330, \end{aligned} \tag{4.3}$$

and in particular,

$$\|\hat{\mathbf{y}} - \mathbf{y}\|_\infty \le 3.523072.$$

Proof. Let us denote the preliminary results of the exact DCT–II using the factorization (4.1) by

$$\begin{aligned} \hat{\mathbf{x}}^{(1)} &:= \sqrt{2}\,T_8(0)\,\mathbf{x},\\ \hat{\mathbf{x}}^{(2)} &:= (\sqrt{2}I_4 \oplus I_4)\,T_8(0,\,1)\hat{\mathbf{x}}^{(1)},\\ \hat{\mathbf{x}}^{(3)} &:= (I_4 \oplus \sqrt{2}I_4)\,T_8(0,\,1,\,0,\,0)\,\hat{\mathbf{x}}^{(2)},\\ \hat{\mathbf{x}}^{(4)} &:= A_8(0,\,1)\,\hat{\mathbf{x}}^{(3)}. \end{aligned}$$

Then we have $\hat{\mathbf{y}} = 2\,C_8^{II}\,\mathbf{x} = B_8\hat{\mathbf{x}}^{(4)}$. Further, let $\mathbf{e}^{(s)} := \mathbf{x}^{(s)} - \tilde{\mathbf{x}}^{(s)}$ $(s \in \{1,2,3,4\})$ denote the error of a single step in the algorithm, where $\mathbf{x}^{(s)}$ are defined in Algorithm 4.1, and where $\tilde{\mathbf{x}}^{(1)} := \sqrt{2}\,T_8(0)\mathbf{x}$, $\tilde{\mathbf{x}}^{(2)} := (\sqrt{2}\,I_4 \oplus I_4)\,T_8(0,1)\mathbf{x}^{(1)}$, $\tilde{\mathbf{x}}^{(3)} := (I_4 \oplus \sqrt{2}\,T_4)\,T_8(0,1,0,0)\mathbf{x}^{(2)}$, $\tilde{\mathbf{x}}^{(4)} := A_8(0,1)\,\mathbf{x}^{(3)}$. Observe that $\mathbf{e}^{(1)} = \mathbf{x}^{(1)} - \tilde{\mathbf{x}}^{(1)} = \mathbf{0}$. Then for the error $\|\hat{\mathbf{y}} - \mathbf{y}\|_2$ we find

$$\begin{aligned} \|\hat{\mathbf{y}} - \mathbf{y}\|_2 &= \|\hat{\mathbf{x}}^{(4)} - \mathbf{x}^{(4)}\|_2 = \|\hat{\mathbf{x}}^{(4)} - \tilde{\mathbf{x}}^{(4)} - \mathbf{e}^{(4)}\|_2\\ &\le \|A_8(0,\,1)\|_2\|\hat{\mathbf{x}}^{(3)} - \mathbf{x}^{(3)}\|_2 + \|\mathbf{e}^{(4)}\|_2\\ &\le \|(I_4 \oplus \sqrt{2}I_4)T_8(0,\,1,\,0,\,0)\|_2\|\hat{\mathbf{x}}^{(2)} - \mathbf{x}^{(2)}\|_2 + \|\mathbf{e}^{(3)}\|_2 + \|\mathbf{e}^{(4)}\|_2\\ &= \sqrt{2}\,\|\mathbf{e}^{(2)}\|_2 + \|\mathbf{e}^{(3)}\|_2 + \|\mathbf{e}^{(4)}\|_2. \end{aligned}$$

With $h(t) := \frac{3}{4} + \sin t + \frac{1}{2}\cos t + \frac{1}{4}(\tan\frac{t}{2})^2$ we find by Theorem 3.1 that

$$\begin{aligned}
\|\mathbf{e}^{(2)}\|_2 &\le (h(\tfrac{\pi}{16}) + h(\tfrac{3\pi}{16}))^{1/2} \approx 1.783877,\\
\|\mathbf{e}^{(3)}\|_2 &\le (h(\tfrac{\pi}{4}) + h(\tfrac{\pi}{8}))^{1/2} \approx 1.859588,\\
\|\mathbf{e}^{(4)}\|_2 &\le h(\tfrac{\pi}{4})^{1/2} \approx 1.361453,
\end{aligned}$$

and

$$\|\hat{\mathbf{y}} - \mathbf{y}\|_2 \le 5.743824.$$

Let us now estimate the componentwise error using the same notations as above. After the first step of Algorithm 4.1 we obtain $\hat{\mathbf{x}}^{(1)} = \mathbf{x}^{(1)}$. After the second step of Algorithm 4.1 we find $\hat{x}_j^{(2)} - x_j^{(2)} = 0$ for $j \in \{0,1,2,3\}$ and by Remark 3.2, (1)

$$\begin{aligned}
|\hat{x}_4^{(2)} - x_4^{(2)}| &\le 1.039638, & |\hat{x}_5^{(2)} - x_5^{(2)}| \le 1.067408,\\
|\hat{x}_6^{(2)} - x_6^{(2)}| &\le 0.777785, & |\hat{x}_7^{(2)} - x_7^{(2)}| \le 0.597545.
\end{aligned}$$

After the third step, we obtain by Remark 3.2, (1) that

$$\begin{aligned}
|\hat{x}_0^{(3)} - x_0^{(3)}| &\le 1.060660, & |\hat{x}_1^{(3)} - x_1^{(3)}| \le 0.853553,\\
|\hat{x}_2^{(3)} - x_2^{(3)}| &\le 1.061396, & |\hat{x}_3^{(3)} - x_3^{(3)}| \le 0.691342,\\
|\hat{x}_4^{(3)} - x_4^{(3)}| &\le 2.107046, & |\hat{x}_5^{(3)} - x_5^{(3)}| \le 2.107046,\\
|\hat{x}_6^{(3)} - x_6^{(3)}| &\le 1.375330, & |\hat{x}_7^{(3)} - x_7^{(3)}| \le 1.375330.
\end{aligned}$$

The multiplication of $A_8(0,1)$ in the fourth step leads to

$$\begin{aligned}
|\hat{x}_j^{(4)} - x_j^{(4)}| &= |\hat{x}_j^{(3)} - x_j^{(3)}|, \qquad j \in \{0,1,2,3,4\},\\
|\hat{x}_5^{(4)} - x_5^{(4)}| &\le \tfrac{1}{\sqrt{2}} \cdot 3.482376 + 1.060660 = 3.523072,\\
|\hat{x}_6^{(4)} - x_6^{(4)}| &\le \tfrac{1}{\sqrt{2}} \cdot 3.482376 + 0.853553 = 3.315965,\\
|\hat{x}_7^{(4)} - x_7^{(4)}| &= |\hat{x}_6^{(3)} - x_6^{(3)}| \le 1.375330.
\end{aligned}$$

Finally, the permutation in the last step provides the componentwise worst case error as given in (4.3). In particular, we see that in two components, the error is smaller than 1 always, and in 3 further components, the error can exceed 1 only slightly. □

In fact, the numerical results imply that Algorithm 4.1 performs really well.

Example 4.3. For given $\mathbf{x} \in \mathbb{Z}^8$, $\mathbf{y}$ denotes the integer DCT–II of $\mathbf{x}$ computed by Algorithm 4.1 and $\hat{\mathbf{y}} = 2\,C_8^{II}\mathbf{x}$ is the exact DCT–II of $\mathbf{x}$ (scaled by 2 and rounded to 3 decimal places). In the following table we give some examples for the performance of the Algorithm 4.1.

$\mathbf{x}$	1	1	2	2	3	3	5	5
$\hat{\mathbf{y}}$	15.556	-8.000	1.307	-1.009	0	0.674	-0.541	1.591
$\mathbf{y}$	16	-8	1	-1	0	0	-1	1

$\mathbf{x}$	121	122	122	127	126	129	120	123
$\hat{\mathbf{y}}$	700.036	-3.992	-11.759	4.257	2.828	-3.607	4.871	-8.302
$\mathbf{y}$	700	-4	-12	5	3	-4	5	-8

$\mathbf{x}$	129	-13	12	45	-23	-69	-133	99
$\hat{\mathbf{y}}$	33.234	187.467	156.260	-115.689	320.319	-28.685	161.058	-60.160
$\mathbf{y}$	33	188	156	-115	320	-29	161	-61

$\mathbf{x}$	22	33	44	55	66	77	88	99
$\hat{\mathbf{y}}$	342.240	-141.731	0	-14.816	0	-4.420	0	-1.115
$\mathbf{y}$	342	-142	0	-15	0	-5	0	-1

For further numerical examples and a numerical consideration of the error distribution we refer to [15].

5 Two–Dimensional Integer DCT–II of Size 8 × 8

The two–dimensional (2–d) DCT–II has important applications in image compression (JPEG, MPEG). Therefore we extend our results of Section 4 to the 2–d integer DCT–II. Let $X \in \mathbb{Z}^{8\times 8}$ be given. Then the 2–d DCT–II *of size* 8×8 of X is defined by $C_8^{II}\,X\,(C_8^{II})^T$. Let $\hat{Y} = (2\,C_8^{II})\,X\,(2\,C_8^{II})^T$. The simple *row–column method* for computing of $\hat{Y}$ is based on the observation

$$\hat{Y} = (2\,C_8^{II}\,X)\,(2\,C_8^{II})^T = \hat{Z}\,(2\,C_8^{II})^T = (2\,C_8^{II}\,\hat{Z}^T)^T$$

with $\hat{Z} := 2\,C_8^{II}\,X$. Now we compute integer approximations of $\hat{Z}$ and $\hat{Y}$ by Algorithm 4.1.

Algorithm 5.1.
Input: $X = (\mathbf{x}_0, \ldots, \mathbf{x}_7) \in \mathbb{Z}^{8\times 8}$.

1. For $k = 0, \ldots, 7$ compute the integer approximation $\mathbf{z}_k$ of $\hat{\mathbf{z}}_k := 2\,C_8^{II}\mathbf{x}_k$ by Algorithm 4.1.
2. Set $Z := (\mathbf{z}_0, \ldots, \mathbf{z}_7)$ and $(\mathbf{u}_0, \ldots, \mathbf{u}_7) := Z^T$.

3. For $k = 0, \ldots, 7$ compute the integer approximation $\mathbf{v}_k$ of $\hat{\mathbf{v}}_k := 2\, C_8^{II} \mathbf{u}_k$ by Algorithm 4.1.
4. Form $Y := (\mathbf{v}_0, \ldots, \mathbf{v}_7)^T$.

Output: $Y \in \mathbb{Z}^{8\times 8}$ integer approximation of $\hat{Y} = 4\, C_8^{II} X\, (C_8^{II})^T$.

Let us consider the (worst case) errors of Algorithm 5.1 estimated in the Frobenius norm and in the maximum norm when the resulting integer matrix Y is compared with the exact (scaled) 2–d DCT–II $\hat{Y}$.

Theorem 5.2. *Let $X \in \mathbb{Z}^{8\times 8}$ be an arbitrary matrix. Using Algorithm 5.1, the resulting integer approximation $Y \in \mathbb{Z}^{8\times 8}$ of $\hat{Y} = 4\, C_8^{II}\, X\, (C_8^{II})^T$ satisfies the error estimate*

$$\|\hat{Y} - Y\|_F \le 48.737963.$$

Further, the componentwise errors can be estimated by the error matrix

$$(|\hat{y}_{jk} - y_{jk}|)_{j,k=0}^{7} \le \begin{pmatrix} 6.718 & 6.187 & 6.287 & 6.186 & 6.718 & 6.186 & 6.287 & 6.186 \\ 7.764 & 7.233 & 7.333 & 7.233 & 7.764 & 7.233 & 7.333 & 7.233 \\ 6.718 & 6.187 & 6.288 & 6.187 & 6.718 & 6.187 & 6.288 & 6.187 \\ 9.180 & 8.649 & 8.749 & 8.649 & 9.180 & 8.649 & 8.749 & 8.649 \\ 6.510 & 5.979 & 6.080 & 5.979 & 6.510 & 5.979 & 6.080 & 5.979 \\ 8.973 & 8.442 & 8.542 & 8.442 & 8.973 & 8.442 & 8.542 & 8.442 \\ 6.348 & 5.817 & 5.918 & 5.817 & 6.348 & 5.817 & 5.918 & 5.817 \\ 7.032 & 6.501 & 6.602 & 6.501 & 7.032 & 6.501 & 6.602 & 6.501 \end{pmatrix},$$

where $\hat{Y} = (\hat{y}_{jk})_{j,k=0}^{7}$ and $Y = (y_{jk})_{j,k=0}^{7}$, and in particular

$$\|\hat{Y} - Y\|_\infty \le 9.179926.$$

Proof. By (4.2), we know that the computed vectors $\mathbf{z}_k$ ($k = 0, \ldots, 7$) in step 1 of Algorithm 5.1 satisfy the estimate

$$\|\mathbf{z}_k - \hat{\mathbf{z}}_k\|_2^2 \le (5.743824)^2.$$

Summing up, this yields

$$\|Z - \hat{Z}\|_F^2 = \sum_{k=0}^{7} \|\mathbf{z}_k - \hat{\mathbf{z}}_k\|_2^2 < 8 \cdot (5.743824)^2$$

with the matrix $\hat{Z} := (\hat{\mathbf{z}}_0, \ldots, \hat{\mathbf{z}}_7)$. Hence,

$$\|Z - \hat{Z}\|_F \le 2\sqrt{2} \cdot 5.743824.$$

Set $(\hat{\mathbf{u}}_0, \ldots, \hat{\mathbf{u}}_7)^T := \hat{Z}^T$ and $\tilde{\mathbf{v}}_k := 2\, C_8^{II} \hat{\mathbf{u}}_k$ $(k = 0, \ldots, 7)$. Further let $\mathbf{v}_k$ $(k = 0, \ldots, 7)$ be the computed vectors in step 3 of Algorithm 5.1. Applying again (4.2), we get

$$\|\mathbf{v}_k - \tilde{\mathbf{v}}_k\|_2^2 \le (5.743824)^2$$

and hence

$$\|Y - \tilde{Y}\|_F \le 2\sqrt{2} \cdot 5.743824$$

with $Y := (\mathbf{v}_0, \ldots, \mathbf{v}_7)^T$ and $\tilde{Y} := (\tilde{\mathbf{v}}_0, \ldots, \tilde{\mathbf{v}}_7)^T = Z\,(2\, C_8^{II})^T$. Since the Frobenius norm is unitarily invariant, we have $\|\tilde{Y} - \hat{Y}\|_F = 2\,\|Z - \hat{Z}\|_F$. The Frobenius norm is consistent such that we can estimate

$$\|Y - \hat{Y}\|_F \le \|Y - \tilde{Y}\|_F + \|\tilde{Y} - \hat{Y}\|_F \le 6\sqrt{2} \cdot 5.743824.$$

Now we consider the componentwise error. By (4.3) we obtain in step 1 of Algorithm 5.1 with $\mathbf{z}_k := (z_{jk})_{j=0}^7$ and $\hat{\mathbf{z}}_k := (\hat{z}_{jk})_{j=0}^7$ the error matrix

$$(|\hat{z}_{jk} - z_{jk}|)_{j,k=0}^7 \le (\mathbf{f}, \mathbf{f}, \ldots, \mathbf{f}) \in \mathbb{R}^{8\times 8}$$

with

$$\begin{aligned} \mathbf{f} &:= (f_0, f_1, \ldots, f_7)^T \\ &= (1.061,\ 2.107,\ 1.061,\ 3.523,\ 0.854,\ 3.316,\ 0.691,\ 1.375)^T. \end{aligned}$$

Here the errors in (4.3) are rounded to 3 decimal places. In the second step, the matrix is transposed only, i.e., with $\mathbf{1} := (1, 1, \ldots, 1)^T \in \mathbb{Z}^8$ we have

$$(|\hat{z}_{kj} - z_{kj}|)_{j,k=0}^7 \le (f_0\mathbf{1}, f_1\mathbf{1}, \ldots, f_7\mathbf{1}).$$

In the third step we find with $\mathbf{v}_k := (v_{jk})_{j=0}^7$ and $\hat{\mathbf{v}}_k := (\hat{v}_{jk})_{j=0}^7$ that

$$\begin{aligned} (|\hat{v}_{jk} - v_{jk}|)_{j,k=0}^7 &\le (\mathbf{f}, \ldots, \mathbf{f}) + 2\,|C_8^{II}|\,(|\hat{z}_{kj} - z_{kj}|)_{j,k=0}^7 \\ &\le \left(\mathbf{f} + 2 f_0\,|C_8^{II}|\,\mathbf{1}, \ldots, \mathbf{f} + 2 f_7\,|C_8^{II}|\mathbf{1}\right), \end{aligned}$$

where

$$|C_8^{II}| := \tfrac{1}{2}\left(\varepsilon_8(j)\,|\cos \tfrac{j(2k+1)\pi}{16}|\right)_{j,k=0}^7.$$

Here the error $2 f_j |C_8^{II}|\mathbf{1}$ is caused by the error of the previous step and $\mathbf{f}$ by the integer DCT–II applied to $\mathbf{u}_k := (z_{kj})_{j=0}^7$. By

$$|C_8^{II}|\mathbf{1} \approx (2.828,\ 2.563,\ 2.613,\ 2.563,\ 2.828,\ 2.563,\ 2.613,\ 2.563)^T$$

and $(|\hat{y}_{jk} - y_{jk}|)_{j,k=0}^{7} = (|\hat{v}_{kj} - v_{kj}|)_{j,k=0}^{7}$, we finally obtain the error matrix as given in the theorem. This completes the proof. □

We want to finish with some numerical results for the 2–d integer DCT–II applying Algorithm 5.1.

Example 5.3. Let $X \in \mathbb{Z}^{8\times 8}$ denote the input matrix, Y is the 2–d integer DCT–II of X computed by Algorithm 5.1, and $\hat{Y} = 4C_8^{II}\, X\, (C_8^{II})^T$ is the exact (scaled) 2–d DCT–II of X, where each entry is rounded to the next integer. For

$$X := \begin{pmatrix} 13 & 14 & 14 & 15 & 14 & 12 & 10 & 10 \\ 14 & 14 & 23 & 32 & 15 & 12 & 9 & 6 \\ 15 & 20 & 21 & 13 & 12 & 14 & 21 & 8 \\ 15 & 16 & 16 & 17 & 18 & 19 & 20 & 12 \\ 15 & 16 & 16 & 16 & 16 & 15 & 15 & 15 \\ 13 & 16 & 18 & 16 & 17 & 15 & 14 & 15 \\ 16 & 19 & 17 & 17 & 16 & 15 & 13 & 12 \\ 16 & 15 & 16 & 16 & 16 & 15 & 12 & 11 \end{pmatrix}$$

we obtain that

$$\hat{Y} = \begin{pmatrix} 487 & 41 & -43 & -9 & -15 & 14 & -7 & -1 \\ -9 & 13 & -18 & -12 & -4 & 21 & -5 & -4 \\ -24 & 23 & -14 & -7 & 18 & 0 & 7 & -10 \\ -12 & -9 & 2 & -7 & 20 & -17 & 6 & -5 \\ -11 & -20 & 13 & 20 & 9 & -10 & 2 & 2 \\ -1 & -12 & 21 & 19 & -17 & -12 & -8 & 8 \\ -12 & -13 & 25 & 7 & -20 & -11 & -2 & 17 \\ -8 & 5 & 24 & 2 & -14 & -7 & -5 & 7 \end{pmatrix}$$

and

$$Y := \begin{pmatrix} 488 & 40 & -43 & -8 & -16 & 14 & -7 & 0 \\ -10 & 15 & -17 & -13 & -2 & 20 & -3 & -4 \\ -24 & 22 & -14 & -8 & 18 & -1 & 7 & -11 \\ -13 & -9 & 2 & -12 & 22 & -18 & 6 & -4 \\ -11 & -19 & 14 & 20 & 9 & -10 & 1 & 3 \\ -1 & -13 & 23 & 19 & -16 & -15 & -8 & 8 \\ -13 & -11 & 25 & 8 & -20 & -9 & -1 & 16 \\ -8 & 4 & 23 & 2 & -13 & -7 & -6 & 8 \end{pmatrix}.$$

The greatest componentwise error occurs in the $(3,3)$ position of Y and is 4.555. □

Finally, we consider the behavior of the errors $\|\hat{Y} - Y\|_\infty := \max_{j,k=0,\ldots,7} |\hat{y}_{jk} - y_{jk}|$ in more detail. As input matrices we use 1000 random matrices in $\mathbb{Z}^{8\times 8}$ with entries in the range $[-128, 127]$. We compute the r–th quantiles for $r = \frac{j}{10}$, $j = 1, \ldots, 10$ for Algorithm 5.1. After sorting the errors of 1000 resulting matrices, the r–th quantile is the smallest value that separates the errors into two parts; $1000\,r$ of the sorted errors are less than or equal to the quantile value, the other $1000\,(1-r)$ errors are greater than the quantile. The quantile for $r = 1.0$ is the maximal error occurring. In the following table the r–th quantiles are rounded to 3 decimal places.

r=0.1	r=0.2	r=0.3	r=0.4	r=0.5	r=0.6	r=0.7	r=0.8	r=0.9	r=1.0
2.333	2.496	2.627	2.751	2.860	2.984	3.125	3.365	3.655	5.654

References

[1] A. R. Calderbank, I. Daubechies, W. Sweldens, B. L. Yeo: *Wavelet transforms that map integers to integers,* Appl. Comput. Harmon. Anal. **5** (1998), 332–369.

[2] W. K. Cham, P. C. Yip: *Integer sinusoidal transforms for image processing,* Internat. J. Electron. **70** (1991), 1015–1030.

[3] Y.-J. Chen, S. Oraintara, T. Q. Nguyen: *Integer discrete cosine transform Int*DCT, Preprint, Univ. Boston, 2000.

[4] Y.-J. Chen, S. Oraintara, T. D. Tran, K. Amaratunga, T. Q. Nguyen: *Multiplierless approximation of transforms using lifting scheme and coordinate descent with adder constraint,* Proc. IEEE Internat. Conf. Acoust. Speech Signal Process., Vol. **3**, 2002, 3136–3139.

[5] L. Z. Cheng, H. Xu, Y. Luo: *Integer discrete cosine transform and its fast algorithm,* Electron. Lett. **37** (2001), 64–65.

[6] I. Daubechies, W. Sweldens: *Factoring wavelet transforms into lifting steps,* J. Fourier Anal. Appl. **4** (1998), 247–269.

[7] K. Komatsu, K. Sezaki: *Reversible discrete cosine transform,* Proc. IEEE Internat. Conf. Acoust. Speech Signal Process., 1998, 1769–1772.

[8] K. Komatsu, K. Sezaki: *2D lossless discrete cosine transform,* Proc. IEEE Internat. Conf. Image Process., 2001, 466–469.

[9] J. Liang, T. D. Tran: *Fast multiplierless approximations of the* DCT *with the lifting scheme,* IEEE Trans. Signal Process. **49** (2001), 3032–3044.

[10] C. Loeffler, A. Lightenberg, G. Moschytz: *Practical fast* 1–*d* DCT *algorithms with* 11 *multiplications,* Proc. IEEE Internat. Conf. Acoust. Speech Signal Process., Vol. **2**, 1989, 988–991.

[11] M. W. Marcellin, M. J. Gormish, A. Bilgin, M. P. Boliek: *An overview of* JPEG–2000, Proc. Data Compression Conf., 2000, 523–541.

[12] W. Philips: *Lossless* DCT *for combined lossy/lossless image coding,* Proc. IEEE Internat. Conf. Image Process., Vol. **3**, 1998, 871–875.

[13] G. Plonka: *A global method for invertible integer* DCT *and integer wavelet algorithms*, Preprint, Univ. Duisburg, 2003.

[14] G. Plonka, M. Tasche: *Split–radix algorithms for discrete trigonometric transforms,* Preprint, Univ. Duisburg, 2002.

[15] G. Plonka, M. Tasche: *Reversible integer* DCT *algorithms,* Preprint, Univ. Duisburg, 2002.

[16] M. Primbs: *Integer* DCT *Algorithms* (in German), Diploma Thesis, Institute of Mathematics, Univ. Duisburg, 2003.

[17] K. R. Rao, P. Yip: *Discrete Cosine Transform: Algorithms, Advantages, Applications*, Academic Press, Boston 1990.

[18] G. Strang: *The discrete cosine transform,* SIAM Rev. **41** (1999), 135–147.

[19] T. D. Tran: *The Bin*DCT: *Fast multiplierless approximation of the* DCT, IEEE Signal Process. Lett. **7** (2000), 141–144.

[20] Y. Zeng, L. Cheng, G. Bi, A. C. Kot: *Integer* DCT*s and fast algorithms,* IEEE Trans. Signal Process. **49** (2001), 2774–2782.

Addresses:

Gerlind Plonka
University of Duisburg
Institute of Mathematics
D–47048 Duisburg
Germany

Manfred Tasche
University of Rostock
Department of Mathematics
D–18051 Rostock
Germany

International Series of Numerical Mathematics
Vol. 145,

Approximation of Density Functions and Reconstruction of the Approximant

Manfred Reimer

Abstract

Polynomials are not as stiff as they are sometimes thought to be. We report on a positive summation method for Appell series with favourable localisation properties, which we gain from the Newman–Shapiro operators which belong to a properly chosen hypersphere. Moreover, the Appell approximants can be calculated from the image under the Radon transform of the approximated function. We discuss the complexity and the stability problem of the resulting reconstruction method for tomography.

1 Appell Series and Summation

For $r \in \mathbb{N}$ we denote by B^r and S^{r-1} the unit ball and the unit sphere in $\mathbb{R}^r$, respectively; $\omega_{r-1}$is the area of S^{r-1}.

For $r+s-1 \neq 0$, the polynomials $V_m^{r,s}$, $m \in \mathbb{N}_0^r$, of Appell and Kampé de Fériet are generated by the equation

$$\frac{1}{(1-2tx+t^2)^{\frac{r+s-1}{2}}} = \sum_{\mu=0}^{\infty} \sum_{|m|=\mu} V_m^{r,s}(x)\, t^m, \tag{1.1}$$

where $|m| = m_1 + \cdots + m_r$. Here $t = (t_1, \ldots, t_r)'$ and $x = (x_1, \ldots, x_r)'$ are varying in $\mathbb{R}^r$ under the restrictions $|x| \leq 1$, $|t| < 1$, and s is called the *(Appell–) index*. For $r \geq 3$ and $s = -1$,

$$|x|^\mu V_m^{r,-1}\Big(\frac{x}{|x|}\Big)$$

is a harmonic homogeneous polynomial of degree $\mu := |m|$, and for $t_r := 0$ the definition (1.1) collapses and takes the form of the defining equation of the polynomials $V_{\bar{m}}^{r-1,s+1}$, $\bar{m} = (m_1, \ldots, m_{r-1})$. For $r \geq 2$ this yields

$$V_{\bar{m},0}^{r,s}(\bar{x}, *) = V_{\bar{m}}^{r-1,s+1}(\bar{x}),$$

where $\bar{x} = (x_1, \ldots, x_{r-1})'$.

In what follows we assume $r \in \mathbb{N} \setminus \{1, 2\}$ and $s \in \mathbb{N}_0 \cup \{-1\}$; in particular, the condition $r + s - 1 \neq 0$ is always satisfied. Then we obtain, with some changes in the notation, the following chain of equalities

$$V_m^{r,s}(x) = V_{m,0}^{r+1,s-1}(x, *) = \cdots = V_{m,0,\ldots,0}^{r+s+1,-1}(x, *, \ldots, *). \tag{1.2}$$

It follows that the restrictions of the $V_m^{r,s}$ onto S^{r+s} are spherical harmonics in $r + s + 1$ variables, which do not depend on the variables $x_{r+1}, \ldots, x_{r+s+1}$. This leads us to introduce the inner products

$$\langle \tilde{F}, \tilde{G} \rangle_{r+s} := \int_{S^{r+s}} \tilde{F}(\tilde{x}) \tilde{G}(\tilde{x}) d\tilde{\omega}(\tilde{x})$$

for functions $\tilde{F}, \tilde{G} \in C(S^{r+s})$, where $\tilde{\omega}$ is the standard measure on S^{r+s}, and

$$[F, G]_{r,s} := \omega_s \int_{B^r} F(x) G(x) \left(1 - |x|^2\right)^{\frac{s-1}{2}} dx$$

for functions $F, G \in C(B^r)$. Using the orthogonality property of the spherical harmonics, we get

$$[V_m^{r,s}, V_n^{r,s}]_{r,s} = \langle V_{m,0,\ldots,0}^{r+s+1,-1}, V_{n,0,\ldots,0}^{r+s+1,-1} \rangle_{r+s} = 0 \tag{1.3}$$

for $|m| \neq |n|$, which is a main result of Appell and Kampé de Fériet, [1].

Now let us define the spaces

$$\mathbf{V}_\mu^{r,s} := span\left\{ V_m^{r,s} : |m| = \mu \right\}$$

for $\mu \in \mathbb{N}_0$. Then

$$\mathbb{P}_\mu^r = \mathbf{V}_0^{r,s} \oplus \mathbf{V}_1^{r,s} \oplus \cdots \oplus \mathbf{V}_\mu^{r,s}$$

is an orthogonal decomposition with respect to the inner product $[\cdot, \cdot]_{r,s}$ of the space $\mathbb{P}_\mu^r$ of all real r–variate polynomials of total degree $\mu \in \mathbb{N}_0$. Moreover, for functions $F \in C(B^r)$ we define $\tilde{F} \in C(S^{r+s})$ by

$$\tilde{F}(x_1, \ldots, x_{r+s+1}) := F(x_1, \ldots, x_r)$$

for $(x_1, \ldots, x_{r+s+1})' \in S^{r+s}$, which implies $(x_1, \ldots, x_r)' \in B^r$. Then the following holds.

If F_ν, $\nu \in \mathbb{N}_0$, is the orthogonal component of F in $\mathbf{V}_\nu^{r,s}$ with respect to $[\,\cdot\,,\,\cdot\,]_{r,s}$, then $\tilde{F}_\nu$ is the orthogonal component of $\tilde{F}$ in the space of spherical harmonics of degree ν with respect to $\langle\,\cdot\,,\,\cdot\,\rangle_{r+s}$. Now let us assume that

$$A = \left(a_{\mu,\nu}\right)_{\mu,\nu\in\mathbb{N}_0}$$

is an infinite subdiagonal summation matrix for the Laplace series on $C(S^{r+s})$ with respect to the maximum norm. Then we obtain

$$\lim_{\mu\to\infty} \sum_{\nu=0}^{\mu} a_{\mu,\nu}\tilde{F}_\nu = \tilde{F}, \tag{1.4}$$

again in the maximum norm of $C(S^{r+s})$, and this implies

$$\lim_{\mu\to\infty} \sum_{\nu=0}^{\mu} a_{\mu,\nu}F_\nu = F, \tag{1.5}$$

now in the maximum norm of $C(B^r)$. In other words, the linear operators

$$L_\mu^A : C(B^r) \to \mathbb{P}_\mu^r,$$

defined by

$$L_\mu^A F := \sum_{\nu=0}^{\mu} a_{\mu,\nu}F_\nu$$

for $F \in C(B^r)$, converge pointwise in the maximum norm as $\mu \to \infty$.

Actually there are suitable summation matrices. For instance we know by results of Kogbetliantz, [3], that the Cesàro method (C, k) defines a proper matrix A, provided the index satisfies

$$k > \frac{r+s-1}{2}.$$

For $k \geq r+s$ the operators L_μ^A are even positive. However, the saturation order is $\mathcal{O}(\frac{1}{\mu})$, which is too poor in applications. So the question arises, is there a better choice of A?

2 What is a Good Choice of the Summation Matrix?

In this section we replace temporarily $r+s+1$ by r, i.e. we consider the unit sphere S^{r-1} in $\mathbb{R}^r$. For even $\mu = 2\nu \in \mathbb{N}_0$, Newman and Shapiro [5] introduced the positive linear operators

$$L_\mu^{(r)} : C(S^{r-1}) \to \mathbb{P}_\mu^r(S^{r-1}) \tag{2.1}$$

given by

$$\left(L_\mu^{(r)}F\right)(x) := \int_{S^{r-1}} F(t)K_\mu^{(r)}(tx)d\omega(t) \tag{2.2}$$

for $F \in C(S^{r-1})$, $x \in S^{r-1}$, where

$$K_\mu^{(r)}(\xi) := c_\nu \cdot \left[\frac{C_{\nu+1}^\lambda(\xi)}{\xi - \eta_{\nu+1}^\lambda}\right]^2, \tag{2.3}$$

with $\lambda := \frac{r-2}{2}$, while $\eta_{\nu+1}^\lambda$ is the greatest zero of $C_{\nu+1}^\lambda$ and the constant c_ν is to be chosen such that

$$L_\mu^{(r)}1 = 1 \tag{2.4}$$

is valid. We complete the definition by putting $L_\mu^{(r)} := L_{\mu-1}^{(r)}$ if μ is odd. Newman and Shapiro introduced these operators to prove Jackson's inequality on the sphere. Actually, a constant k_r exists such that

$$\|F - L_\mu^{(r)}F\|_\infty \leq k_r \cdot \omega_1\left(F, \frac{1}{\mu}\right) \tag{2.5}$$

holds, where $\omega_1(F, \phi)$ is the modulus of continuity of the first order with respect to the geodetic distance on S^{r-1}. Recently we improved this result, replacing in (2.5) the modulus of continuity by the modulus of continuity of the second order. The proof takes full advantage of the particular choice of the kernel, which allows us to estimate the contribution of the kernel to the surface integral from outside of a small cap by the contribution arising just from the interior of this cap itself. For details we refer to our original paper, [8].

It follows that, possibly with some new constant k_r, the inequalities

$$\|F - L_\mu^{(r)}F\|_\infty \leq k_r \cdot \|F^{(j)}\|_\infty \cdot \frac{1}{\mu^j} \tag{2.6}$$

are valid for $j \in \{1, 2\}$ and $F \in C^{(j)}(S^{r-1})$, which is the space consisting of all $F \in C(S^{r-1})$ with bounded pure directional derivatives of order j. We remark that by a well known result of Korovkin, the equations (2.6) cannot hold for a positive operator simultaneously for $j = 1, 2$ and 3, which says that $\mathcal{O}(\frac{1}{\mu^j})$ is the exact saturation order of the Newman–Shapiro operators, which cannot be improved upon by any other sequence of positive operators.

Moreover, the inequality (2.5) holds even if $L_\mu^{(r)}$ is replaced by a discrete version $\hat{L}_\mu^{(r)}$, defined as follows. For fixed μ let (2.2) be evaluated by a positive quadrature with nodes $t_1, \dots, t_M$ and weights $A_1, \dots, A_M$, which is exact of degree 2μ. For instance we could use a product Gauß quadrature. Then we obtain an operator defined by

$$\left(\hat{L}_\mu^{(r)} F\right)(x) := \sum_{j=1}^{M} A_j F(t_j) K_\mu^{(r)}(t_j x)$$

for $F \in C(S^{r-1})$, $x \in S^{r-1}$, instead of (2.2). It has the form of a hyperinterpolation operator, is in general not a projection, but satisfies

$$\|F - \hat{L}_\mu^{(r)} F\|_\infty \le \hat{k}_r \cdot \omega_1\left(F, \frac{1}{\mu}\right) \tag{2.7}$$

instead of (2.5), posssibly again with some new constant $\hat{k}_r$. In this sense the first order estimates are stable against discretisation, which is surprising in view of the irregularities which must occur in the distribution of the nodes and weights. Actually, to prove (2.7) an intrinsic investigation is needed, which takes advantage of the particular distribution of the weights in a positive quadrature, described by the *regularity condition* of Sloan and Womersley, [10]. Fortunately, we could prove that the regularity condition is always satisfied, see [7].

Now we replace r by $r+s+1$ again, and ask, how can we use the operators $L_\mu^{(r+s+1)}$, which act on $C(S^{r+s})$, to get results for $C(B^r)$? Of course, we could apply the operators to the function $\tilde{F} \in C(S^{r+s})$ defined by $F \in C(B^r)$ in Section 1, but this would be at the price of increasing the complexity of the problem.

Fortunately the $L_\mu^{(r)}$ can be represented by some infinite subdiagonal summation matrix $A^{(r)}$, which we call the *Newman–Shapiro matrix.* It is defined by the expansion

$$K_\mu^{(r)} = \sum_{\kappa=0}^{\mu} a_{\mu,\nu}^{(r)} G_\kappa^r,$$

where, for $\kappa \in \mathbb{N}_0$,

$$G_\kappa^r = \frac{2\kappa + r - 2}{(r-2)\omega_{r-1}} \cdot C_\kappa^\lambda$$

is the reproducing kernel function of the space of spherical harmonics on S^{r-1} of degree κ. Using the formula of *Christoffel–Darboux* and the *linearisation formulae* of *Rogers and Ramanujan* we were able to prove that the matrix $A^{(r)}$ is nonnegative, [8].

Now let F and $\tilde{F}$ again be as in Section 1, but let $A := A^{(r+s+1)}$ be the Newman–Shapiro matrix belonging to $r+s+1$ dimensions. Then we obtain

$$\begin{aligned}\left(L_\mu^{(r+s+1)}\tilde{F}\right)(x_1,\ldots,x_{r+s+1}) &= \sum_{\kappa=0}^{\mu} a_{\mu,\kappa}^{(r+s+1)}\tilde{F}_\kappa(x_1,\ldots,x_{r+s+1}) \\ &= \sum_{\kappa=0}^{\mu} a_{\mu,\kappa}^{(r+s+1)}F_\kappa(x_1,\ldots,x_{r+s+1}) \\ &= \left(L_\mu^A F\right)(x_1,\ldots,x_r)\end{aligned}$$

for $(x_1,\ldots,x_{r+s+1})' \in S^{r+s}$, i.e. $(x_1,\ldots,x_r)' \in B^r$. It follows that this choice of A provides L_μ^A with the following, most favourable properties.

- The L_μ^A are positive operators $C(B^r) \to \mathbb{P}_\mu^r$.
- $\|F - L_\mu^A F\|_{B^r} = \|\tilde{F} - L_\mu^{(r+s+1)}\tilde{F}\|_{S^{r+s}} \leq k_r\tilde{\omega}_j(\tilde{F}, \frac{1}{\mu})$ holds for $j = 1, 2$, where $\tilde{\omega}_j(\,\cdot\,,\,\cdot\,)$ is the modulus of continuity of order j with respect to the geodetic distance on S^{r+s}.
- $\|F - L_\mu^A F\|_{B^r} = \mathcal{O}(\frac{1}{\mu^j})$ holds for $j = 1, 2$ under suitable smoothness assumptions on F, i.e.
- the saturation order of the L_μ^A is best possible.
- There is no need to evaluate $L_\mu^{(r+s+1)}\tilde{F}$ instead of $L_\mu^A F$, i.e. there is no artificial increase of the order of complexity.

Finally, we remark that the kernels $K_\mu^{(r)}$ and $K_\mu^{(r+s+1)}$ are very steep at the point $\xi = 1$. Hence they provide the Newman–Shapiro operators with a very *strong localisation* property.

3 How Can the Approximants be Gained from the Images Under the Radon Transform?

In what follows we consider the cylinder $Z^r := [-1, 1] \times S^{r-1}$ and let

$$R_0 : \; C(B^r) \to C(Z^r)$$

be the Radon transform, which we *normalize* by defining

$$R := R_0/R_0 1, \tag{3.1}$$

which is such that RF is free of a physical dimension.

Now we assume that R_0F is known, and we want to know how the approximant $L_\mu^A F$ to $F \in C(B^r)$ can be calculated from R_0F or RF. We outline the main ideas, only, and refer to our monograph [9] for details.

First it is important to know that

$$\mathbf{V}_\nu^{r,s} = span\{G_\nu^{r+s+1}(a\cdot)|\, a \in S^{r-1}\}$$

holds for $\nu \in \mathbb{N}_0$, for, by a result of Davison and Grünbaum, [2],

$$\left[RG_\nu^{r+s+1}(a\cdot)\right](\sigma,t) = \tilde{G}_\nu^{r+s+1}(\sigma)\cdot G_\nu^{r+s+1}(at)$$

holds for $a \in S^{r-1}$ and $(\sigma,t) \in Z^r$, where $\tilde{G}_\nu^{r+s+1} := G_\nu^{r+s+1}/G_\nu^{r+s+1}(1)$. It follows that the components F_ν of F in $\mathbf{V}_\mu^{r,s}$ satisfy

$$\left(RF_\nu\right)(\sigma,t) = \tilde{G}_\nu^{r+s+1}(\sigma)\cdot F_\nu(t). \tag{3.2}$$

Now, assuming that (1.5) holds in the uniform norm, we get by the help of (3.2)

$$\sum_{\nu=0}^{\infty} a_{\mu,\nu}\cdot \tilde{G}_\nu^{r+s+1}(\sigma)F_\nu(t) = \left(RF\right)(\sigma,t), \tag{3.3}$$

where the right side is known for all $(\sigma,t) \in Z^r$. Actually, this enables us to reconstruct the F_ν, and hence to calculate $L_\mu^A F$ from RF. In the case $s = 1$ the result has a particularly simple structure, given by

$$\left(L_\mu^A F\right)(x) = \int_{Z^r}\left(R_0F\right)(\sigma,t)K_\mu^A(\sigma,tx)d(\sigma,t) \tag{3.4}$$

for $x \in B^r$, where the kernel is given by

$$K_\mu^A(\sigma,\tau) := \sum_{\nu=0}^{\mu}\lambda_\nu a_{\mu,\nu}C_\nu^{\frac{r}{2}}(\sigma)C_\nu^{\frac{r}{2}}(\tau) \tag{3.5}$$

for $(\sigma,\tau) \in [-1,1]^2$, with the constants

$$\lambda_\nu = \frac{2\nu + r}{\omega_{r-1}^2}.$$

So far, our problem is solved. In particular, for $A = A^{(r+s+1)}$, the $L_\mu^A F$ possess the most desirable properties discussed at the end of Section 2. Numerical images processed by Maier, [4], confirm the high quality of the operators presented.

4 Complexity and Stability

The evaluation of (3.4) by means of a product Gauß quadrature requires $\mathcal{O}(\mu^r)$ evaluations of R_0F. The costs can be reduced considerably by omitting all hyperplanes $\{\sigma t + u | u \perp t\}$ which do not intersect a given neighbourhood of x. In this case the costs are at most of the order caused by the evaluation of the inversion formula of Lorentz and Radon.

In the Newman–Shapiro case, the approximation order is best possible. But in spite of this, a stability problem exists, if the evaluation of R_0F or RF, respectively, is not performed exactly. Suppose, for example, that RF is replaced in practice by some approximation G which can be measured in the supremum norm $\|G\|_\infty$ or in the weighted 2–norm defined by

$$\|G\|_2^2 = \gamma_r \int_{Z^r} [G(\sigma,t)]^2 (1-\sigma^2)^{\frac{r-1}{2}} d(\sigma,t),$$

where γ_r is chosen such that $\|1\|_2 = 1$ is valid. Then we evaluate

$$\left(M_\mu^A G\right)(x) := \Omega_{r-1} \int_{Z^r} G(\sigma,t) K_\mu^A(\sigma, tx)(1-\sigma^2)^{\frac{r-1}{2}} d(\sigma,t),$$

instead of

$$\left(L_\mu^A F\right)(x) = \left(M_\mu^A RF\right)(x).$$

Here we used the definition of R together with

$$\left(R_0 1\right)(\sigma,t) = \Omega_{r-1}(1-\sigma^2)^{\frac{r-1}{2}}.$$

For $p \in \{2,\infty\}$, the error can be estimated in the form

$$\|M_\mu^A G - L_\mu^A F\|_\infty \le \|M_\mu^A\|_{p,\infty} \cdot \|G - RF\|_p, \tag{4.1}$$

where

$$\|M_\mu^A\|_{\infty,\infty} \le \|M_\mu^A\|_{2,\infty}$$

holds in view of the choice of γ_r. Under rather general assumptions on the summation matrix A, we can prove the estimate

$$\|M_\mu^A\|_{2,\infty} \le \sqrt{\frac{2}{r(r+1)}} \cdot \frac{(\mu+r+1)^r}{(r-2)!}. \tag{4.2}$$

It is valid, in particular, for $A = A^{(r+2)}$, i.e. in the Newman–Shapiro case where the index is $s = 1$. Note that the bound of (4.2) is growing as $\mu \to \infty$ at the same rate as $\dim \mathbb{P}_\mu^r$, while the absolute constant decreases rapidly with r. For details we refer to our monograph [9], which is to appear.

References

[1] P. Appell, J. Kampé de Fériet: *Fonctions Hypergéometriques et Hypersphériques. Polynomes d'Hermite,* Gauthier–Villars, Paris, 1926.

[2] M. E. Davison, F. A. Grünbaum: *Tomographic reconstruction with arbitrary directions,* Comm. Pure Appl. Math. **34** (1983), 428–448.

[3] E. Kogbetliantz: *Recherches sur la sommabilité de séries ultrasphériques par la méthode de moyennes arithmétiques,* J. Math. Pure Appl. Ser. 9, **3** (1924), 107–187.

[4] U. Maier: *Tomographic reconstruction using Cesàro means and Newman–Shapiro operators.* In: *Algorithms for Approximation IV,* J. Levesley, I. J. Anderson, J. C. Mason (eds.), 478–485, University of Huddersfield 2002.

[5] D. J. Newman, H. S. Shapiro: *Jackson's theorem in higher dimensions.* In: *Über Approximationstheorie*, 208–219, Birkhäuser, Basel 1964.

[6] M. Reimer: *Discretized Newman–Shapiro operators and Jackson's inequality on the sphere,* Result. Math. **36** (1999), 331–341.

[7] M. Reimer: *Hyperinterpolation on the sphere at the minimal projection order,* J. Approx. Theory **104** (2000), 272–286.

[8] M. Reimer: *Generalized hyperinterpolation on the sphere and the Newman–Shapiro operators,* Constr. Approx. **18** (2002), 183–204.

[9] M. Reimer: *Multivariate Polynomials in Approximation,* to appear in ISNM, Birkhäuser, Basel.

[10] I. H. Sloan, R. S. Womersley: *Constructive polynomial approximation on the sphere,* J. Approx. Theory **103** (2000), 91–98.

Address:

Manfred Reimer
Fachbereich Mathematik
Universität Dortmund
D–44221 Dortmund
Germany

International Series of Numerical Mathematics
Vol. 145,

On Some Multivariate Quadratic Spline Quasi–Interpolants on Bounded Domains

Paul Sablonnière

Dedicated to Manfred Reimer on the occasion of his 70th birthday

Abstract

We study some C^1 quadratic (or d–quadratic) spline quasi–interpolants on bounded domains $\Omega \subset \mathbb{R}^d, d = 1, 2, 3$. These operators are of the form $Qf(x) = \sum_{k \in K(\Omega)} \mu_k(f) B_k(x)$, where $K(\Omega)$ is the set of indices of B–splines B_k whose support is included in the domain Ω and $\mu_k(f)$ is a discrete linear functional based on values of f in a neighbourhood of $x_k \in supp(B_k)$. The data points x_j are vertices of a uniform or nonuniform partition of the domain Ω where the function f is to be approximated. Beyond the simplicity of their evaluation, these operators are uniformly bounded independently of the given partition and they provide the best approximation order to smooth functions.

Introduction and Notations

The aim of this paper is to study some C^1 quadratic (or d–quadratic) spline *discrete quasi–interpolant* (dQIs) on bounded domains $\Omega \subset \mathbb{R}^d, d = 1, 2, 3$. These operators are of the form $Qf(x) = \sum_{k \in K(\Omega)} \mu_k(f) B_k(x)$, where $K(\Omega)$ is the set of indices of B–splines B_k whose support is included in the domain Ω and $\mu_k(f)$ is a *discrete linear functional* $\sum_{i \in I(r)} \lambda_k(i) f(x_{i+k})$, with $I(r) = \mathbb{Z}^d \cap [-r, r]^d$ for $r \in \mathbb{N}$ fixed (and small). The data points x_j are vertices of a uniform or nonuniform partition of the domain Ω where the function f is to be approximated.

Such operators have been widely studied in recent years (see, e.g., [4], [6]–[11], [13], [21], [27]–[31], [38]), but in general, except in the univariate or multivariate tensor–product cases, they are defined on the *whole space* $\mathbb{R}^d$: here we restrict our study to *bounded domains* and to C^1 *quadratic spline* dQIs. Their main interest lies in the fact that they provide approximants having the best approximation order and small norms while being easy to compute. They are particularly useful as initial approximants at the first step of a multiresolution analysis.

First, we study univariate dQIs on uniform and non–uniform meshes of a bounded interval of the real line (Section 1) or on bounded rectangles of the plane with a uniform or non–uniform criss–cross triangulation (Section 2). We use quadratic B–splines whose Bernstein–Bézier (abbr. BB)–coefficients are given in technical reports [32] and which extend previous results given in [12]. In Section 3, we study a bivariate blending sum of two univariate dQIs of Section 1 on a rectangular domain. Finally, in Section 4, we study a trivariate blending sum of a univariate dQI (Section 1) and of the bivariate dQI described in Section 2. For blending and tensor product operators, see, e.g., [2], [3], [14], [17]–[19], [26].

For all these operators, we prove that the sup–norms are *bounded independently of the given partition* of the domain. Using a classical result of approximation theory: $\|f - Sf\| \leq (1 + \|S\|)d(f, \mathcal{S}_2)$ (see, e.g., [15], chapter 5), and the fact that the dQI S is exact on the space $\Pi_2 \in \mathcal{S}_2$ of quadratic polynomials, we obtain $f - Sf = O(h^3)$ for f smooth enough, where h is the maximum of diameters of the elements (segments, triangles, rectangles, prisms) of the partition of the domain. A detailed study of approximation error estimates will be published elsewhere.

1 Quadratic Spline dQIs on a Bounded Interval

Let $X = \{x_0, x_1, \ldots, x_n\}$ be a partition of a bounded interval $I = [a, b]$, with $x_0 = a$ and $x_n = b$. For $1 \leq i \leq n$, let $h_i = x_i - x_{i-1}$ be the length of the subinterval $I_i = [x_{i-1}, x_i]$. Let $\mathcal{S}_2(X)$ be the $n+2$–dimensional space of C^1 quadratic splines on this partition. A basis of this space is formed by quadratic B–splines $\{B_i, 0 \leq i \leq n+1\}$. Define the set of evaluation points

$$\Theta_n = \{\theta_0 = x_0,\ \ \theta_i = \frac{1}{2}(x_{i-1} + x_i),\ \ for\ \ 1 \leq i \leq n,\ \ \theta_{n+1} = x_n\}$$

The simplest dQI associated with Θ_n is the *Schoenberg–Marsden* operator (see, e.g., [22], [35]):

$$S_1 f := \sum_{i=0}^{n+1} f(\theta_i) B_i$$

This operator is exact on Π_1. Moreover $S_1 e_2 = e_2 + \sum_{i=1}^{n} \frac{1}{4} h_i^2 B_i$.

Therefore, it is easy to verify that the *differential quasi-interpolant*:

$$\tilde{S}_2 f = f(x_0)B_0 + \sum_{i=1}^{n}[f(\theta_i) - \frac{1}{8}h_i^2 f''(\theta_i)]B_i + f(x_n)B_{n+1}$$

is exact on Π_2. Then, we can replace the second derivatives by second order divided differences based on *three consecutive points* $\{\theta_{i-1}, \theta_i, \theta_{i+1}\}$ andwe obtain (see also [1], [20]) the unique dQI of type

$$S_2 f = f(x_0)B_0 + \sum_{i=1}^{n}\mu_i(f)B_i + f(x_n)B_{n+1}$$

whose coefficient functionals are of the form

$$\mu_i(f) = a_i f(\theta_{i-1}) + b_i f(\theta_i) + c_i f(\theta_{i+1}), \quad 1 \le i \le n$$

and which is exact on the space Π_2 of quadratic polynomials. Using the B-spline expansions of monomials $e_k(x) = x^k$ for $k = 0, 1, 2$ (see, e.g., [22], [35])

$$e_0 = \sum B_i, \quad e_1 = \sum \theta_i B_i, \quad e_2 = \sum \theta_i^{(2)} B_i,$$

where $\theta_0^{(2)} = x_0^2$, $\theta_i^{(2)} = x_{i-1}x_i$, and $\theta_{n+1}^{(2)} = x_n^2$, writing the equations

$$S_2 e_k = e_k \ \ for \ \ 0 \le k \le 2,$$

and using the following notations, with the convention $h_0 = h_{n+1} = 0$

$$\sigma_i = \frac{h_i}{h_{i-1} + h_i}, \quad \sigma_i' = \frac{h_{i-1}}{h_{i-1} + h_i} = 1 - \sigma_i,$$

we finally obtain, for $1 \le i \le n$:

$$a_i = -\frac{\sigma_i^2 \sigma_{i+1}'}{\sigma_i + \sigma_{i+1}'}, \quad b_i = 1 + \sigma_i \sigma_{i+1}', \quad c_i = -\frac{\sigma_i (\sigma_{i+1}')^2}{\sigma_i + \sigma_{i+1}'}.$$

Defining the basic (or fundamental) functions of S_2 by

$$\begin{aligned}
\tilde{B}_0 &= B_0 + a_1 B_1, \\
\tilde{B}_i &= c_{i-1}B_{i-1} + b_i B_i + a_{i+1}B_{i+1}, \ 1 \le i \le n, \\
\tilde{B}_{n+1} &= c_n B_n + B_{n+1},
\end{aligned}$$

we can express S_2f in the following form

$$S_2f = \sum_{i=0}^{n+1} f(\theta_i)\tilde{B}_i.$$

In [23], Marsden proved the existence of a unique *Lagrange interpolant* Lf in $\mathcal{S}_2(X)$ satisfying $Lf(\theta_i) = f(\theta_i)$ for $0 \le i \le n+1$. He also proved the following

Theorem 1.1. *For f bounded on I and for any partition X of I, the Chebyshev norm of the Lagrange operator L is uniformly bounded by 2.*

Here is a similar result for the dQI S_2 defined above:

Theorem 1.2. *For f bounded on I and for any partition X of I, the Chebyshev norm of the dQI S_2 is uniformly bounded by 3.*

Proof. From the expression of S_2, it is clear that

$$\|S_2\|_\infty \le max\{|a_i| + |b_i| + |c_i|, 1 \le i \le n\}$$

Now, from the following inequalities, which are valid for any partition X of I

$$2\sigma_i^2\sigma'_{i+1} \le \sigma_i + \sigma'_{i+1}, \;\; 2\sigma_i(\sigma'_{i+1})^2 \le \sigma_i + \sigma'_{i+1},$$

we deduce that

$$|a_i| \le \frac{1}{2}, \;\; |b_i| \le 2, \;\; |c_i| \le \frac{1}{2}$$

for $1 \le i \le n$, whence the desired result. □

Remark 1. This result can be slightly improved by using a convenient upper bound of the Lebesgue function $\Lambda_2 = \sum_{i=0}^{n+1} |\tilde{B}_i|$ of S_2. Each function $|\tilde{B}_i|$ being bounded above by the continuous quadratic spline $\overline{B}_i$ whose BB–coefficients are absolute values of those of $\tilde{B}_i$, we obtain $\Lambda_2 \le \overline{\Lambda}_2 = \sum_{i=0}^{n+1} \overline{B}_i$, and it can be proved that $\|S_2\|_\infty \le \max \overline{\Lambda}_2 \le 2.5$.

Remark 2. For uniform partitions, we can compute a smaller bound of the norm by using the Lebesgue function and we obtain $\|S_2\|_\infty \le 1.66$.

Remark 3. In [1], we also studied another dQI defined on an infinite partition $X = \{x_i, i \in \mathbb{Z}\}$ of the real line by

$$S_3f = \sum_{i \in \mathbb{Z}} \nu_i(f)B_i,$$

where the discrete coefficient functionals ν_i are based on subsets of *five consecutive points* $\Theta_i = \{\theta_j = \frac{1}{2}(x_{j-1} + x_j),\ \ i-2 \le j \le i+2\}$:

$$\nu_i(f) = a_i f(\theta_{i-2}) + b_i f(\theta_{i-1}) + c_i f(\theta_i) + d_i f(\theta_{i+1}) + e_i f(\theta_{i+2})$$

and we proved that the set of coefficients:

$$a_i = -\frac{h_i^2}{\eta_{i-1}(\eta_{i-1} + \eta_{i+1})},\quad c_i = 1 + \frac{h_i^2}{\eta_{i-1}\eta_{i+1}},\quad e_i = -\frac{h_i^2}{\eta_{i+1}(\eta_{i-1} + \eta_{i+1})}$$

with $\eta_i = h_{i-1} + 2h_i + h_{i+1}$ and $b_i = d_i = 0$, is *almost optimal* in the sense that it minimizes the sum

$$Z_i = |a_i| + |b_i| + |c_i| + |d_i| + |e_i|$$

under the three constraints insuring that S_3 is exact on Π_2:

$$a_i + b_i + c_i + d_i + e_i = 1,\quad \theta_{i-2}a_i + \theta_{i-1}b_i + \theta_i c_i + \theta_{i+1}d_i + \theta_{i+2}e_i = \theta_i,$$

$$\theta_{i-2}^2 a_i + \theta_{i-1}^2 b_i + \theta_i^2 c_i + \theta_{i+1}^2 d_i + \theta_{i+2}^2 e_i = x_{i-1}x_i$$

This result is only true under some additional constraints on the partition. Assuming they are satisfied, it is easy to see that for all indices i we have

$$Z_i = 1 + \frac{2h_i^2}{\eta_{i-1}\eta_{i+1}} \le 3$$

therefore the sup–norm of S_3 is also uniformly bounded

$$\|S_3\|_\infty \le 3.$$

Details and further results on other types of dQIs will be given in [20].

2 Quadratic Spline dQIs on a Bounded Rectangle

In this section, we study some C^1 quadratic spline dQIs on a nonuniform criss–cross triangulation of a rectangular domain. More specifically, let $\Omega = [a_1, b_1] \times [a_2, b_2]$ be a rectangle decomposed into mn subrectangles by the two partitions

$$X_m = \{x_i,\ \ 0 \le i \le m\},\ \ Y_n = \{y_j,\ \ 0 \le j \le n\}$$

respectively of the segments $I = [a_1, b_1] = [x_0, x_m]$ and $J = [a_2, b_2] = [y_0, y_n]$.

For $1 \le i \le m$ and $1 \le j \le n$, we set $h_i = x_i - x_{i-1}$, $k_j = y_j - y_{j-1}$, $I_i = [x_{i-1}, x_i]$, $J_j = [y_{j-1}, y_j]$, $s_i = \frac{1}{2}(x_{i-1} + x_i)$ and $t_j = \frac{1}{2}(y_{j-1} + y_j)$. Moreover $s_0 = x_0, s_{m+1} = x_m, t_0 = y_0, t_{n+1} = y_n$.

In this section and the next one, we use the following notations:

$$\sigma_i = \frac{h_i}{h_{i-1} + h_i}, \quad \sigma_i' = \frac{h_{i-1}}{h_{i-1} + h_i} = 1 - \sigma_i,$$

$$\tau_j = \frac{k_j}{k_{j-1} + k_j}, \quad \tau_j' = \frac{k_{j-1}}{k_{j-1} + k_j} = 1 - \tau_j,$$

for $1 \le i \le m$ and $1 \le j \le n$, with the convention $h_0 = h_{m+1} = k_0 = k_{n+1} = 0$.

$$a_i = -\frac{\sigma_i^2 \sigma_{i+1}'}{\sigma_i + \sigma_{i+1}'}, \quad b_i = 1 + \sigma_i \sigma_{i+1}', \quad c_i = -\frac{\sigma_i (\sigma_{i+1}')^2}{\sigma_i + \sigma_{i+1}'},$$

$$\bar{a}_j = \frac{\tau_j^2 \tau_{j+1}'}{\tau_j + \tau_{j+1}'}, \quad \bar{b}_j = 1 + \tau_j \tau_{j+1}', \quad \bar{c}_j = -\frac{\tau_j (\tau_{j+1}')^2}{\tau_j + \tau_{j+1}'}.$$

for $0 \le i \le m+1$ and $0 \le j \le n+1$.

Let $\mathcal{K}_{mn} = \{(i,j) : 0 \le i \le m+1, \ \ 0 \le j \le n+1\}$, then the data sites are the mn intersection points of diagonals in subrectangles $\Omega_{ij} = I_i \times J_j$, the $2(m+n)$ midpoints of the four edges and the four vertices of Ω, i.e. the $(m+2)(n+2)$ points of the following set

$$\mathcal{D}_{mn} := \{M_{ij} = (s_i, t_j), \ \ (i,j) \in \mathcal{K}_{mn}\}.$$

As in Section 2, the simplest dQI is the bivariate Schoenberg–Marsden operator:

$$S_1 f = \sum_{(i,j) \in \mathcal{K}_{mn}} f(M_{ij}) B_{ij}$$

where

$$\mathcal{B}_{mn} := \{B_{ij}, 0 \le i \le m+1, \ \ 0 \le j \le n+1\}$$

is the collection of $(m+2)(n+2)$ B–splines (or generalized box–splines) generating the space $\mathcal{S}_2(\mathcal{T}_{mn})$ of all C^1 piecewise quadratic functions on the criss–cross triangulation $\mathcal{T}_{mn}$ associated with the partition $X_m \times Y_n$ of the domain Ω (see, e.g., [12], [13]). There are mn *inner B–splines* associated with the set of indices

$$\hat{\mathcal{K}}_{mn} = \{(i,j), 1 \le i \le m, 1 \le j \le n\}$$

whose restrictions to the boundary Γ of Ω are equal to zero. We add $2m+2n+4$ *boundary B–splines* whose restrictions to Γ are univariate quadratic B–splines. Their set of indices is

$$\tilde{\mathcal{K}}_{mn} := \{(i,0),(i,n+1), 0 \le i \le m+1;\ (0,j),(m+1,j),\ \ 0 \le j \le n+1\}$$

The BB–coefficients of inner B–splines whose indices are in $\{(i,j), 2 \le i \le m-1, 2 \le j \le n-1\}$ are given in [27] and [28]. The other ones can be found in the technical report [32]. The B–splines are positive and form a partition of unity (blending system). The boundary B–splines are *linearly independent* as the univariate ones. But the inner B–splines are *linearly dependent*, the dependence relationship being:

$$\sum_{(i,j)\in\hat{\mathcal{K}}_{mn}} (-1)^{i+j} h_i k_j B_{ij} = 0$$

It is well known that S_1 is exact on bilinear polynomials, i.e.,

$$S_1 e_{rs} = e_{rs} \ \ for\ \ 0 \le r,s \le 1$$

Moreover, one has

$$S_1 e_{20} = e_{20} + \frac{1}{4} \sum_{(i,j)\in\mathcal{K}_{mn}} h_i^2 B_{ij}, \ \ S_1 e_{02} = e_{02} + \frac{1}{4} \sum_{(i,j)\in\mathcal{K}_{mn}} k_j^2 B_{ij}$$

Therefore it is easy to verify that the following *Differential Quasi–Interpolant* (abbr. DQI)

$$\tilde{S}_2 f = \sum_{(i,j)\in\mathcal{K}_{mn}} \nu_{ij}(f) B_{ij}$$

where

$$\nu_{ij}(f) = f(M_{ij}) - \frac{1}{8}[h_i^2 \partial_1^2 f(M_{ij}) + k_j^2 \partial_2^2 f(M_{ij})]$$

is exact on Π_2. In the uniform case, we find again

$$\nu_{ij}(f) = f(M_{ij}) - \frac{h^2}{8}[\partial_1^2 f(M_{ij}) + \partial_2^2 f(M_{ij})] = f(M_{ij}) - \frac{h^2}{8}\Delta f(M_{ij})$$

whose study is detailed in Chapter 6 of [27].

Substituting divided differences based on M_{ij} and its four neighbouring points for partial derivatives, we obtain the following dQI, which is also exact on Π_2:

$$S_2 f = \sum_{(i,j)\in\hat{\mathcal{K}}_{mn}} \mu_{ij}(f) B_{ij}$$

where the coefficient functionals are given by

$$\mu_{ij}(f) = (b_i + \bar{b}_j - 1)f(M_{ij}) + a_i f(M_{i-1,j}) + c_i f(M_{i+1,j}) \\ + \bar{a}_j f(M_{i,j-1}) + \bar{c}_j f(M_{i,j+1}).$$

Theorem 2.1. *The sup–norm of S_2 is uniformly bounded independently of the partition $\mathcal{T}_{mn}$ of the domain:*

$$\|S_2\|_\infty \leq 5.$$

Proof. The B–splines forming a partition of unity, one has, for $\|f\|_\infty \leq 1$

$$\|S_2\|_\infty \leq \max_{(i,j)\in\mathcal{K}_{mn}} \{|a_i| + |c_i| + |\bar{a}_j| + |\bar{c}_j| + |b_i + \bar{b}_j - 1|\}.$$

As in the proof of Theorem 1.2, we obtain

$$|a_i|,\ |c_i|,\ |\bar{a}_j|,\ |\bar{c}_j| \ \leq \frac{1}{2}.$$

Moreover,

$$|b_i + \bar{b}_j - 1| = 1 + \sigma_i \sigma'_{i+1} + \tau_j \tau'_{j+1} \leq 3,$$

whence the result. □

Remark. For a uniform partition, we obtain $\|S_2\|_\infty \leq 2.4$, but this bound is not optimal and can still be slightly reduced.

3 A Biquadratic Blending Sum of Univariate dQIs

In this section, we study a biquadratic dQI on a rectangular domain $\Omega = [a_1, b_1] \times [a_2, b_2]$ which is a blending sum of bivariate extensions of quadratic spline dQIs of Section 1. We use the same notations as in Section 1 for the domain Ω, the partitions of $I = [a_1, b_1]$, $J = [a_2, b_2]$ and data sites. The partition considered on Ω is the tensor product of partitions of I and J.

We use the two sets of univariate B–splines

$$\{B_i(x), 0 \leq i \leq m+1\},\ \ \{B_j(y), 0 \leq j \leq n+1\}.$$

and the two sets of univariate fundamental functions introduced in Section 2:

$$\{\tilde{B}_i(x), 0 \leq i \leq m+1\},\ \ \{\tilde{B}_j(y), 0 \leq j \leq n+1\}$$

The associated *extended bivariate* dQIs are, respectively, (see, e.g., [14] for bivariate extensions of univariate operators)

$$P_1 f(x,y) := \sum_{i=0}^{m+1} f(s_i,y) B_i(x), \quad P_2 f(x,y) := \sum_{i=0}^{m+1} f(s_i,y) \tilde{B}_i(x),$$

$$Q_1 f(x,y) := \sum_{j=0}^{n+1} f(x,t_j) B_j(y), \quad Q_2 f(x,y) := \sum_{j=0}^{n+1} f(x,t_j) \tilde{B}_j(y).$$

We also recall the expressions of the operators P_2 and Q_2 in terms of B–splines:

$$P_2 f(x,y) := \sum_{i=0}^{m+1} \{a_i f(s_{i-1},y) + b_i f(s_i,y) + c_i f(s_{i+1},y)\} B_i(x) B_j(y),$$

$$Q_2 f(x,y) := \sum_{j=0}^{n+1} \{\bar{a}_j f(x,t_{j-1}) + \bar{b}_j f(x,t_j) + \bar{c}_j f(x,t_{j+1})\} B_i(x) B_j(y),$$

where the coefficients are the same as those defined in Section 1 above.

The bivariate dQI considered in this section is now defined as the blending sum

$$R := P_1 Q_2 + P_2 Q_1 - P_1 Q_1,$$

and it can be written in the following form

$$Rf(x,y) = \sum_{(i,j)\in\mathcal{K}_{mn}} f(M_{ij}) \overline{B}_{ij}(x,y)$$

where the biquadratic fundamental functions are defined by

$$\overline{B}_{ij}(x,y) := B_i(x)\tilde{B}_j(y) + \tilde{B}_i(x) B_j(y) - B_i(x) B_j(y)$$

In terms of tensor–product B–splines $B_{ij}(x,y) = B_i(x)B_j(y)$, we have:

$$Rf(x,y) = \sum_{(i,j)\in\mathcal{K}_{mn}} \mu_{ij}(f) B_{ij}(x,y),$$

where the coefficient functionals are given by

$$\begin{aligned} \mu_{ij}(f) \; &:= \; a_i f(M_{i-1,j}) + c_i f(M_{i+1,j}) \\ &\quad + \bar{a}_j f(M_{i,j-1}) + \bar{c}_j f(M_{i,j+1}) + (b_i + \bar{b}_j - 1) f(M_{ij}). \end{aligned}$$

Theorem 3.1. *The operator R is exact on the 8-dimensional subspace $(\Pi_1[x] \otimes \Pi_2[y]) \oplus (\Pi_2[x] \otimes \Pi_1[y])$ of biquadratic polynomials generated by monomials $\{1, x, y, x^2, xy, y^2, x^2y, xy^2\}$. Moreover, its sup-norm is bounded from above independently of the nonuniform partition $X_m \otimes Y_n$ of the domain Ω,*

$$\|R\|_\infty \leq 5.$$

Proof. For any polynomial $r(x,y) = p(x)q(y) \in \Pi_{22}$, we have for $k = 1, 2$, respectively,

$$P_k r(x,y) = q(y) P_k p(x), \quad Q_k r(x,y) = p(x) Q_k q(y).$$

Therefore

$$Rr(x,y) = P_1 p(x) Q_2 q(y) + P_2 p(x) Q_1 q(y) - P_1 p(x) Q_1 q(y).$$

As $P_k p = p$ (resp. $Q_k q = q$) for $p \in \Pi_k$ (resp. for $q \in \Pi_k$), we obtain $Rr = r$ for all monomials in Π_{22}, except x^2y^2, whence the result.

Now, for any function f uniformly bounded above by 1 in Ω, we have

$$|Rf(x,y)| \leq \max_{(i,j) \in \mathcal{K}_{mn}} \{|a_i| + |c_i| + |\overline{a}_j| + |\overline{c}_j| + |b_i + \overline{b}_j - 1|\},$$

and the result follows as in the proof of Theorem 2.1. □

4 A Trivariate Blending Sum of Univariate and Bivariate Quadratic dQIs

In this section, we study a trivariate dQI on a parallelepiped $\Omega = [a_1, b_1] \times [a_2, b_2] \times [a_3, b_3]$ which is a blending sum of trivariate extensions of univariate and bivariate dQIs as seen in Sections 2 and 3. We consider the three partitions

$$X_m := \{x_i, \ 0 \leq i \leq m\}, \ Y_n = \{y_j, \ 0 \leq j \leq n\}, \ Z_p := \{z_k, \ 0 \leq k \leq p\}$$

of the segments $I = [a_1, b_1] = [x_0, x_m]$, $J = [a_2, b_2] = [y_0, y_n]$ and $K = [a_3, b_3] = [z_0, z_p]$, respectively. For the projection $\Omega' = [a_1, b_1] \times [a_2, b_2]$ of Ω on the $xy-plane$, the notations are those of Section 2. For the projection $\Omega'' = [a_3, b_3]$ of Ω on the $z - axis$, we use the following notations, for $1 \leq k \leq p$:

$$l_k = z_k - z_{k-1}, \ K_k = [z_{k-1}, z_k], \ u_k = \frac{1}{2}(z_{k-1} + z_k),$$

with $u_0 = z_0$ and $u_{p+1} = z_p$. For mesh ratios of subintervals, we set

$$\omega_k = \frac{l_k}{l_{k-1} + l_k}, \ \omega_k' = \frac{l_{k-1}}{l_{k-1} + l_k} = 1 - \omega_k$$

for $1 \le k \le p$, with $l_0 = l_{p+1} = 0$ (all these ratios lie between 0 and 1).

Let $\mathcal{K} = \mathcal{K}_{mnp} = \{(i,j,k),\ \ 0 \le i \le m+1,\ 0 \le j \le n+1,\ 0 \le k \le p+1\}$, then the set of data sites is

$$\mathcal{D} = \mathcal{D}_{mnp} = \{N_{ijk} = (x_i, y_j, z_k), (i,j,k) \in \mathcal{K}_{mnp}\}.$$

The partition of Ω considered here is the tensor product of partitions on Ω' and Ω'', i.e., a partition into *vertical prisms with triangular horizontal sections.*

Setting $\mathcal{K}'_{mn} = \{(i,j), 0 \le i \le m+1,\ 0 \le j \le n+1\}$, we consider the bivariate B–splines and fundamental splines on $\Omega' = [a_1, b_1] \times [a_2, b_2]$ defined in Section 2 above:

$$\{B_{ij}(x,y),\ \ (i,j) \in \mathcal{K}'_{mn}\}, \textit{and}\ \ \{\tilde{B}_{ij}(x,y),\ \ (i,j) \in \mathcal{K}'_{mn}\}$$

and the univariate B–splines and fundamental splines on $[a_3, b_3]$ defined in Section 1:

$$\{B_k(z),\ \ 0 \le k \le p+1\}\ \ \textit{and}\ \ \{\tilde{B}_k(z),\ \ 0 \le k \le p+1\}.$$

The extended trivariate dQIs that we need for the construction are the following

$$P_1 f(x,y,z) := \sum_{(i,j)\in\mathcal{K}'_{mn}} f(s_i, t_j, z) B_{ij}(x,y),$$

$$P_2 f(x,y,z) := \sum_{(i,j)\in\mathcal{K}'_{mn}} f(s_i, t_j, z) \tilde{B}_{ij}(x,y),$$

$$Q_1 f(x,y,z) := \sum_{k=0}^{p+1} f(x,y,u_k) B_k(z), \quad Q_2 f(x,y,z) := \sum_{k=0}^{p+1} f(x,y,u_k) \tilde{B}_k(z).$$

For the sake of clarity, we give the expressions of P_2 and Q_2 in terms of B–splines:

$$P_2 f(x,y,z) = \sum_{(i,j)\in\mathcal{K}'_{mn}} \mu_{ij}(f) B_{ij}(x,y),$$

where

$$\mu_{ij}(f) = a_i f(s_{i-1}, t_j, z) + c_i f(s_{i+1}, t_j, z)) + \overline{a}_j f(s_i, t_{j-1}, z) + \overline{c}_j f(s_i, t_{j+1}, z)$$
$$+ (b_i + \overline{b}_j - 1) f(s_i, t_j, z);$$

$$Q_2 f(x,y,z) := \sum_{k=0}^{p+1} \{\hat{a}_k f(x,y,u_{k-1}) + \hat{b}_k f(x,y,u_k) + \hat{c}_k f(x,y,u_{k+1})\} B_k(z),$$

$$\hat{a}_k = -\frac{\omega_k^2 \omega'_{k+1}}{\omega_k + \omega'_{k+1}}, \quad \hat{b}_k = 1 + \omega_k \omega'_{k+1}, \quad \hat{c}_k = -\frac{\omega_k (\omega'_{k+1})^2}{\omega_k + \omega'_{k+1}}\ .$$

We now define the trivariate blending sum

$$R = P_1 Q_2 + P_2 Q_1 - P_1 Q_1 \,.$$

Setting

$$\overline{B}_{ijk}(x,y,z) = B_{ij}(x,y)\tilde{B}_k(z) + \tilde{B}_{ij}(x,y)B_k(z) - B_{ij}(x,y)B_k(z)$$

we obtain

$$Rf = \sum_{(i,j,k)\in\mathcal{K}_{mnp}} f(N_{ijk})\overline{B}_{ijk}.$$

In terms of tensor product B–splines $B_{ijk} = B_{ij}B_k$, one has

$$Rf = \sum_{(i,j,k)\in\mathcal{K}_{mnp}} \nu_{ijk}(f)B_{ijk}$$

where $\nu_{ijk}(f)$ is based on the 7 neighbours of N_{ijk} in $\mathbb{R}^3$:

$$\nu_{ijk}(f) = \hat{a}_k f(N_{i,j,k-1}) + \hat{c}_k f(N_{i,j,k+1}) + a_i f(N_{i-1,j,k}) + c_i f(N_{i+1,j,k})$$
$$+ \overline{a}_j f(N_{i,j-1,k}) + \overline{c}_j f(N_{i,j+1,k}) + (b_i + \overline{b}_j + \hat{c}_k - 1) f(N_{ijk}).$$

Theorem 4.1. *The operator R is exact on the 15–dimensional subspace $(\Pi_1[x,y]\otimes\Pi_2[z])\oplus(\Pi_2[x,y]\otimes\Pi_1[z])$ of $\Pi_2[x,y]\otimes\Pi_2[z]$, generated by monomials $\{1,x,y,z,x^2,y^2,z^2,xy,xz,yz,x^2z,xyz,y^2z,xz^2,yz^2\}$. Moreover, its sup–norm is bounded from above independently of the nonuniform partition of the domain Ω*

$$\|R\|_\infty \le 8.$$

Proof. The proof is similar to those of previous Theorems 1.2, 2.1, 3.1. □

5 Some Applications

Without going into details, we cite a number of possible applications of the preceding results. Let Q be any of the previous operators.

1) *Approximate Integration.*
It is possible to approximate $\int_\Omega f$ by $\int_\Omega Qf$, giving rise to several interesting quadrature formulas in $\mathbb{R}^d$, especially for $d = 2,3$. We plan to develop this point elsewhere.

2) *Approximate Differentiation: Spectral Methods.*
In the same way, one can approximate the first (partial) derivatives of f by those of Qf at the data sites. We thus obtain differentiation matrices which can be also used for second derivatives and for spectral methods (see, e.g., [16], [37]).

3) *Approximation of Zeros of Polynomials.*
We have tested the approximation of the Legendre polynomial $f(x) = P_8(x)$ on the interval $[-1, 1]$ by the dQI $S_2 f$ of Section 1 based on Chebyshev points with $n = 32$. We obtain $\|f - S_2 f\|_\infty \approx 0.0034$ and quite good approximations of the 8 roots of f in the interval. There is practically no difference between the approximation $S_2 f$ and the Marsden interpolant of f which needs the solution of a linear system of $n + 2$ equations.

For bivariate or trivariate functions, the advantage of using dQIs over interpolants is still bigger since one avoids the solution of large linear systems. Moreover, at least in the bivariate case, one can use the nice properties of piecewise quadratic surfaces (see, e.g., the results given by M. J. D. Powell in [25]).

Acknowledgements. The author thanks D. Barrera, M. Pasadas and the Departamento de Matemática Aplicada of the University of Granada (Spain) for their kind invitation which allowed the realization of this paper.

References

[1] D. Barrera, M. J. Ibanẽz, P. Sablonnière: *Near–best quasi-interpolants on uniform and nonuniform partitions in one and two dimensions.* Prépublication IRMAR 02–52, 2002. Submitted for publication in Proceedings of the International Conference on *Curves and Surfaces* (Saint–Malo, July 2002).

[2] G. Baszenski: *n–th order polynomial spline blending*, In: *Multivariate Approximation Theory III*, W. Schempp, K. Zeller (eds.), ISNM, Volume **75**, 35–46, Birkhäuser, Basel 1985.

[3] L. Beutel, H. Gonska: *Simultaneous approximation by tensor product operators*, Technical report SM–DU–504, Universität Duisburg, 2001.

[4] B. D. Bojanov, H. A. Hakopian, A. A. Sahakian: *Spline Functions and Multivariate Interpolation*, Kluwer, Dordrecht 1993.

[5] C. de Boor: *A Practical Guide to Splines*, Springer, New York 2001. (revised edition).

[6] C. de Boor: *Splines as linear combinations of B-splines*, In: *Approximation Theory II*, G.G. Lorentz et al. (eds.), 1–47, Academic Press, New York 1976.

[7] C. de Boor: *Quasiinterpolants and approximation power of multivariate splines*, In: *Computation of Curves and Surfaces*, W. Dahmen, M. Gasca, C. A. Micchelli (eds.), 313–345, Kluwer, Dordrecht 1990.

[8] C. de Boor, K. Höllig, S. Riemenschneider: *Box-splines*, Springer, New York 1993.

[9] C. de Boor, G. Fix: *Spline approximation by quasi-interpolants*, J. Approx. Theory **8** (1973), 19–45.

[10] G. Chen, C. K. Chui, M. J. Lai: *Construction of real-time spline quasi-interpolation schemes*, CAT Report **107**, Texas A&M University, 1986.

[11] C. K. Chui, M. J. Lai: *A multivariate analog of Marsden's identity and a quasi-interpolation scheme*, Constr. Approx. **3** (1987), 111–122.

[12] C. K. Chui, L. L. Schumaker, R. H. Wang: *On spaces of piecewise polynomials with boundary conditions III. Type II triangulations*, In: *Canadian Mathematical Society Conference Proceedings*, Volume **3** (1983), 67–80, American Mathematical Society.

[13] C. K. Chui: *Multivariate Splines*, CBMS–NSF Regional Conference Series in Applied Mathematics, vol. 54, SIAM, Philadelphia 1988.

[14] F. J. Delvos, W. Schempp: *Boolean Methods in Interpolation and Approximation*, Longman Scientific & Technical 1989.

[15] R. A. DeVore, G. G. Lorentz: *Constructive Approximation*, Springer, Berlin 1993.

[16] B. Fornberg: *A Practical Guide to Pseudospectral Methods*, Cambridge University Press 1996.

[17] H. Gonska: *Degree of simultaneous approximation of bivariate functions by Gordon operators*, J. Approx. Theory **62** (1990), 170–191.

[18] H. Gonska: *Simultaneous approximation by generalized n–th order blending operators*, In: ISNM Volume **90**, 173–180, Birkhäuser, Basel 1989.

[19] W. J. Gordon: *Distributive lattices and approximation of multivariate functions*, In: *Approximation with Special Emphasis on Spline Functions*, I. J. Schoenberg (ed.), 223–277, Academic Press, New York 1969.

[20] M. J. Ibañez–Pérez: *Cuasi–interpolantes spline discretos con norma casi minima: teoria y aplicaciones.* Tesis doctoral, Universidad de Granada, 2003 (In preparation).

[21] T. Lyche, L. L. Schumaker: *Local spline approximation*, J. Approx. Theory **15** (1975), 294–325.

[22] J. M. Marsden, I. J. Schoenberg: *An identity for spline functions with applications to variation diminishing spline approximation.* J. Approx. Theory **3** (1970), 7–49.

[23] J. M. Marsden: *Operator norm bounds and error bounds for quadratic spline interpolation*, In: *Approximation Theory*, Banach Center Publications, vol. **4** (1979), 159–175.

[24] G. Nürnberger: *Approximation by Spline Functions*, Springer, Berlin 1989.

[25] M. J. D. Powell: *Piecewise quadratic surface fitting in contour plotting*, In: *Software for Numerical Mathematics*, D. J. Evans (ed.), 253–271. Academic Press, London 1974.

[26] Rong–Qing Jia: *Approximation by multivariate splines: an application of boolean methods*, In: *Numerical Methods of Approximation Theory*, Volume **9**, D. Braess, L. L. Schumaker (eds.), ISNM Volume **105**, 117–134, Birkhäuser, Basel 1992.

[27] P. Sablonnière: *Bases de Bernstein et approximants spline.* Thèse de doctorat, Université de Lille, 1982.

[28] P. Sablonnière: *Bernstein–Bézier methods for the construction of bivariate spline approximants.* Comput. Aided Geom. Design **2** (1985), 29–36.

[29] P. Sablonnière: *Quasi–interpolants associated with H–splines on a three–direction mesh.* J. Comput. & Appl. Math. **66** (1996), 433–442.

[30] P. Sablonnière: *Quasi–interpolantes sobre particiones uniformes*, Meeting in Approximation Theory, Ubeda, Spain, July 2000. Prépublication IRMAR 00–38, 2000.

[31] P. Sablonnière: *H–splines and quasi–interpolants on a three directional mesh.* Prépublication IRMAR 02–11, 2002. To appear in Proceedings IDoMAT 2001.

[32] P. Sablonnière: *BB–coefficients of basic bivariate quadratic splines on uniform and nonuniform criss–cross triangulations of a rectangular domain.* Prépublications IRMAR, 2002 (to appear).

[33] P. Sablonnière, F. Jeeawock–Zedek: *Hermite and Lagrange interpolation by quadratic splines on nonuniform criss–cross triangulations.* In: *Curves and Surfaces*, P. J. Laurent, A. Le Méhauté, L. L. Schumaker (eds.), 445–452, A.K. Peters, Wellesley 1991.

[34] I. J. Schoenberg: *Cardinal Spline Interpolation*, CBMS–NSF Regional Conference Series in Applied Mathematics, vol. 12, SIAM, Philadelphia 1973.

[35] I. J. Schoenberg: *Selected Papers*, Volumes 1 and 2, edited by C. de Boor, Birkhäuser, Boston 1988.

[36] L. L. Schumaker: *Spline Functions: Basic Theory*, John Wiley & Sons, New York 1981.

[37] L. N. Trefethen: *Spectral Methods in Matlab,* SIAM , Philadelphia 2000.

[38] J. Ward: *Polynomial reproducing formulas and the commutator of a locally supported spline*, In: *Topics in multivariate approximation*, C. K. Chui, L. L. Schumaker, F. Utreras (eds.), 255–263, Academic Press, Boston 1987.

Address:

Paul Sablonnière
INSA de Rennes
20, avenue des Buttes de Coësmes, CS 14315
F–35043 Rennes cedex
France

International Series of Numerical Mathematics
Vol. 145, ©2003 Birkhäuser Verlag Basel/Switzerland

How to Generate Smoother Refinable Functions from Given Ones

Tomas Sauer

Dedicated to Prof. M. Reimer on the occasion of his 70th birthday

Abstract

We consider the question of how and to what extent the simple process of smoothing refinable functions by creating an additional zero at -1 of the associated symbol can be transferred to the multivariate case.

1 Introduction

Let $\Gamma = \mathbb{Z}^s$ denote the grid of integer vectors with s components. Recall that a function $\phi : \mathbb{R}^s \to \mathbb{R}$ is called *refinable* if there exists a so-called *mask* a belinging to $\ell_{00}(\mathbb{Z}^s)$, the space of finitely supported sequences over $\mathbb{Z}^s$, such that

$$\phi = \sum_{\alpha\in\Gamma} a(\alpha)\,\phi\,(2\cdot - \alpha)\,. \tag{1.1}$$

To keep notation simple, we only refer to the most basic case of *dyadic scaling* here in the introduction as well as in the univariate case.

Usually, refinable functions are given only *implicitly* as solutions of a *refinement equation* (1.1), so all that is really known about the function ϕ is the mask that determines it via the refinement equation. Therefore, there has been an enormous amount of work explaining how to read properties of the refinable function ϕ off the mask. Clearly, this is only possible under additional assumptions on the mask, since in general not any mask gives rise to a refinable function at all, nor does the solution of the refinement equation or the refinement equation a given function solves have to be unique.

On the other hand, from a constructive point of view, a worthwhile thing might be to obtain masks which do give rise not only to a refinable function, but to a refinable function with additional properties like smoothness. In our context here, smoothness will refer to the existence of derivatives of a certain order. In particular, to use knowledge obtained before, one might ask how to transform a smooth solution of a refinement equation into a smoother one.

The paper is organized as follows. After recalling the univariate case, we introduce the notation and concepts needed for the multivariate case. Theorem 5.1 then points out the relation between a convergent subdivision scheme (and therefore its associated refinable function) and its associated "derivative" scheme. Based on this result, we then show that multivariate smoothing requires additional properties of the derivative. This concept is finally illustrated by considering the simplest example of bivariate refinable functions with dyadic scaling.

2 The Univariate Case

In the univariate case, where $\Gamma = \mathbb{Z}$, there exists a simple process of "smoothing" refinable functions in order to obtain ones of higher regularity, namely, *convolving* the function with the characteristic function χ of the unit interval. Hence, starting out with a refinable function ϕ, one defines

$$\psi = \phi * \chi = \int_0^1 \phi(\cdot - t)\, dt. \tag{2.1}$$

It is well–known that ψ is a refinable function again; indeed, it is easy to see that the mask a_ψ is generated from the mask for a_ϕ by setting

$$a_\psi^*(z) = \frac{z+1}{2}\, a_\phi^*(z), \qquad \text{where} \qquad a^*(z) = \sum_{j\in\Gamma} a(j)\, z^j$$

denotes the *symbol* associated to the mask a. That ψ has indeed one higher order of differentiability than ϕ has follows readily from taking derivative of both sides of (2.1) which gives $\psi' = \phi(\cdot - 1) - \phi$, hence the derivative ψ' has at least the same regularity as ϕ.

It is a nice observation that starting this process "from the scratch", beginning with the simplest refinable function χ, one obtains after k such smoothing steps the *cardinal B–spline* of order k, that is, the B–spline with knots at the integers $0, \ldots, k+1$, $k \in \mathbb{N}$.

The more interesting fact, however, is that the above simple smoothing process generates *all* refinable functions of a given differentiability order provided that the function is stable. Recall that a function $\phi \in L_p(\mathrm{I\!R})$ is *stable* if there exists constants $A, B > 0$ such that

$$A \|c\|_p \leq \|\phi * c\|_p \leq B \|c\|_p, \qquad \phi * c := \sum_{j\in\Gamma} \phi(\cdot - j)\, c(j). \tag{2.2}$$

It should be remarked that if a compactly supported function ϕ belongs to $L_p(\mathrm{I\!R}) \cap L_q(\mathrm{I\!R})$, then it is p–stable if and only if it is q–stable – the norm in the definition (2.2) of stability is not essential. Let us record the smoothing relation for continuously differentiable univariate functions in the following theorem, cf. [1, 4] and the references therein.

Theorem 2.1. *Let $\psi \in C(\mathrm{I\!R})$ be a stable solution of the refinement equation*

$$\psi = \sum_{j\in\Gamma} a_\psi(j)\, \psi(2 \cdot -j). \tag{2.3}$$

Then $\psi \in C^{k+1}(\mathrm{I\!R})$ if and only if there exists a stable function $\phi \in C^k(\mathrm{I\!R})$ satisfying the refinement equation

$$\phi = \sum_{j\in\Gamma} a_\phi(j)\, \phi(2 \cdot -j). \tag{2.4}$$

where

$$a_\psi^*(z) = \frac{z+1}{2}\, a_\phi^*(z). \tag{2.5}$$

The proof of this result, which can be done for example by the methods provided in [7, 8] for the even more general situation of vector valued functions and matrix valued masks, relies on the fact that the "smoothing process" in (2.1) as well its "inversion" preserve stability. That is, starting with a stable function ϕ one obtains a stable function ψ and vice versa.

There is, however, another way of looking at the factorization property (2.5), namely, in terms of *subdivision operators.* Recall that the stationary subdivision operator S_a acts on sequences $c : \Gamma \to \mathrm{I\!R}$ from the sequence space $\ell(\Gamma)$ by setting

$$S_a\, c = \sum_{k\in\Gamma} a(\cdot - 2k)\, c(k).$$

The idea behind the action of these operators is that the sequence $S_a^r\, c$ is considered as a discrete function defined on the refined grid $\Gamma_r := 2^{-r}\Gamma$. With this picture in mind, we can say that the subdivision scheme is L_p*–convergent* provided there exists, for any initial sequence $c \in \ell_p(\Gamma)$, a *limit function* $f_c \in$

$L_p(\mathbb{R})$, $1 \le p < \infty$ or $f_c \in C_u(\mathbb{R})$, $p = \infty$, where C_n denotes the set of uniformly continuous and uniformly bounded functions on $\mathbb{R}$, such that

$$\lim_{r\to\infty} \left\| f_c - (\chi * S_a^r c)\,(2^r\cdot) \right\|_p = 0.$$

Here, the characteristic function can be replaced by any *test function*, cf. [3], and convergence is independent of the actual choice of the test function. To exclude trivialities like $a = 0$, one usually also demands that there exists at least one $c \in \ell_p(\Gamma)$ such that $f_c \neq 0$. Note that the subdivision operator is closely related to the refinement operator T_a defined as

$$T_a f = \sum_{j\in\Gamma} a(j)\, f\,(2\cdot - j)\,;$$

in fact, they are almost (up to the scaling factor 2) *adjoints* for the bilinear operator $\langle f, c\rangle = f * c$, i.e., $\langle T_a f, c\rangle = \langle f, S_a c\rangle\,(2\cdot)$. To express (2.5) in terms of subdivision operators, we need to recall the (backward) difference operator Δ which is defined, for any $c \in \ell(\Gamma)$, as $\Delta c = c(\cdot - 1) - c$. Taking into account that in terms of symbols we have the identities

$$(\Delta c)^*\,(z) = (z-1)\,c^*(z) \qquad \text{and} \qquad (S_a c)^*\,(z) = a^*(z)\,c^*\left(z^2\right),$$

the latter one following from the *slantedness* of the subdivision operator, we obtain the following result.

Theorem 2.2. *Let $a, b \in \ell_{00}(\Gamma)$. Then $a^*(z) = (z+1)\,b^*(z)$ if and only if $\Delta S_a = S_b \Delta$.*

Therefore, the process of smoothing a refinable function can also be considered as "smoothing" a subdivision process by adding an additional "zero at -1" to the symbol of this subdivision scheme. This is the concept whose extension to the multivariate case we are going to give a closer look as smoothing "by convolution" extends more or less trivially.

3 The Multivariate Case – Basic Definitions

We now ask the question how much of the results from the preceding section can be transferred to the multivariate case. To that end, we first introduce some more terminology. A matrix $M \in \mathbb{Z}^{s\times s}$ is called *expanding* or a *dilation matrix* if all its eigenvalues are greater than one in modulus, or, equivalently, if $\|M^{-r}\| \to 0$ as $r \to \infty$. Clearly, M is a dilation matrix if and only if its transpose $\overline{M} := M^T$ is also a dilation matrix. The coarse grid we now consider is $\Gamma = \Gamma_0 := \mathbb{Z}^s$ while the finer grids now take the form $\Gamma_r := M^r\,\Gamma$. The univariate case thus corresponds to the choice $M = 2I$ which is the most widely investigated situation also in the multivariate case. Given a finitely supported sequence $a \in \ell_{00}(\Gamma)$, the associated subdivision operator is now defined as

$$S_a\, c = \sum_{\alpha\in\Gamma} a\,(\cdot - M\alpha)\; c(\alpha), \qquad c \in \ell_p(\Gamma); \tag{3.1}$$

as long as a is finitely supported, it is easily seen that this operator is continuous. Note that so far we only consider the *scalar* case and not yet *vector subdivision operators* that iterate matrix valued masks on vector- or even matrix–valued sequences.

We denote by $W_p\,(\mathbb{R}^s)$ the spaces $L_p\,(\mathbb{R}^s)$ if $1 \le p < \infty$ and the space $C_u\,(\mathbb{R}^s)$ of uniformly bounded uniformly continuous real valued functions on $\mathbb{R}^s$, endowed with their canonical norms. A function $\phi \in W_p\,(\mathbb{R}^s)$ is called *refinable* with respect to a if

$$\phi = T_a\,\phi = \sum_{\alpha\in\Gamma} a(\alpha)\,\phi\,(M\,\cdot - \alpha)\,, \tag{3.2}$$

and it is called *stable* if there are constants $A, B > 0$ such that

$$A\,\|c\|_{\ell_p} \le \|\phi * c\|_{W_p} \le B\,\|c\|_{\ell_p}\,, \qquad \phi * c := \sum_{\alpha\in\Gamma} \phi(\cdot - \alpha)\, c(\alpha). \tag{3.3}$$

Denoting by χ the characteristic function of the unit cube $[0,1]^s$, the subdivision operator is said to converge if for any $c \in \ell_p(\Gamma)$ there exists a function $f_c \in W_p\,(\mathbb{R}^s)$ such that

$$\lim_{r\to\infty} \|f_c - (\chi * S_a^r c)\,(M^r\cdot)\|_p = 0$$

and that $f_c \neq 0$ for at least one $c \in \ell_p(\Gamma)$. It is well–known from [3] that the convergence of the subdivision scheme implies that $f_c = \phi * c$ and that the existence of a *stable* solution of the refinement equation $\phi = T_a\phi$ implies the convergence of the subdivision scheme.

Things become different from the univariate case once we start differencing the subdivision scheme. To that end, we introduce the *partial backwards difference operators* δ_j, $j = 1, \ldots, s$, which are defined as

$$\delta_j\, c = c(\cdot - e_j) - c, \qquad j = 1, \ldots, s, \qquad c \in \ell(\Gamma),$$

and stack them into a discrete (backwards) gradient

$$\Delta = [\delta_j \ : \ j = 1, \dots, s] = \begin{bmatrix} \delta_1 \\ \vdots \\ \delta_s \end{bmatrix}.$$

As has been pointed out in [1, 4, 6, 11, 12], the factorization property still has the innocently looking form $\Delta S_a = S_b \Delta$, but now $b \in \ell_{00}^{s \times s}(\Gamma)$ is a *matrix valued* mask, that is, S_b becomes a vector subdivision scheme and b does not depend on a in such a simple way as in the univariate case, it is not even unique. Note that it *formally* makes no difference in the definition (3.1) of the subdivision operator S_a if the mask consists of scalars or of $N \times L$ matrices as long as the subdivision operator now acts on $\ell_p^{N \times L}(\Gamma)$, the space of $N \times L$ matrix valued sequences, where one usually takes $L = 1$, which is no restriction as it just would mean acting on the otherwise L columns of a matrix separately. Despite of the formal similarity of the subdivision operators, vector subdivision nevertheless behaves very different from scalar subdivision in many cases even in the univariate case. For an extensive treatment of these effects appearing in the context of convergence and smoothness analysis, the reader is referred to [7, 8].

There is an algebraic interpretation of the identity

$$\Delta S_a = S_b \Delta \tag{3.4}$$

which was already mentioned in [1] for $M = 2I$ and which has been explored in [6,10–12]. This interpretation involves the notion of a *quotient ideal*; recall that an *ideal* I in a commutative ring R with unit 1 is a subset of R which is closed under addition and which has the property that $I \cdot R = I$. The two rings which we will consider in the context of this paper will either be the complex coefficient *algebraic polynomials*

$$\Pi = \mathbb{C}[z] = \mathbb{C}[z_1, \dots, z_s]$$

or the *Laurent polynomials*

$$\Lambda = \mathbb{C}\left[z, z^{-1}\right] = \mathbb{C}\left[z_1, \dots, z_s, z_1^{-1}, \dots, z_s^{-1}\right].$$

Recall that by Hilbert's Basissatz any ideal in Π or Λ is *finitely generated* which means that for any such ideal I there exists a finite subset $\mathcal{F}$ of Π or Λ, respectively, such that

$$I = \langle \mathcal{F} \rangle = \left\{ \sum_{f \in \mathcal{F}} g_f\, f \ : \ g_f \in R \right\}, \qquad R = \Pi, \Lambda.$$

Given two ideals $I, J \subset R$, their *quotient ideal* or *colon ideal*, cf. [2], is defined as

$$I : J = \{ f \in R \ : \ f \cdot J \subseteq I \}.$$

Moreover, for a given matrix $A \in \mathbb{Z}^{s \times s}$, we denote by $\left\langle z^A - 1 \right\rangle$ the ideal generated by the Laurent polynomials $z^{a_j} - 1$, $j = 1, \dots, s$, where $a_1, \dots, a_s$ are the column vectors of the matrix A. Here it suffices to consider *polynomial ideals* as long as $\left\{ z^A - 1 \right\} \subset \Pi$, or, equivalently, $A \in \mathbb{N}_0^{s \times s}$, but one has to consider Laurent polynomial ideals as soon as A has negative entries. For differences between these two rings and concepts, let me just refer to [10].

Taking symbol representation of (3.4) yields the identity

$$(z_j - 1)\, a^*(z) = \sum_{k=1}^{s} b_{jk}^*(z)\, (z^{m_k} - 1), \qquad j = 1, \dots, s, \tag{3.5}$$

where $m_1, \dots, m_s$ are the columns of M, hence the matrix valued (Laurent) polynomial $b^*(z)$ is the *representation* of the symbol a^* with respect to the quotient ideal $\left\langle z^M - 1 \right\rangle : \langle z - 1 \rangle$. In other words, the quotient ideal containment

$$a^* \in \left\langle z^M - 1 \right\rangle : \langle z - 1 \rangle \tag{3.6}$$

is the analog of the factorization property (2.5) in the multivariate case.

4 Vector Subdivision

The identity (3.4) shows that in order to understand the nature of smoothing of subdivision schemes and thus of the associated refinable functions it is necessary to consider vector subdivision schemes. It can be shown, cf. [3,6,12], that an W_p–convergent vector subdivision scheme iterating $N \times N$ matrix masks b on N–vectors yields the existence of a refinable $N \times N$ *matrix* function ψ such that

$$\psi = T_a \psi = \sum_{\alpha \in \Gamma} \psi\,(M \cdot - \alpha)\, b(\alpha) \qquad \text{and} \qquad f_c = \psi * c = \sum_{\alpha \in \Gamma} \psi(\cdot - \alpha)\, c(\alpha).$$

Note, however, that since ψ is a matrix valued function and since b and c are matrix and vector valued sequences, respectively, the order in the products matters now. In coincidence with (3.3), we say that the matrix function ψ is *stable* if there exist numbers $A, B > 0$ such that

$$A \|c\|_{\ell_p^N} \leq \|\psi * c\|_{W_p^N} \leq B \|c\|_{\ell_p^N}. \qquad (4.1)$$

The details on the p–norms for matrix and vector valued sequences and functions will be omitted here, they can be found e.g. in [12].

Let $E = M[0,1)^s \cap \Gamma$ denote the canonical set of representers of the quotient group $\Gamma/M\Gamma$. Summing up all coefficients that belong to the same of these equivalence classes into

$$b_\epsilon := \sum_{\alpha \in \Gamma} b(\epsilon - M\alpha),$$

a fundamental quantity related to the vector subdivision scheme is

$$n_b := \dim \left(\bigcap_{\epsilon \in E} \left\{ y \in \mathbb{R}^N \; : \; b_\epsilon y = y \right\} \right),$$

the dimension of the joint eigenspace of the matrices b_ϵ, $\epsilon \in E$ with respect to the eigenvalue 1, which is also the dimension of the space of *constant* eigenvectors of S_b with respect to the eigenvalue 1. The vector subdivision scheme based on the mask b is said to be *of full rank* if $n_b = N$; in this case we also have that the rank of the matrix function ψ is N as well. In fact, this paper will focus almost exclusively on full rank schemes.

5 Smooth Stable Refinable Functions

In this section we want to exploit the relationship between smoothness of a stable refinable function and vector subdivision schemes. We begin with the smoothing argument. To that end, we need to introduce the notion of an *isotropic* matrix whose powers have uniformly bounded condition numbers, that is, there exists a constant C such that

$$\kappa(M^r) = \|M^r\| \, \|M^{-r}\| < C, \qquad r \in \mathbb{N}_0.$$

Obviously, any multiply of the identity matrix, in particular $M = 2I$, has this property. Moreover, it is easily seen that M is isotropic if and only if $\overline{M} = M^T$ is isotropic.

Theorem 5.1. *Let M be isotropic, let ϕ be a stable solution of the refinement equation $\phi = T_a \phi$ and suppose that the scalar subdivision scheme S_a and the s–vector full rank subdivision scheme S_b are related by $\Delta S_a = S_b \Delta$. If, in addition, the limit function ψ of $\overline{M}^r S_b^r$, $r \to \infty$, exists and has a k–th derivative in $W_p^{s\times s}(\mathbb{R}^s)$, then ϕ has a $(k+1)$st derivative.*

In the case that M is a scalar multiple of the identity matrix, like $M = 2I$, the above convergence condition could also be seen as the *divided difference* factorization $\Delta S_a = \frac{1}{2} S_b \Delta$, cf. [4], which we know from (2.5) already, together with the "usual" convergence of the subdivision scheme S_b. In general, we only request the "algebraic condition" $\Delta S_a = S_b \Delta$ here, but then we need to demand that a "normalized" or "rescaled" version of the subdivision operator has to converge.

Proof of Theorem 5.1: We first note that the stability of ϕ implies the convergence of $S_a^r c$ to the limit function $\phi * c$ for any initial sequence c and any p–norm. What remains to show is that the *gradient* of an approximating sequence also converges in norm which means that the gradient of ϕ must exist and coincide with this limit. To do so, we choose a multiindex $\gamma \in \mathbb{N}^s$ such that $\gamma_j > 1$, $j = 1, \ldots, s$, and let

$$N_\gamma(x) = \prod_{j=1}^s \underbrace{(\chi * \cdots * \chi)}_{\gamma_j}(x_j)$$

denote the tensor product B–spline of order γ which has the property that

$$\nabla N_\gamma = \left[\frac{\partial}{\partial x_j} N_\gamma \; : \; j = 1, \ldots, s\right] = -\left[\delta_j N_{\gamma - e_j} \; : \; j = 1, \ldots, s\right]$$

so that for any $c \in \ell(\Gamma)$

$$\nabla (N_\gamma * c) = -\begin{bmatrix} N_{\gamma-e_1} & & \\ & \ddots & \\ & & N_{\gamma-e_s} \end{bmatrix} * (\Delta c) =: N'_\gamma * \Delta c, \qquad (5.1)$$

where N'_γ is an $s \times s$ diagonal matrix valued function. Taking also into account that for any matrix A and any function $f \in C^1(\mathbb{R}^s)$ we have the identity $\nabla f(A\cdot) = \left(A^T \nabla f\right)(A\cdot)$, it follows that

$$\begin{aligned} \nabla (N_\gamma * S_a^r c)(M^r \cdot) &= \overline{M}^r \left(N'_\gamma * \Delta S_a^r c\right)(M^r \cdot) = \overline{M}^r \left(N'_\gamma * S_b^r \Delta c\right)(M^r \cdot) \\ &= \left(\left(\overline{M}^r N'_\gamma \overline{M}^{-r}\right) * \overline{M}^r S_b^r \Delta c\right)(M^r \cdot). \end{aligned}$$

We claim that the matrix functions $F_r = \overline{M}^r N'_\gamma \overline{M}^{-r}$, $r \in \mathbb{N}$, are uniformly stable. To that end, we choose any $c \in \ell_p(\Gamma)$ and observe that due to

$$\overline{M}^{-r} F_r * c = N'_\gamma * \left(\overline{M}^{-r} c\right)$$

we have, with A, B being the stability constants of N'_γ, that

$$\|F_r * c\|_p \leq \left\|\overline{M}^r\right\|_p \left\|\overline{M}^{-r} F_r * c\right\|_p \leq B \left\|\overline{M}^{-r}\right\|_p \left\|\overline{M}^r\right\|_p \|c\|_p$$

as well as

$$A \|c\|_p \leq \left\|\overline{M}^r\right\|_p \left\|N'_\gamma * \overline{M}^{-r} c\right\|_p \leq \left\|\overline{M}^r\right\|_p \left\|\overline{M}^{-r}\right\|_p \|F_r * c\|_p,$$

which yields, since M and thus also $\overline{M}$ are isotropic, that

$$\frac{A}{C} \|c\|_p \leq \|F_r * c\|_p \leq (BC) \|c\|_p, \qquad r \in \mathbb{N}.$$

Consequently, the sequence F_r, $r \in \mathbb{N}$, forms a sequence of matrix valued test functions and a literal repetition of the arguments from [3] shows that $\nabla (N_\gamma * S_a^r c)(M^r \cdot)$ converges to a function which thus must be the gradient of ϕ. $\square$

It might appear unnecessarily eccentric at first view to consider the convergence of this sequence $S_r := \overline{M}^r S_b^r$ mapping $\ell(\Gamma)$ to $\ell(\Gamma)$ and relating $S_r c$ to the grid $M^{-r}\Gamma$, but in fact this approach is a very natural one. Let $\psi \in W_p(\mathbb{R}^s)^{s \times s}$ be the canonical *matrix valued* limit function obtained as the limit $r \to \infty$ of $(\chi * S_r \delta I)(M^r \cdot)$, where I simply denotes the $s \times s$ identity matrix and $\delta \in \ell(\Gamma)$ the Dirac sequence $\delta(\alpha) = \delta_{\alpha,0}$, $\alpha \in \Gamma$. Then, for $x \in \mathbb{R}^s$,

$$\psi(x) = \lim_{r \to \infty} \overline{M}^r (\chi * S_b^r \delta I)(M^r x) = \overline{M} \lim_{r \to \infty} \overline{M}^{r-1} \left(\chi * S_b^{r-1} S_b \delta I\right) \left(M^{r-1} M x\right)$$

from which it follows that

$$\psi(x) = M^T (\phi * S_b \delta I)(M \cdot) = M^T \sum_{\alpha \in \Gamma} \psi(M \cdot - \alpha)\, b(\alpha) = M^T T_a \psi,$$

which is precisely the form a refinement equation for the gradient of a scalar refinable function has to take: just differentiate the refinement equation $\phi = T_a \phi$ from (3.2) to come to this conclusion.

Remark 5.2. *If $M = \lambda I$, $\lambda > 1$ is a simple multiple of the identity matrix, then Theorem 5.1 can be interpreted just as in the univariate case: The subdivision operator S_b is defined via the "divided difference" identity $\Delta S_a = \lambda^{-1} S_b \Delta$*

and then "conventional" convergence of the subdivision operator is to be considered.

Based on Theorem 5.1, we can now reconsider the smoothing methods from the univariate case, by trying the following strategy: begin with a refinable function ψ for a re–scaled vector subdivision scheme S_b, and define S_a by the multivariate analog of multiplication of the factor $(z+1)/2$; if this holds, then ψ relates to the gradient of the refinable function ϕ for a and thus ϕ is a smoothened version of ψ.

Obviously, this strategy is different from the straightforward method for $M = 2I$ of smoothing a function by convolving it with the characteristic function of the unit cube, if only as it involves *matrix valued* functions. Note, however, that convolution of a refinable function with the characteristic function increases the total degree of the symbol a^* by s and thus leads to a significant increase in the support size as soon as $s > 1$. As an example, note that the minimally supported masks of a certain given approximation order have a significantly smaller total degree than the auto-convolution of characteristic functions already for $s = 2$ and $M = 2I$ as was pointed out in [10].

6 Redundancies and Syzygies

In the multivariate case, the "smoothing" (3.6) requests containment in a quotient ideal and that the "factorized" symbol b^* is a *representation* of a^* with respect to this quotient ideal as in (3.5). Let us give this latter identity a closer inspection. In fact, multiplying the jth equation in the system (3.5) by $(z_k - 1)$ and subtracting the $(z_j - 1)$–fold of the kth equation, we obtain that

$$0 = \sum_{\ell=1}^{s} \Big((z_k - 1)\, b_{j\ell}^*(z) - (z_j - 1)\, b_{k\ell}^*(z) \Big) \left(z^{m_\ell} - 1\right). \tag{6.1}$$

A telescoping summation readily yields that (6.1) holds for any pair $1 \le j \ne k \le s$ if and only if it holds for the special choice $k = j+1$ and $j = 1, \ldots, s$. Defining the (column) vectors

$$\left[z^M - 1\right] := \left[z^{m_j} - 1 \,:\, j = 1, \ldots, s\right],$$

and

$$[z-1]_- := \left[(-1)^j\,(z-1) \,:\, j = 1, \ldots, s\right],$$

equation (6.1) takes the simple form

$$0 = [z-1]_-^T \; b^*(z) \; \left[z^M - 1\right]. \tag{6.2}$$

Recall that for a Laurent polynomial vector $p \in \Lambda^n$ a *syzygy* $s \in \Lambda^n$ is a "linear dependency" relation $0 = s^T p$. The set of all syzygies of p forms a module $\mathcal{S}(p)$ which is called the *module of syzygies* of p. Noting that (6.2) *characterizes* all vector subdivision masks that allow for a "smoothing", we can rephrase the existence of a mask a such that $\Delta S_a = S_b \Delta$ in the following result.

Theorem 6.1. *For a mask* $b \in \ell_{00}^{s\times s}(\Gamma)$ *there exists a mask* $a \in \ell_{00}(\Gamma)$ *such that* $\Delta S_a = S_b \Delta$ *if and only if*

$$\left(b^*(z)\right)^T \; [z-1]_- \in \mathcal{S}\left(z^M - 1\right). \tag{6.3}$$

The condition (6.3) *becomes void if and only if* $s = 1$.

Syzygies and Gröbner bases for syzygies (in the context of Gröbner bases on modules in general) have been introduced and studied in [9]. These methods can be used to construct a generating set (again called a *basis*) of the module of syzygies and thus to parameterize all the matrix valued masks that permit a smoothing operation.

7 A Simple Example

We only consider an example with $s = 2$ and $M = 2I$ here; this example is particularly simple since it consists only of a scalar renormalization of the subdivision operator S_b. Moreover, we want the diagonal elements of b^* to coincide, that is, $b_{11}^*(z) = b_{22}^*(z)$. For simplicity of notation, we will use $z = (x, y)$ instead of $z = (z_1, z_2)$ in this paragraph. Now, the problem is to determine the two additional symbols b_{12}^* and b_{21}^* such that (6.3) is satisfied. In fact, (3.5) simply becomes

$$\begin{aligned} a^*(z) &= (x-1)^{-1}\left(b_{11}^*(z)\left(x^2-1\right) + b_{12}^*(z)\left(y^2-1\right)\right) && (7.1)\\ &= (y-1)^{-1}\left(b_{21}^*(z)\left(x^2-1\right) + b_{22}^*(z)\left(y^2-1\right)\right). && (7.2)\end{aligned}$$

From that it follows that $b_{12}^*(z) = (x-1)\, c_{12}^*(z)$ as well as $b_{21}^*(z) = (y-1)\, c_{21}^*(z)$ and thus, since $b_{11}^* = b_{22}^*$,

$$\begin{aligned} 0 &= b_{11}^*(z)\,(x+1-y-1) + \left(y^2-1\right)\, c_{12}^*(z) - \left(x^2-1\right)\, c_{21}^*(z) \\ &= b_{11}^*(z)\,(x-y) + \left(y^2-1\right)\, c_{12}^*(z) - \left(x^2-1\right)\, c_{21}^*(z) \end{aligned}$$

Obviously, this requires that $(x-y)\,b_{11}^*(z) \in \langle z^2-1\rangle$ and then c_{12}^* and c_{21}^* are the "coefficients" in the representation of b_{11}^* with respect to this ideal. Indeed, this is the case if $b_{11}^*(z) \in \langle z^2-1\rangle : \langle z-1\rangle$ as this latter quotient ideal consists of all polynomials that vanish at $(-1,-1)$, $(1,-1)$, $(-1,1)$, while the factor $(x-y)$ takes care of the point $(1,1)$ which completes the other three points to the variety of the ideal $\langle z^2-1\rangle$. Since

$$\left\langle z^2-1\right\rangle : \langle z-1\rangle = \left\langle z^2-1\right\rangle + \langle (x+1)(y+1)\rangle,$$

it follows that b_{11}^* must be of the form

$$b_{11}^*(z) = f(z)\,(x+1)(y+1) + g_1(z)\left(x^2-1\right) + g_2(z)\left(y^2-1\right) \tag{7.3}$$

with appropriate polynomials f, g_1, g_2. Because of the identity

$$(x-y)(x+1)(y+1) = (y+1)\left(x^2-1\right) - (x+1)\left(y^2-1\right)$$

which yields

$$\begin{aligned}(x-y)\,b_{11}^*(z) &= \Big((y+1)\,f(z) + (x-y)\,g_1(z)\Big)\left(x^2-1\right)\\ &\quad -\Big((x+1)\,f(z) - (x-y)\,g_2(z)\Big)\left(y^2-1\right),\end{aligned}$$

we are left with the canonical (though not necessarily unique) choices $c_{12}^*(z) = (x+1)\,f(z) - (x-y)\,g_2(z)$ and $c_{21}^*(z) = (y+1)\,f(z) + (x-y)\,g_1(z)$, that is,

$$\begin{aligned}b_{12}^*(z) &= \left(x^2-1\right) f(z) - (x-1)(x-y)\,g_2(z)\\ b_{21}^*(z) &= \left(y^2-1\right) f(z) + (y-1)(x-y)\,g_1(z).\end{aligned}$$

Substituting this and (7.3) into either (7.1) or (7.2), we finally get that the "smoothened" symbol a^* takes the form

$$a^*(z) = \frac{x+y}{2}(x+1)(y+1)\,f(z) + (x+1)^2(x-1)\,g_1(z) + (y+1)^2(y-1)\,g_2(z).$$

If, for additional simplicity, we choose $g_1 = g_2 = 0$, then we get the family of smoothened functions

$$a^*(z) = \frac{x+y}{2}(x+1)(y+1)\,f(z); \tag{7.4}$$

for example, setting $f = \frac{1}{2}$, that is $b_{11} = b_{22}$ correspond to the characteristic function χ, then we get that

$$b^*(z) = \frac{1}{2}\begin{bmatrix} (x+1)(y+1) & (x^2-1) \\ (y^2-1) & (x+1)(y+1) \end{bmatrix},$$

and a corresponds to "the" three directions box spline.

Slightly more interesting, the selection $f(z) = (x+y)/2$ corresponding to the matrix mask

$$b^*(z) = \frac{1}{4}\begin{bmatrix} (x+1)(y+1)(x+y) & (x^2-1)\,(x+y) \\ (y^2-1)\,(x+y) & (x+1)(y+1)(x+y) \end{bmatrix}$$

leads to

$$a^*(z) = \frac{(x+y)^2}{4}(x+1)(y+1),$$

featuring another *three directions box spline*, this time with a duplicated direction, however.

The latter example shows that smoothing is indeed an intricate procedure as the smoothed scalar function is not yet differentiable; so, we can conclude that the refinable function ψ is not yet continuous and this effect is caused by the *off–diagonal* elements of b^* which are enforced by the redundancy in the syzygies.

References

[1] A. S. Cavaretta, W. Dahmen, C. A. Micchelli: *Stationary Subdivision,* Memoirs Amer. Math. Soc. **93**, Providence, R. I., 1991.

[2] D. Cox, J. Little, D. O'Shea: *Ideals, Varieties and Algorithms*, Springer, New York–Berlin–Heidelberg–Tokyo, 1996.

[3] W. Dahmen, C. A. Micchelli: *Biorthogonal wavelet expansions*, Constr. Approx. **13** (1997), 294–328.

[4] N. Dyn: *Subdivision schemes in Computer Aided Geometric Design*, In: *Advances in Numerical Analysis – Volume II, Wavelets, Subdivision Algorithms and Radial Basis Functions*, W. Light (ed.), 36–104, Clarendon, Oxford, 1992.

[5] K. Jetter, G. Plonka: *A survey on L_2–approximation order from shift–invariant spaces*, In: *Multivariate Approximation and Applications*, N. Dyn, D. Leviatan, D. Levin, A. Pinkus (eds.), 73–111, Cambridge University Press, Cambridge, 2001.

[6] V. Latour, J. Müller, W. Nickel: *Stationary subdivision for general scaling matrices*, Math. Z. **227** (1998), 645–661.

[7] C. A. Micchelli, T. Sauer: *Regularity of multiwavelets*, Advances Comput. Math. **7** (1997), 455–545.

[8] C. A. Micchelli, T. Sauer: *On vector subdivision*, Math. Z. **229** (1998), 621–674.

[9] H. M. Möller, F. Mora: *New constructive methods in classical ideal theory*, J. Alg. **100** (1986), 138–178.

[10] H. M. Möller, T. Sauer: *Multivariate refinable functions of high approximation order via quotient ideals of Laurent polynomials*, Advances Comput. Math., to appear.

[11] T. Sauer: *Polynomial interpolation, ideals and approximation order of refinable functions*, Proc. Amer. Math. Soc. **130** (2002), 3335–3347.

[12] T. Sauer: *Subdivisión vectorial estacionaria: ideales cociente, diferencias y potencia de aproximación*, Rev. R. Acad. Cienc. Serie A. Mat. **96** (2002), 257–277.

Address:

Tomas Sauer
Lehrstuhl für Numerische Mathematik
Justus–Liebig–Universität Gießen
Heinrich–Buff–Ring 44
D–34592 Gießen
Germany

International Series of Numerical Mathematics
Vol. 145,

Reconstructing Multivariate Functions from Large Data Sets

Holger Wendland

Dedicated to Professor Manfred Reimer on the occasion of his 70th birthday

Abstract

In this paper we discuss an efficient method for reconstructing multivariate functions from large sets of scattered data. Our method is based on (conditionally) positive definite kernels and a partition of unity. It is tailored to solve such reconstruction problems efficiently even if the space dimension is high and the number of data points is huge. We generalize the classical interpolation approach to dealing with other information than point evaluations. Finally, we demonstrate the generality of our approach by providing three different examples.

Introduction

The reconstruction of multivariate functions from multivariate data sets is one of the major problems in numerical analysis and approximation theory. In particular, if the data set is huge and consists of unorganized and unstructured points, special methods have to be developed for an efficient reconstruction. The simplest situation is given by a set of *data sites* $X = \{x_1, \dots, x_N\} \subseteq \Omega \subseteq \mathbb{R}^d$ situated in a bounded and open region $\Omega \subseteq \mathbb{R}^d$ and a set of *data values* $f_1, \dots, f_N \in \mathbb{R}$, which are assumed to stem from a function $f : \Omega \to \mathbb{R}$, i.e. we might assume $f(x_j) = f_j$, $1 \le j \le N$. The only assumption on the data sites throughout this paper will be that of being pairwise distinct.

Then, we are interested in finding a *good* approximant $s_{f,X}$ to f. The term *good* is in general determined by describing how fast $s_{f,X}$ converges to f if the data sites X fill out the region Ω. If the data values are exact, interpolation is a good choice for the reconstruction process. If they contain noise, smoothing techniques are necessary.

A more general approximation problem can be described as follows. Suppose the unknown function f comes from a (complete) normed linear space $\mathcal{F}$ and the information is provided in terms of $\lambda_1(f), \ldots, \lambda_N(f)$, where $\Lambda = \{\lambda_j\}_{j=1}^N \subseteq \mathcal{F}^*$ is a set of linearly independent continuous functionals on $\mathcal{F}$. Then, a reconstruction of f has to be based upon these values. For example, an interpolating function $s_{f,\Lambda}$ now has to satisfy $\lambda_j(s_{f,\Lambda}) = \lambda_j(f)$, $1 \le j \le N$.

The rest of this paper is organized as follows. In the next section we shortly review the relevant material on (conditionally) positive definite kernels. In the second section we describe the partition of unity approach, review old and establish new error estimates, and extend the approach to the case of generalized interpolation. In the final section we turn to three different examples, collocation for partial differential equations, coupling of fluid–structure–interaction problems, and the reconstruction of surfaces from point clouds.

1 Conditionally Positive Definite Kernels

The most flexible approach to scattered data problems is based upon a (conditionally) positive definite kernel. In the sequel we will always assume that $\Omega \subseteq \mathbb{R}^d$ is an open and bounded set.

Definition 1.1. *Let $\mathcal{P}$ be a finite dimensional subspace of $C(\Omega)$. A symmetric function $\Phi : \Omega \times \Omega \to \mathbb{R}$ is called* conditionally positive definite with respect to *$\mathcal{P}$ if for any N distinct points $x_1, \ldots, x_N \in \Omega$ and any $\alpha \in \mathbb{R}^N \setminus \{0\}$ satisfying $\sum_j \alpha_j p(x_j) = 0$ for all $p \in \mathcal{P}$ we have*

$$\sum_{j,k=1}^N \alpha_j \alpha_k \Phi(x_j, x_k) > 0.$$

If $\mathcal{P} = \{0\}$ we call Φ a positive definite kernel. Moreover, if $\mathcal{P} = \pi_{m-1}(\mathbb{R}^d)$ is the set of all d–variate polynomials of degree at most $m-1$, Φ is also called conditionally positive definite of order m.

The most prominent examples of positive definite kernels are *radial basis functions.* Here, Φ is defined by $\Phi(x, y) = \phi(\|x - y\|_2)$ with a univariate function $\phi : [0, \infty) \to \mathbb{R}$, and assumed to be (conditionally) positive definite on all of $\mathbb{R}^d$. Examples are the Gaussians $\phi(r) = e^{-\alpha r^2}$, $\alpha > 0$, the thin–plate spline $\phi(r) = r^2 \log r$, the cubic $\phi(r) = r^3$, the multiquadric $\phi(r) = \sqrt{r^2 + c^2}$, $c > 0$, the inverse multiquadric $\phi(r) = \frac{1}{\sqrt{r^2+c^2}}$ and the compactly supported functions like $\phi(r) = (1-r)_+^4(4r+1)$. All these functions are (conditionally) positive definite of some order m. The minimal order m and more details can for example be found in [12].

If the data contains no noise, interpolation is an appropriate reconstruction tool. It is well–known that a conditionally positive definite kernel allows an interpolant of the form

$$s_{f,X} = \sum_{j=1}^{N} \alpha_j \Phi(\cdot, x_j) + \sum_{k=1}^{Q} \beta_k p_k,$$

where $p_1, \ldots, p_Q$ form a basis for $\mathcal{P}$. The coefficients $\alpha \in \mathbb{R}^N$ and $\beta \in \mathbb{R}^Q$ are determined by the interpolation conditions

$$s(x_i) = f(x_i), \qquad 1 \le i \le N, \tag{1.1}$$

together with the *natural boundary conditions*

$$\sum_{j=1}^{N} p_k(x_j) = 0, \qquad 1 \le k \le Q. \tag{1.2}$$

Moreover, the coefficients and hence the interpolant are uniquely determined by (1.1) and (1.2) if the data set X is $\mathcal{P}$–unisolvent, which means that the zero function is the only function from $\mathcal{P}$ that vanishes on X.

Associated to every conditionally positive definite kernel comes a natural function space $\mathcal{N}_\Phi$, which carries a semi–inner product with null space $\mathcal{P}$ and which is complete with respect to this semi–inner product. The benefit of this *native space* is that much is known on the approximation error of the interpolant if the data stems from a target function from $\mathcal{N}_\Phi$. To be more precise, if we measure the data density by

$$h_{X,\Omega} := \sup_{x \in \Omega} \min_{x_j \in X} \|x - x_j\|_2$$

and $f \in \mathcal{N}_\Phi$, then

$$|f(x) - s_{f,X}(x)| \le F_\Phi(h_{X,\Omega}) |f|_{\mathcal{N}_\Phi}.$$

The function F_Φ depends only on the kernel Φ, in particular on its smoothness. For the aforementioned kernels the function F_Φ can for example be found in [10]. We will also discuss some of them in more detail later on. Native spaces themselves have been thoroughly investigated by Madych–Nelson [7, 8] and Schaback [11].

The native space or more precisely its dual $\mathcal{N}_\Phi^*$, plays also an important role in case of generalized interpolation. The following result is an obvious consequence of the fact that the functionals are assumed to be linearly independent and are defined on a reproducing kernel Hilbert space (cf. [15]).

Proposition 1.2. *Let $\Phi : \Omega \times \Omega \to \mathbb{R}$ be a positive definite kernel. Suppose $\lambda_1, \ldots, \lambda_N \in \mathcal{N}_\Phi^*$ are linearly independent. Then there exists a unique function $s_{f,\Lambda}$ of the form*

$$s_{f,\Lambda} = \sum_{j=1}^{N} \alpha_j (I \otimes \lambda_j)\Phi,$$

satisfying $\lambda_j(s_{f,\Lambda}) = \lambda_j(f)$, $1 \le j \le N$.

Here we have used the notation $(\mu \otimes \lambda)\Phi$ to indicate that μ acts on Φ with respect to Φ's first argument, while λ acts with respect to its second argument. Moreover, I denotes the identity, meaning that there is no action with respect to this argument.

Obviously, Proposition 1.2 is equivalent to the fact that the generalized interpolation matrix with entries $(\lambda_i \otimes \lambda_j)\Phi$ is positive definite.

2 Partition of Unity Methods

If the number of data sites is huge, fast solution and fast evaluation techniques for the (generalized) interpolation problem are necessary. In particular, if the kernel is globally supported, the interpolation matrix is dense and a direct solver would need $\mathcal{O}(N^3)$ time and space for computing the solution and another $\mathcal{O}(N)$ time for each evaluation. To remedy this problem, *fast evaluation* methods based on *multipole expansion* of the kernel have successfully been developed, see for example the overview article [2]. Despite the fact that these methods do a really good job, they still have some drawbacks. First of all, if different kernels are considered, a multipole expansion is necessary for each of them. Second, coding the fast solver is a delicate task, since not only the underlying space has to be subdivided into a sequence of nested patches, but also for each patch the multipole expansion has to be taken care of. Third, interpolation is sacrificed for efficiency. The resulting functions are rather approximants than interpolants. While this might be less problematic in case of data with noise, there are other situations (as we will soon see) where interpolation is a crucial requirement.

Here, we want to introduce another method based upon a partition of unity. The idea of using a partition of unity combined with radial basis functions goes back to Franke [6]. Moreover, such methods have recently also attracted attention in the context of meshless methods for the numerical solution of partial differential equations, see for example [1]. To this end, we cover our region Ω with a finite number of locally overlapping patches Ω_j, $1 \le j \le M$. We assume that these patches satisfy certain regularity assumptions, for example all of them should satisfy a uniform cone condition with the same angle and radius. Moreover, it should be possible to answer the following geometric queries in constant time. If a patch Ω_j is given, what are the data sites $X_j = \{x \in X : x \in \Omega_j\} = \Omega_j \cap X$? If a point $x \in \Omega$ is given, what are the patches Ω_j that contain x? To answer these questions in constant time it is necessary that each point $x \in \Omega$ is contained only in a small number of patches, say K of them with K independent of N. Moreover, the number of points in each X_j has also to be a small number, say L, when compared to N. Again L should be independent of N.

For such a covering, we compute the local interpolants s_{f,X_j} to f and form a global interpolant by setting

$$s = \sum_{j=1}^{M} w_j s_{f,X_j}. \tag{2.1}$$

The functions w_j are chosen to form a partition of unity with respect to the covering $\{\Omega_j\}$, i.e. w_j is continuous, nonnegative, with support in Ω_j, and all of them sum up to one: $\sum_{j=1}^{M} w_j(x) = 1$, $x \in \Omega$.

Theorem 2.1. *Under the aforementioned assumptions the partition of unity approximant s from (2.1) can be computed in $\mathcal{O}(N)$ time and evaluated in $\mathcal{O}(1)$ time. Moreover, if the local interpolants satisfy*

$$|f(x) - s_{f,X_j}(x)| \le C_f F(h_{X_j,\Omega_j}), \quad x \in \Omega_j,\ 1 \le j \le M,$$

with a constant C_f depending only on f but not on j and a nondecreasing function $F : [0,\infty) \to [0,\infty)$, then the global approximant enjoys the error estimate

$$|f(x) - s(x)| \le C_f F(h), \qquad x \in \Omega,$$

with $h = \max h_{X_j,\Omega_j}$.

Proof. The computation of the global approximant includes the computation of $\mathcal{O}(M) = \mathcal{O}(N)$ local interpolants. By assumption, for every x_j, there are at most $K = \mathcal{O}(1)$ patches containing x_j and they are found in constant time. Hence, inserting j into lists associated to these patches, allows us to store the indices of the data sites for each patch in $\mathcal{O}(N)$ time. Since there are at most $L = \mathcal{O}(1)$ points in X_j, setting up and solving the local linear system can be performed in $\mathcal{O}(L^3) = \mathcal{O}(1)$ time. This gives the computational complexity of

$\mathcal{O}(N)$ in the preprocessing phase. Moreover, using again the assumption that, for every $x \in \Omega$, we can find all patches containing x in constant time and the number of these patches is bounded by $K = \mathcal{O}(1)$, we see that evaluation can be done in $\mathcal{O}(1)$ time as well.

For the error estimate we use the fact that the $\{w_j\}$ are supported in Ω_j to derive

$$\begin{aligned} |f(x) - s(x)| &\leq \sum_{j:x\in\Omega_j} w_j(x)|f(x) - s_{f,X_j}(x)| \\ &\leq C_f \sum_{j:x\in\Omega_j} w_j(x)F(h_{X_j,\Omega_j}) \\ &\leq C_f F(h) \sum_{j=1}^{N} w_j(x), \end{aligned}$$

so that the partition of unity property of the $\{w_j\}$ gives the desired result. □

Obviously, this result holds for any kind of local approximation process. It is even possible to use a *completely different* approximation on each patch.

In [13] it was shown that the additional assumptions on the local processes in Theorem 2.1 are satisfied if the target function f comes from the native space $\mathcal{N}_\Phi$ of the kernel in use. If all the local patches satisfy a uniform cone condition with angle θ and radius r, the constant C_f can be chosen to be $C_f = C_{\theta,r,\Phi}|f|_{\mathcal{N}_\Phi}$. The paper [13] contains also more details on how the data structures have to be chosen so that the geometric queries can be answered efficiently.

Recent results on *escaping the native space* (cf. [9, 4]) allow some less restrictive convergence results. Suppose the kernel Φ is given by $\Phi(x,y) = \phi(x-y)$ and $\phi : \mathbb{R}^d \to \mathbb{R}$ possesses a generalized Fourier transform $\widehat{\phi} : \mathbb{R}^d \setminus \{0\} \to (0,\infty)$ satisfying

$$c_1\|\omega\|_2^{-2\beta} \leq \widehat{\phi}(\omega) \leq c_2\|\omega\|_2^{-2\beta}, \qquad \|\omega\|_2 \geq r_0 > 0, \tag{2.2}$$

with $\beta > d/2$. For example, the thin–plate splines and the compactly supported functions of minimal degree enjoy such a behaviour. Then, the function F in Theorem 2.1 can be chosen to be $F(h) = h^{\beta-d/2}$ for functions from the native space which is now a Sobolev or a Beppo Levi space. But in this situation we have even more. To formulate our next result, we introduce the notation C_0^k for the space of all k times continuously differentiable functions, which vanish together with all their derivatives at infinity. This space is equipped with the norm $\|f\|_{C_0^k} = \max_{|\alpha|\leq k} \|D^\alpha f\|_{L_\infty(\mathbb{R}^d)}$. Moreover, we denote the usual Sobolev space of order k on $\mathbb{R}^d$ by W_2^k. For functions from $W_2^k \cap C_0^k$ we define the norm $\|f\|_{C_0^k \cap W_2^k}$ by the maximum of the C_0^k and the Sobolev norm.

Finally, we say that a set X is quasi–uniform on Ω with constant c_q if $q_X \leq h_{X,\Omega} \leq c_q q_X$. Here, $q_X = \frac{1}{2} \min_{i \neq j} \|x_i - x_j\|_2$ is the separation distance of X.

Theorem 2.2. *Suppose ϕ is a function satisfying (2.2) with $\beta \geq k > d/2$. Suppose that all the local point sets X_j are quasi–uniform with the same constant $c_q > 0$. Then we have for every $f \in C_0^k \cap W_2^k$ that*

$$\|f - s\|_{L_\infty}(\Omega) \leq Ch^{k-\frac{d}{2}} \|f\|_{C_0^k \cap W_2^k}$$

with $h = \max_j h_{X_j,\Omega_j}$.

Proof. In [9] it has been proven that our assumptions lead to the local error estimates

$$\|f - s_{f,X_j}\|_{L_\infty(\Omega_j)} \leq Ch_{X_j,\Omega_j}^{k-d/2} \rho_{X_j,\Omega_j}^{\beta-k} \|f\|_{C_0^k \cap W_2^k}, \tag{2.3}$$

where ρ_{X_j,Ω_j} is the local mesh ratio defined by $\rho_{X_j,\Omega_j} = h_{X_j,\Omega_j}/q_{X_j}$. Moreover, the constant C in (2.3) depends on the region Ω_j precisely in the same way as the constant C does in the classical radial basis function estimates, i.e. on the angle and radius of the cone condition. Since we assumed all patches to satisfy the same cone condition, the constant C in (2.3) is indeed independent of j. Finally, the mesh ratio can be bounded by the quasi–uniformity constant c_q, because all local sets are assumed to be quasi–uniform with the same constant. Hence, (2.3) becomes

$$\|f - s_{f,X_j}\|_{L_\infty(\Omega_j)} \leq ch_{X_j,\Omega_j}^{k-d/2} \|f\|_{C_0^k \cap W_2^k},$$

with a constant $c > 0$ independent of j so that Theorem 2.1 gives the desired result. □

The assumption on all X_j to be quasi–uniform with respect to the same constant is not a severe one. Quasi–uniformity is destroyed be nearly coalescing points. Such points are always critical and should be removed from the data set in a preprocessing step.

The assumption on f to be from C_0^k can also be weakend, at least in the case of thin–plate spline interpolation (see [4]).

We end this section by discussing well–posedness of the generalized interpolation problem. For simplicity, we restrict ourselves to the case of a positive definite kernel Φ and functionals from the dual of the native space, which have a one point support. Later on, we will relax this last condition.

Hence, let us assume that $\Lambda = \{\lambda_1, \ldots, \lambda_N\} \subseteq \mathcal{N}_\Phi^*$ is a set of N linearly independent, compactly supported functionals with support $\text{supp}(\lambda_j) = \{x_j\} \subseteq \Omega$. As before, we define $\Lambda_j = \{\lambda_k \in \Lambda : x_k \in \Omega_j\}$ and solve local problems based on these local sets. But to make sure that the global function satisfies the generalized interpolation conditions, we have to be careful. To this end, we assume that the weight functions $\{w_j\}$ are contained in $\mathcal{N}_\Phi$ and define new functionals $\mu_k^{(j)} \in \mathcal{N}_\Phi^*$ by

$$\mu_k^{(j)}(f) = \lambda_k(w_j f), \qquad \lambda_k \in \Lambda_j, 1 \le j \le M. \tag{2.4}$$

If for each j the functionals $\{\mu_k^{(j)} : \lambda_k \in \Lambda_j\}$ are linearly independent, we can form the local approximants by

$$s_{f,\Lambda_j} = \sum_{\lambda_i \in \Lambda_j} \alpha_i^{(j)} (I \otimes \mu_i^{(j)}) \Phi, \tag{2.5}$$

and determine the coefficients by

$$\mu_k^{(j)}(s_{f,\Lambda_j}) = \sum_{\lambda_i \in \Lambda_j} \alpha_i^{(j)} (\mu_k^{(j)} \otimes \mu_i^{(j)}) \Phi = \lambda_k(f) w_j(x_k), \qquad \lambda_k \in \Lambda_j. \tag{2.6}$$

Proposition 1.2 together with our assumptions guarantees a unique solution to this system for every j. If we finally form the global generalized interpolant by

$$s = \sum_{j=1}^{M} w_j s_{f,\Lambda_j}, \tag{2.7}$$

we find

$$\begin{aligned} \lambda_k(s) &= \sum_{j=1}^{M} \lambda_k(w_j s_{f,\Lambda_j}) = \sum_{j:\lambda_k \in \Lambda_j} \mu_k^{(j)}(s_{f,\Lambda_j}) = \sum_{j:x_k \in \Omega_j} \lambda_k(f) w_j(x_k) \\ &= \sum_{j=1}^{M} \lambda_k(f) w_j(x_k) = \lambda_k(f) \end{aligned}$$

for every $1 \le k \le N$. Hence we have established the following result.

Theorem 2.3. *Suppose $\Phi : \Omega \times \Omega \to \mathbb{R}$ is a positive definite kernel and $\lambda_1, \ldots, \lambda_N \in \mathcal{N}_\Phi^*$ are linearly independent functionals with supp$(\lambda_j) = \{x_j\}$. Let $\Lambda_j := \{\lambda_k : x_k \in \Omega_j\}$ and define the functionals $\mu_k^{(j)}$ as in (2.4). If the functionals $\{\mu_k^{(j)} : \lambda_k \in \Lambda_j\}$ are linearly independent for $1 \le j \le M$, the local systems (2.6) have a unique solution of the form (2.5), and the global generalized interpolant s from (2.7) satisfies $\lambda_k(s) = \lambda_k(f)$, $1 \le k \le N$.*

3 Examples

In this section we discuss some possible applications of the partition of unity method. Throughout this section we will assume that the weight functions are positive on all of Ω_j.

We start with an obvious application of Theorem 2.3 by showing that Hermite interpolation is always well–posed. Hence, our method can for example be used for solving partial differential equations numerically by collocation. In the sequel $\alpha \in \mathbb{N}_0^d$ denotes a multiindex with length $|\alpha| = \alpha_1 + \ldots + \alpha_d$ and D^α denotes the usual derivative

$$D^\alpha = \frac{\partial^{|\alpha|}}{\partial x^\alpha} = \frac{\partial^{|\alpha|}}{\partial \chi_1^{\alpha_1} \cdots \partial \chi_d^{\alpha_d}},$$

where $x = (\chi_1, \ldots \chi_d)^T \in \mathbb{R}^d$. A typical boundary value problem might be to find the solution u of

$$\begin{aligned} Lu &= f \qquad \text{on } \Omega, \\ u &= g \qquad \text{on } \partial\Omega, \end{aligned}$$

with a linear partial differential operator L having nonvanishing coefficients and given right hand sides f and g. To solve this system numerically by collocation, we choose two sets of points $X = \{x_1, \ldots, x_N\} \subseteq \Omega$ and $Y = \{y_1, \ldots, y_M\} \subseteq \partial\Omega$ and define the functionals λ_j by

$$\lambda_j(u) = \begin{cases} Lu(x_j) & \text{for } 1 \le j \le N, \\ u(y_{j-N}) & \text{for } N+1 \le j \le N+M. \end{cases}$$

It follows from the next more general result that the associated discrete system has a unique solution.

Corollary 3.1. *Suppose Φ is a positive definite kernel of the form $\Phi(x,y) = \phi(x-y)$ with $\phi \in C^{2k}(\mathbb{R}^d)$. Suppose $X = \{x_1, \ldots, x_N\}$ are distinct points in $\Omega \subseteq \mathbb{R}^d$ and functionals λ_j are defined by $\lambda_j(f) := D^{\alpha^{(j)}} f(x_j)$, $f \in \mathcal{N}_\Phi$, where $|\alpha^{(j)}| \le k$ for all $1 \le j \le N$. Then these functionals, as well as the associated functionals $\{\mu_k^{(j)} : \lambda_k \in \Lambda_j\}$ are linearly independent. There exists a function s of the form (2.7) satisfying*

$$\lambda_j(s) = \lambda_j(f), \qquad 1 \le j \le N.$$

Proof. It is known (cf. [11]) that our assumptions on the kernel Φ imply $\mathcal{N}_\Phi \subset C^k(\mathbb{R}^d)$. Hence, all functionals are well–defined on $\mathcal{N}_\Phi$. Moreover, from the results in [15] on Hermite interpolation by radial basis functions it is clear that all the λ_j are linearly independent. This can be shown by constructing compactly supported functions $g_\ell \in C^\infty(\mathbb{R}^d)$ with

$$\lambda_k(g_\ell) = D^{(\alpha^{(k)})} g_\ell(x_k) = \delta_{\ell,k}.$$

It remains to show that all the sets $\{\mu_k^{(j)} : x_k \in \Omega_j\}$ consist of linearly independent functionals. To see this, we have to dwell a little more on the construction of the aforementioned functions g_ℓ. One way to construct them is to choose a compactly supported function $f_0 \in C^\infty(\mathbb{R}^d)$ with $f_0(0) = 1$ and support in the ball of radius $\epsilon < q_X$ around zero and to set

$$g_\ell(x) = \frac{(x - x_\ell)^{\alpha^{(\ell)}}}{\alpha^\ell!} f_0(x - x_\ell).$$

The choice of the support of f_0 ensures $\lambda_k(g_\ell) = 0$ for $k \neq \ell$ and the multivariate version of Leibniz' rule gives $\lambda_\ell(g_\ell) = f_0(0) = 1$. But the same arguments yield $\lambda_k(g_\ell w_j) = 0$ if $k \neq \ell$ and $\lambda_\ell(g_\ell w_j) = f_0(0) w_j(x_\ell) = w_j(x_\ell) > 0$ for $x_\ell \in \Omega_j$. Hence,

$$0 = \sum_{x_k \in \Omega_j} c_k \mu_k^{(j)}(f) = \sum_{x_k \in \Omega_j} c_k \lambda_k(w_j f), \qquad f \in \mathcal{N}_\Phi,$$

leads to $0 = c_k w_j(x_k)$ and $c_k = 0$ if f is specialized to be g_k. □

The conditions can even be relaxed in the sense that double points x_j are allowed as long as the associated multiindices differ. Numerical examples and a convergence analysis of this method are the subject of a follow–up paper.

Our second example deals with the problem of fluid–structure–interaction. In the simulation of large aircraft during flight, engineers use different models for describing different physical observations. In particular, they have one model for describing the structural integrity of the plane and one model for the aerodynamical behaviour. A coupling of these models makes it necessary to exchange data between them. For example, displacement vectors are calculated on the structural grid and have to be transported to the computional dynamics (CFD) mesh. On the other hand, the CFD model is used to calculate pressure or forces, and these are needed on the structural mesh. Moreover, certain additional conditions like conservation of energy and exact reproduction of rigid body motions have to be taken into account. Hence, interpolation is crucial.

One way of ensuring the conservation of energy and forces is to interpolate the displacement vectors and to use the transpose of the interpolation matrix for transporting the forces (see [3]). Another possibility is to use the conservation as an additional interpolation condition. For example, conservation of forces means

$$\sum_{j=1}^{N} f(x_j) = \sum_{j=1}^{n} f(y_j)$$

if $X = \{x_1, \dots, x_N\}$ and $Y = \{y_1, \dots, y_n\}$ denote the nodes of the two different models.

Theorem 3.2. *Suppose $X = \{x_1, \dots, x_N\} \subseteq \Omega$ and $Y = \{y_1, \dots, y_n\}$ are two sets of distinct points. Let $X_j = X \cap \Omega_j = \{x_1^{(j)}, \dots, x_{N_j}^{(j)}\}$ and $Y_j = Y \cap \Omega_j = \{y_1^{(j)}, \dots, y_{n_j}^{(j)}\}$, $1 \le j \le M$. Suppose $\emptyset \neq Y_j \not\subseteq X_j$. Define for each $1 \le j \le M$ the functionals $\lambda_k^{(j)}(f) = f(x_k^{(j)})$, $1 \le k \le N_j$, and*

$$\lambda_{N_j+1}^{(j)}(f) = \sum_{i=1}^{n_j} w_j(y_i^{(j)}) f(y_i^{(j)}).$$

Then there exists a unique function s_j of the form $s_j = \sum_{k=1}^{N_j+1} \alpha_k^{(j)} (I \otimes \lambda_k^{(j)}) \Phi$ satisfying $s_j(x_k^{(j)}) = f(x_k^{(j)})$, $1 \le k \le N_j$, and

$$\lambda_{N_j+1}^{(j)}(s_j) = \sum_{k=1}^{N_j} w_j(x_k^{(j)}) f(x_k^{(j)}).$$

Moreover, the global function $s = \sum_{j=1}^{M} w_j s_j$ satisfies $s(x_k) = f(x_k)$, $1 \le k \le N$ and

$$\sum_{k=1}^{N} f(x_k) = \sum_{i=1}^{n} s(y_i).$$

Proof. We know that the local systems have a unique solution of the prescribed form, if the linear functionals $\lambda_k^{(j)}$, $1 \le k \le N_j + 1$, are linearly independent. This follows immediately from the fact that Y_j contains at least one point different from X_j. It is also clear that the global interpolant satisfies the interpolation conditions $s(x_k) = f(x_k)$, $1 \le k \le N$. Finally, we see that

$$\sum_{i=1}^{n} s(y_i) \quad = \quad \sum_{i=1}^{n} \sum_{j=1}^{M} s_j(y_i) w_j(y_i)$$

$$= \sum_{j=1}^{M}\sum_{i=1}^{n_j} s_j(y_i^{(j)})w_j(y_i^{(j)})$$
$$= \sum_{j=1}^{M}\sum_{k=1}^{N_j} f(x_k^{(j)})w_j(x_k^{(j)})$$
$$= \sum_{k=1}^{N} f(x_k),$$

which follows from the local conservation conditions. □

Note that this result differs from Theorem 2.3 in such a way that not all the functionals λ_j have support in one point. Moreover, once again, we actually have two sets of functionals $\{\lambda_j\}$ and $\{\mu_j\}$ and our interpolation conditions are rather of the form

$$\lambda_j(s) = \mu_j(f), \qquad 1 \le j \le N.$$

It should be clear by now how additional conditions like discretely conserving energy can be added to the partition of unity interpolant. Again, more details will be provided in a follow–up paper.

Our final example deals with the reconstruction of a $3D$ surface $\mathcal{S}$ from an unorganized $3D$–point cloud $X = \{x_1, \dots, x_N\} \subseteq \mathbb{R}^3$. The idea here is to assume that $\mathcal{S}$ can be represented as an implicit surface $\mathcal{S} = \{x \in \mathbb{R}^3 : F(x) = 0\}$ and we want to approximate the unknown function $F : \mathbb{R}^3 \to \mathbb{R}$ by a partition of unity interpolant. Obviously, we need additional information since the interpolation conditions $s(x_j) = 0$ would lead to the zero function. The key idea is to add off–surface points to the data set and to assign a positive value to a data site outside the surface and a negative value to a data site inside the surface. These additional data points are constructed along (estimated) surface normals. To this end first a surface normal in (every) point x_j is constructed. We actually do not need a normal for every point and in regions where a normal cannot be estimated, we simply stick to the surface points alone. Then, the normals are consistently oriented, and finally off–surface points are added. After computing the partition of unity interpolant, the implicit representation has to be polygonized, which can for example be done by the *marching cubes algorithm*. Figure 1 shows the Stanford bunny, which is a noiseless data set of moderate size, as an example. On the left the original bunny is shown while on the right hand side our reconstruction is presented. As one can clearly see, all the details of the surface are recovered. For more details, we refer the reader to [5] and [14].

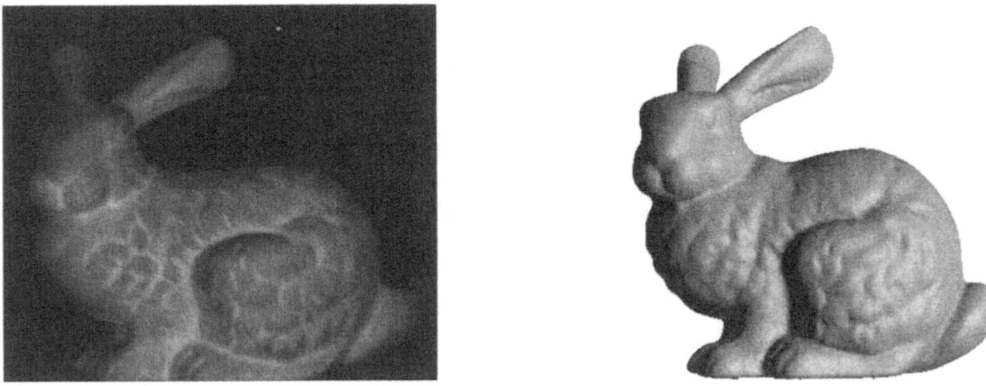

Figure 1: The Stanford bunny.

Acknowledgement. I am grateful to Stanford Computer Graphics Laboratory for the bunny data set and the original bunny image.

References

[1] I. Babuska, J. M. Melenk: *The partition of unity method*, Int. J. Numer. Methods Eng. **40** (1997), 727–758.

[2] R. K. Beatson, L. Greengard: *A short course on fast multipole methods.* In *Wavelets, multilevel methods and elliptic PDEs. 7th EPSRC numerical analysis summer school*, University of Leicester, Leicester, GB, July 8–19, 1996, M. Ainsworth et al. (eds.), Clarendon Press, Oxford, 1997, pp. 1–37.

[3] A. Beckert, H. Wendland: *Multivariate interpolation for fluid–structure–interaction problems using radial basis functions.* Aerosp. Sci. Technol. **5** (2001), 125–134.

[4] R. Brownlee, W. Light: *Approximation orders for interpolation by surface splines to rough functions.* Preprint, Leicester, 2002.

[5] J. C. Carr, R. K. Beatson, J. B. Cherrie, T. J. Mitchell, W. R. Fright, B. C. McCallum, T. R. Evans: *Reconstruction and representation of 3D objects with radial basis functions.* In *Computer Graphics Proceedings, Annual Conference Series* (2001), Addison Wesley, pp. 67–76.

[6] R. Franke: *Smooth interpolation of scattered data by local thin plate splines*, Comput. Math. Appl. **8** (1982), 273–281.

[7] W. R. Madych, S. A. Nelson: *Multivariate interpolation and conditionally positive definite functions*, Approximation Theory Appl. **4** (1988), 77–89.

[8] W. R. Madych, S. A. Nelson: *Multivariate interpolation and conditionally positive definite functions* II, Math. Comput. **54** (1990), 211–230.

[9] F. J. Narcowich, J. D. Ward: *Scattered–data interpolation on* $\mathbb{R}^n$*: Error estimates for radial basis and band–limited functions.* preprint 2002.

[10] R. Schaback: *Error estimates and condition number for radial basis function interpolation.* Adv. Comput. Math. **3** (1995), 251–264.

[11] R. Schaback: *Native Hilbert spaces for radial basis functions* I. In *New Developments in Approximation Theory.* 2nd international Dortmund meeting (IDoMAT) '98, Germany, February 23–27, 1998, M. W. Müller et al. (eds.), *Int. Ser. Numer. Math.* **132**, Birkhäuser, Basel 1999, pp. 255–282.

[12] R. Schaback, H. Wendland: *Characterization and construction of radial basis functions.* In *Multivariate Approximation with Applications*, N. Dyn, D. Leviatan, D. Levin, A. Pinkus (eds.), Cambridge University Press, Cambridge 2001, pp. 1–24.

[13] H. Wendland: *Fast evaluation of radial basis functions: Methods based on partition of unity.* In *Approximation Theory* X*: Wavelets, Splines, and Applications*, C. K. Chui, L. L. Schumaker, J. Stöckler (eds.), Vanderbilt University Press, pp. 473–483.

[14] H. Wendland: *Surface reconstruction from unorganized points.* Preprint, Göttingen 2002.

[15] Z. Wu: *Hermite–Birkhoff interpolation of scattered data by radial basis functions.* Approximation Theory Appl. **8** (1992), 1–10.

Address:

Holger Wendland
Institut für Numerische und Angewandte Mathematik
Universität Göttingen
Lotzestr. 16–18
D–37083 Göttingen
Germany

International Series of Numerical Mathematics
Vol. 145,

Three Scale Versus Matrix Refinement Equations

Georg Zimmermann

Dedicated to Professor Manfred Reimer on the occasion of his 70th birthday

Abstract

We show under what conditions three scale refinement equations are equivalent to matrix refinement equations of a certain structure, and how this equivalence can be used in the modification of refinement masks.

1 Introduction

A good example for a matrix refinement equation is given by the piecewise linear splines in the plane on the four directional grid, see, *e.g.*, [1, 6]. Its main advantage compared to a piecewise linear box spline on the three directional grid is the extra symmetry, but this is paid for by the disadvantage of having to use two functions which are jointly refinable only. This generates the matrix refinement equation. Another nice property is the fact that together, the $\mathbb{Z}^2$–translates of these two functions interpolate on the grid $\mathbb{Z}^2 \cup (\mathbb{Z}^2 + (\frac{1}{2}, \frac{1}{2}))$. Furthermore, it is worth noting that these two functions satisfy

$$\varphi_2(x,y) = \varphi_1(x+y-1, x-y)\,, \tag{1.1}$$

i.e., the second function is a rotated, dilated, and translated version of the first, and we essentially are using one basic function only. Its graph is a four sided, square pyramid.

We are interested in replacing this pyramid by some smoother function, preserving as much structure as possible.

Notation. Because of (1.1), we will use the dilation matrix

$$M = \begin{pmatrix} 1 & 1 \\ 1 & -1 \end{pmatrix} \quad \text{with} \quad M^2 = 2\,I = \begin{pmatrix} 2 & 0 \\ 0 & 2 \end{pmatrix} \quad \text{and thus} \quad M^{-1} = \tfrac{1}{2}\,M\,.$$

We shall frequently need the vectors

$$e_1 = \begin{pmatrix} 1 \\ 0 \end{pmatrix}, \quad e_2 = \begin{pmatrix} 0 \\ 1 \end{pmatrix}, \quad e = \begin{pmatrix} 1 \\ 1 \end{pmatrix},$$

and it is worth noting that

$$M\,e_1 = e\,, \qquad M\,e = 2\,e_1\,.$$

A function $\varphi : \mathbb{R}^2 \to \mathbb{C}$ satifies a *three scale refinement equation*, if there exist coefficient sequences $(h_1[\alpha])_{\alpha\in\mathbb{Z}^2}$ and $(h_2[\alpha])_{\alpha\in\mathbb{Z}^2}$ such that

$$\varphi(x) = \sum_{\alpha\in\mathbb{Z}^2} h_1[\alpha]\,\varphi(Mx-\alpha) + \sum_{\alpha\in\mathbb{Z}^2} h_2[\alpha]\,\varphi(M^2x-\alpha)\,. \tag{1.2}$$

For an integrable solution φ with $\widehat{\varphi}(0) \neq 0$ to exist, we necessarily have

$$1 = \tfrac{1}{2} \sum_{\alpha\in\mathbb{Z}^2} h_1[\alpha] + \tfrac{1}{4} \sum_{\alpha\in\mathbb{Z}^2} h_2[\alpha]\,, \tag{1.3}$$

which follows immediately from taking the integral of (1.2) over $\mathbb{R}^2$.

On the other hand, we are interested in vector valued functions $\Phi = \begin{pmatrix} \varphi_1 \\ \varphi_2 \end{pmatrix} :$ $\mathbb{R}^2 \to \mathbb{C}^2$ satisfying a matrix refinement equation

$$\Phi(x) = \sum_{\alpha\in\mathbb{Z}^2} A[\alpha]\,\Phi(2x-\alpha) \quad \text{with} \quad \varphi_2 = \varphi_1(M\cdot-e_1)\,, \tag{1.4}$$

where $(A[\alpha])_{\alpha\in\mathbb{Z}^2}$ is a sequence of matrix coefficients in $\mathbf{M}_{2\times 2}(\mathbb{C})$.

We shall show that the second row of the matrix refinement equation in (1.4) is actually a three scale refinement equation for φ_2, and we want to study conditions ensuring that we can switch back and forth between (1.2) and (1.4).

We shall always assume that all masks are finitely supported and that all functions have compact support.

2 Equivalence of the Two Types of Scaling Equations

First, we study consequences of (1.4) on the entries of the matrix mask, where we write, as usual, $A[\alpha] = (a_{i,j}[\alpha])_{i,j=1}^2$. We also show under which conditions (1.2) and (1.4) are equivalent.

Theorem 2.1. (i) *Assume that the matrix refinement equation* (1.4) *has a nontrivial solution with the property that the family* $\{\varphi_1(\cdot-\alpha), \varphi_2(\cdot-\alpha) : \alpha \in \mathbb{Z}^2\}$ *is linearly independent. Then the matrix mask satisfies*

$$a_{1,1}[\alpha] = \begin{cases} \sum_\beta a_{2,1}[\beta+e]\, a_{2,1}[\alpha-M\beta+e]\,, & \alpha \in M\mathbb{Z}^2\,, \\ \Big(\sum_\beta a_{2,1}[\beta+e]\, a_{2,1}[\alpha-M\beta+e]\Big) + a_{2,2}[M^{-1}(\alpha+e_1)]\,, & \textit{else,} \end{cases}$$

and

$$a_{1,2}[\alpha] = \quad \sum_\beta a_{2,1}[\beta+e]\, a_{2,2}[\alpha-M\beta+e]\,. \tag{2.1}$$

Furthermore, $\varphi := \varphi_1$ *satisfies the three scale refinement equation* (1.2) *with*

$$h_1[\alpha] = a_{2,1}[\alpha+e]$$

$$\textit{and} \qquad h_2[\alpha] = \begin{cases} 0\,, & \alpha \in M\mathbb{Z}^2\,, \\ a_{2,2}[M^{-1}(\alpha+e_1)]\,, & \textit{else.} \end{cases}$$

(ii) *Conversely, the three scale refinement equation* (1.2) *can be rewritten as a matrix refinement equation of the form* (1.4) *with* $\varphi_1 := \varphi$ *if and only if its mask satisfies*

$$h_2[\alpha] = 0 \quad \forall\, \alpha \in M\mathbb{Z}^2\,, \tag{2.2}$$

and then the matrix mask is given by

$$A[\alpha] = \begin{pmatrix} \Big(\sum_\beta h_1[\beta]\, h_1[\alpha-M\beta]\Big) + h_2[\alpha] & \sum_\beta h_1[\beta]\, h_2[M\alpha-M^2\beta+e_1] \\ h_1[\alpha-e] & h_2[M\alpha-e_1] \end{pmatrix}.$$

Proof. (i) Let Φ be a solution of (1.4). Then $\varphi = \varphi_1$ satisfies

$$\begin{aligned} \varphi(x) = \varphi_1(x) &= \sum_{\alpha\in\mathbb{Z}^2} a_{1,1}[\alpha]\, \varphi_1(M^2x-\alpha) + \sum_{\alpha\in\mathbb{Z}^2} a_{1,2}[\alpha]\, \varphi_2(M^2x-\alpha) \\ &= \sum_{\alpha\in\mathbb{Z}^2} a_{1,1}[\alpha]\, \varphi(M^2x-\alpha) + \sum_{\alpha\in\mathbb{Z}^2} a_{1,2}[\alpha]\, \varphi(M(M^2x-\alpha)-e_1) \\ &= \sum_{\alpha\in\mathbb{Z}^2} a_{1,1}[\alpha]\, \varphi(M^2x-\alpha) + \\ &\qquad + \sum_{\alpha\in M\mathbb{Z}^2+e_1} a_{1,2}[M^{-1}(\alpha-e_1)]\, \varphi(M^3x-\alpha)\,. \end{aligned}$$

On the other hand, we have

$$\begin{aligned}\varphi(x) &= \varphi_2(M^{-1}x+e/2)\\ &= \sum_{\alpha\in\mathbb{Z}^2} a_{2,1}[\alpha]\,\varphi_1(M^2(M^{-1}x+e/2)-\alpha) + \\ &\qquad + \sum_{\alpha\in\mathbb{Z}^2} a_{2,2}[\alpha]\,\varphi_2(M^2(M^{-1}x+e/2)-\alpha)\\ &= \sum_{\alpha\in\mathbb{Z}^2} a_{2,1}[\alpha]\,\varphi_1(Mx+e-\alpha) + \sum_{\alpha\in\mathbb{Z}^2} a_{2,2}[\alpha]\,\varphi_2(Mx+e-\alpha)\\ &= \sum_{\alpha\in\mathbb{Z}^2} a_{2,1}[\alpha]\,\varphi(Mx+e-\alpha) + \sum_{\alpha\in\mathbb{Z}^2} a_{2,2}[\alpha]\,\varphi(M(Mx+e-\alpha)-e_1)\\ &= \sum_{\alpha\in\mathbb{Z}^2} a_{2,1}[\alpha]\,\varphi(Mx+e-\alpha) + \sum_{\alpha\in\mathbb{Z}^2} a_{2,2}[\alpha]\,\varphi(M^2x+e_1-M\alpha)\\ &= \sum_{\alpha\in\mathbb{Z}^2} a_{2,1}[\alpha+e]\,\varphi(Mx-\alpha) + \sum_{\alpha\in M\mathbb{Z}^2-e_1} a_{2,2}[M^{-1}(\alpha+e_1)]\,\varphi(M^2x-\alpha)\,.\end{aligned}$$

This is the three scale refinement equation for φ. To be able to compare this representation with the previous one, we iterate it in the sense that we replace $\varphi(Mx-\alpha)$ by the whole expression, which yields

$$\begin{aligned}\varphi(x) = &\sum_{\alpha\in\mathbb{Z}^2} a_{2,1}[\alpha+e]\Big(\sum_{\beta\in\mathbb{Z}^2} a_{2,1}[\beta+e]\,\varphi(M(Mx-\alpha)-\beta)\\ &\qquad + \sum_{\beta\in M\mathbb{Z}^2-e_1} a_{2,2}[M^{-1}(\beta+e_1)]\,\varphi(M^2(Mx-\alpha)-\beta)\Big)\\ &+ \sum_{\alpha\in M\mathbb{Z}^2-e_1} a_{2,2}[M^{-1}(\alpha+e_1)]\,\varphi(M^2x-\alpha)\\ = &\sum_{\alpha\in\mathbb{Z}^2} a_{2,1}[\alpha+e]\sum_{\beta\in\mathbb{Z}^2} a_{2,1}[\beta+e]\,\varphi(M^2x-(\beta+M\alpha))\\ &+ \sum_{\alpha\in\mathbb{Z}^2} a_{2,1}[\alpha+e]\sum_{\beta\in M\mathbb{Z}^2-e_1} a_{2,2}[M^{-1}(\beta+e_1)]\,\varphi(M^3x-(\beta+M^2\alpha))\\ &+ \sum_{\alpha\in M\mathbb{Z}^2-e_1} a_{2,2}[M^{-1}(\alpha+e_1)]\,\varphi(M^2x-\alpha)\\ = &\sum_{\alpha\in\mathbb{Z}^2} a_{2,1}[\alpha+e]\sum_{\gamma\in\mathbb{Z}^2} a_{2,1}[\gamma-M\alpha+e]\,\varphi(M^2x-\gamma)\\ &+ \sum_{\alpha\in\mathbb{Z}^2} a_{2,1}[\alpha+e]\sum_{\gamma\in M\mathbb{Z}^2-e_1} a_{2,2}[M^{-1}(\gamma-M^2\alpha+e_1)]\,\varphi(M^3x-\gamma)\\ &+ \sum_{\alpha\in M\mathbb{Z}^2-e_1} a_{2,2}[M^{-1}(\alpha+e_1)]\,\varphi(M^2x-\alpha)\end{aligned}$$

$$= \sum_{\alpha\in\mathbb{Z}^2} \Big(\sum_{\beta\in\mathbb{Z}^2} a_{2,1}[\beta+e]\, a_{2,1}[\alpha-M\beta+e] \Big)\, \varphi(M^2x-\alpha) + \sum_{\alpha\in M\mathbb{Z}^2-e_1} a_{2,2}[M^{-1}(\alpha+e_1)]\, \varphi(M^2x-\alpha)$$
$$+ \sum_{\alpha\in M\mathbb{Z}^2-e_1} \Big(\sum_{\beta\in\mathbb{Z}^2} a_{2,1}[\beta+e]\, a_{2,2}[M^{-1}\alpha-M\beta+e/2] \Big)\, \varphi(M^3x-\alpha)\,.$$

Now we have obtained two representations of φ, whose coefficients have to be identical because of the assumed linear independence. This yields (2.1).

(ii) Let φ be a solution of (1.2), and assume that it can be rewritten in the form (1.4) with $\varphi_1 = \varphi$ and $\varphi_2 = \varphi(M\ \cdot\ -e_1)$. Then we can write

$$\begin{aligned}\varphi_2(x) &= \varphi(Mx-e_1)\\ &= \sum_{\beta\in\mathbb{Z}^2} h_1[\beta]\, \varphi(M(Mx-e_1)-\beta) + \sum_{\beta\in\mathbb{Z}^2} h_2[\beta]\, \varphi(M^2(Mx-e_1)-\beta)\\ &= \sum_{\beta\in\mathbb{Z}^2} h_1[\beta]\, \varphi(M^2x-(Me_1+\beta)) + \sum_{\beta\in\mathbb{Z}^2} h_2[\beta]\, \varphi(M^3x-(M^2e_1+\beta))\,.\end{aligned}$$

Because of the terms M^2x and M^3x in the respective arguments, we have to express the first sum using the translates of $\varphi_1(M^2x)$, and the second sum using those of $\varphi_2(M^2x)$. Thus, using $\varphi(x) = \varphi_2(M^{-1}(x+e_1))$, we obtain

$$\begin{aligned}\varphi_2(x) &= \sum_{\beta\in\mathbb{Z}^2} h_1[\beta]\, \varphi_1(M^2x-(Me_1+\beta)) +\\ &\qquad + \sum_{\beta\in\mathbb{Z}^2} h_2[\beta]\, \varphi_2(M^{-1}(M^3x-(M^2e_1+\beta)+e_1))\\ &= \sum_{\alpha\in\mathbb{Z}^2} h_1[\alpha-Me_1]\, \varphi_1(M^2x-\alpha) + \sum_{\beta\in\mathbb{Z}^2} h_2[\beta]\, \varphi_2(M^2x-M^{-1}(\beta+e_1))\,.\end{aligned}$$

The last sum is of the desired form if and only if h_2 satisfies

$$h_2[\beta] = 0 \text{ for all } \beta \text{ with } M^{-1}(\beta+e_1) \notin \mathbb{Z}^2\,.$$

Since $M^{-1}(\beta+e_1) \notin \mathbb{Z}^2$ is equivalent to $\beta \in M\mathbb{Z}^2$, this is just (2.2).

Now assume that φ is a solution of (1.2) and that (2.2) is satisfied. Letting again $\varphi_1 = \varphi$ and $\varphi_2 = \varphi(M\ \cdot\ -e_1)$, the above argument shows that

$$\varphi_2(x) = \sum_{\alpha} a_{2,1}[\alpha]\, \varphi_1(M^2x-\alpha) + \sum_{\alpha} a_{2,2}[\alpha]\, \varphi_2(M^2x-\alpha)$$

with $a_{2,1}[\alpha] = h_1[\alpha - e]$ and $a_{2,2}[\alpha] = h_2[M\alpha - e_1]$. Furthermore, we have

$$\begin{aligned}
\varphi_1(x) &= \varphi(x) \\
&= \sum_\alpha h_1[\alpha]\,\varphi(Mx-\alpha) + \sum_\alpha h_2[\alpha]\,\varphi(M^2x-\alpha) \\
&= \sum_\alpha h_1[\alpha]\Big(\sum_\beta h_1[\beta]\,\varphi(M(Mx-\alpha)-\beta) + \sum_\beta h_2[\beta]\,\varphi(M^2(Mx-\alpha)-\beta)\Big) \\
&\qquad\qquad + \sum_\alpha h_2[\alpha]\,\varphi(M^2x-\alpha) \\
&= \sum_\alpha h_1[\alpha]\Big(\sum_\gamma h_1[\gamma-M\alpha]\,\varphi(M^2x-\gamma) + \sum_\gamma h_2[\gamma-M^2\alpha]\,\varphi(M^3x-\gamma)\Big) \\
&\qquad\qquad + \sum_\gamma h_2[\gamma]\,\varphi(M^2x-\gamma) \\
&= \sum_\gamma \Big(\Big(\sum_\alpha h_1[\alpha]\,h_1[\gamma-M\alpha]\Big) + h_2[\gamma]\Big)\varphi(M^2x-\gamma) \\
&\qquad + \sum_\gamma \Big(\sum_\alpha h_1[\alpha]\,h_2[\gamma-M^2\alpha]\Big)\varphi(M^3x-\gamma)\,.
\end{aligned}$$

Because of (2.2), we may replace in the last sum γ by $M\gamma + e_1$ and obtain

$$\begin{aligned}
\varphi_1(x) &= \sum_\gamma \Big(\Big(\sum_\alpha h_1[\alpha]\,h_1[\gamma-M\alpha]\Big) + h_2[\gamma]\Big)\varphi(M^2x-\gamma) \\
&\qquad + \sum_\gamma \Big(\sum_\alpha h_1[\alpha]\,h_2[M\gamma+e_1-M^2\alpha]\Big)\varphi(M^3x-M\gamma-e_1) \\
&= \sum_\gamma \Big(\Big(\sum_\alpha h_1[\alpha]\,h_1[\gamma-M\alpha]\Big) + h_2[\gamma]\Big)\varphi_1(M^2x-\gamma) \\
&\qquad + \sum_\gamma \Big(\sum_\alpha h_1[\alpha]\,h_2[M\gamma+e_1-M^2\alpha]\Big)\varphi_2(M^2x-\gamma)\,,
\end{aligned}$$

which yields the claim. □

Remark. In [3], it has been shown that any poly–scale refinement equation can be rewritten as a matrix refinement equation with the same scaling matrix. The crucial difference in our setup, though, is that the three scale refinement equation (1.2) uses scaling matrix M, while the matrix refinement equation (1.4) uses $M^2 = 2\,I$.

3 Interpolation

Now we show which conditions on the masks ensure the interpolation property.

Lemma 3.1. *Assume that φ satisfies a three scale refinement equation* (1.2) *with the transcription condition* (2.2). *Then φ interpolates on the lattice $M^{-1}\mathbb{Z}^2$ if and only if the mask satisfies*

$$h_1 = \delta\,,$$

i.e., $h_1[\alpha] = \delta_{0,\alpha} = 1$. *Note that for the matrix mask, this implies*

$$A[\alpha] = \begin{pmatrix} \delta_{0,\alpha} + h_2[\alpha] & h_2[M\alpha{+}e_1] \\ \delta_{e,\alpha} & h_2[M\alpha{-}e_1] \end{pmatrix}.$$

Proof. Interpolation on $M^{-1}\mathbb{Z}^2 = \mathbb{Z}^2 \cup (\mathbb{Z}^2{+}e/2)$ means for $\alpha \in \mathbb{Z}^2$

$$\begin{aligned}\varphi(\alpha) = \delta_{0,\alpha} &= \sum_\beta h_1[\beta]\,\varphi(M\alpha{-}\beta) + \sum_\beta h_2[\beta]\,\varphi(M^2\alpha{-}\beta)\\ &= h_1[M\alpha] + h_2[M^2\alpha]\end{aligned}$$

and

$$\begin{aligned}\varphi(\alpha + e/2) = 0 &= \sum_\beta h_1[\beta]\,\varphi(M\alpha{+}e_1{-}\beta) + \sum_\beta h_2[\beta]\,\varphi(M^2\alpha{+}e{-}\beta)\\ &= h_1[M\alpha{+}e_1] + h_2[M^2\alpha{+}e]\,.\end{aligned}$$

Now condition (2.2) yields $h_2[M^2\alpha] = 0$ and $h_2[M^2\alpha{+}e] = h_2[M(M\alpha{+}e_1)] = 0$, so the interpolation property implies $h_1 = \delta$.

The converse follows from Proposition 3.10 in [2] and condition (2.2). □

4 Application

The above results enable us to work with the well known fast algorithms for the two scale case, but nevertheless use the three scale refinement equation to easily modify the mask without losing symmetries. We want to take this approach to try and smoothen the pyramidal function of the Conti-Jetter scheme, where we have Φ satisfying (1.4) with

$$\mathbf{A} = \begin{pmatrix} & \vdots & \vdots & \vdots & \vdots & \vdots & \\ \cdots & \mathbf{O} & \mathbf{O} & \mathbf{O} & \mathbf{O} & \mathbf{O} & \cdots \\ \cdots & \mathbf{O} & \mathbf{O} & \begin{pmatrix} \frac{1}{2} & 0 \\ 0 & \frac{1}{2} \end{pmatrix} & \begin{pmatrix} 0 & 0 \\ 1 & \frac{1}{2} \end{pmatrix} & \mathbf{O} & \cdots \\ \cdots & \mathbf{O} & \begin{pmatrix} \frac{1}{2} & \frac{1}{2} \\ 0 & 0 \end{pmatrix} & \begin{pmatrix} 1 & \frac{1}{2} \\ 0 & \frac{1}{2} \end{pmatrix} & \begin{pmatrix} \frac{1}{2} & 0 \\ 0 & \frac{1}{2} \end{pmatrix} & \mathbf{O} & \cdots \\ \cdots & \mathbf{O} & \begin{pmatrix} 0 & \frac{1}{2} \\ 0 & 0 \end{pmatrix} & \begin{pmatrix} \frac{1}{2} & \frac{1}{2} \\ 0 & 0 \end{pmatrix} & \mathbf{O} & \mathbf{O} & \cdots \\ \cdots & \mathbf{O} & \mathbf{O} & \mathbf{O} & \mathbf{O} & \mathbf{O} & \cdots \\ & \vdots & \vdots & \vdots & \vdots & \vdots & \end{pmatrix},$$

so $\varphi = \varphi_1$ satisfies (1.2) with $h_1 = \delta$ and

$$h_2 = \begin{pmatrix} & \vdots & \vdots & \vdots & \vdots & \vdots & \\ \cdots & 0 & 0 & 0 & 0 & 0 & \cdots \\ \cdots & 0 & 0 & \frac{1}{2} & 0 & 0 & \cdots \\ \cdots & 0 & \frac{1}{2} & 0 & \frac{1}{2} & 0 & \cdots \\ \cdots & 0 & 0 & \frac{1}{2} & 0 & 0 & \cdots \\ \cdots & 0 & 0 & 0 & 0 & 0 & \cdots \\ & \vdots & \vdots & \vdots & \vdots & \vdots & \end{pmatrix}.$$

For the modification, we remind the reader of the four point scheme, a modification of the mask for the triangle function in the univariate case [4,5]. The mask for the triangle function is

$$h = (\ \cdots\ 0\ 0\ 0\ \tfrac{1}{2}\ 1\ \tfrac{1}{2}\ 0\ 0\ 0\ \cdots\ ).$$

The most simple modification preserving symmetry and interpolation as well as allowing for polynomial reproduction is the mask

$$h = (\ \cdots\ 0\ \vartheta\ 0\ \tfrac{1}{2}{+}\omega\ 1\ \tfrac{1}{2}{+}\omega\ 0\ \vartheta\ 0\ \cdots\ ), \tag{4.1}$$

and polynomial reproduction up to degree 1 requires $\vartheta = -\omega$. The authors show in [5] that this mask generates a $\mathbf{C}^1$–function for $\omega \in \left]0, \frac{1}{8}\right[$, and $\omega = \frac{1}{16}$ actually yields polynomial reproduction up to degree 3.

Analogously, we make for h_2 the "Ansatz"

$$h_2 = \begin{pmatrix} & \vdots & \vdots & \vdots & \vdots & \vdots & \vdots & \vdots & \\ \cdots & 0 & \varrho & 0 & \vartheta & 0 & \varrho & 0 & \cdots \\ \cdots & \varrho & 0 & \lambda & 0 & \lambda & 0 & \varrho & \cdots \\ \cdots & 0 & \lambda & 0 & \frac{1}{2}+\omega & 0 & \lambda & 0 & \cdots \\ \cdots & \vartheta & 0 & \frac{1}{2}+\omega & 0 & \frac{1}{2}+\omega & 0 & \vartheta & \cdots \\ \cdots & 0 & \lambda & 0 & \frac{1}{2}+\omega & 0 & \lambda & 0 & \cdots \\ \cdots & \varrho & 0 & \lambda & 0 & \lambda & 0 & \varrho & \cdots \\ \cdots & 0 & \varrho & 0 & \vartheta & 0 & \varrho & 0 & \cdots \\ & \vdots & \vdots & \vdots & \vdots & \vdots & \vdots & \vdots & \end{pmatrix} . \tag{4.2}$$

The necessary condition for convergence, *i.e.*, the existence of a common left eigenvector of the matrix submasks, is equivalent to the condition

$$\omega = -\vartheta - 2\,\lambda - 2\,\varrho\,,$$

and this yields polynomial reproduction up to total degree 1, by the algorithm given in [7]. With this algorithm, it is also easy to see that reproduction of polynomials up to total degree 3 is equivalent to the conditions

$$\omega = 2\,\varrho + \tfrac{1}{16}\,, \quad \vartheta = -2\,\varrho - \tfrac{1}{16}\,, \quad \lambda = -\varrho\,.$$

Figures 1 and 2 show φ for the cases $\varrho = 0$ and $\varrho = \frac{1}{32}$, respectively.

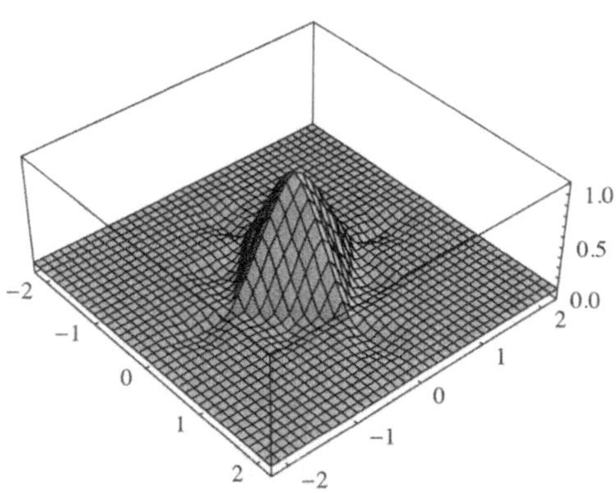

Figure 1.

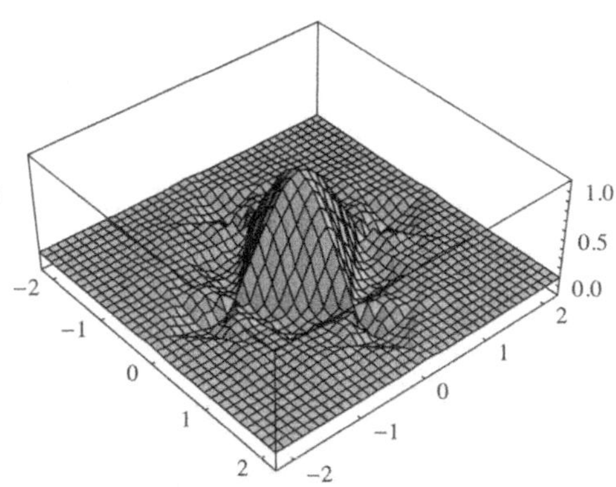

Figure 2.

It is not clear yet whether there exist values for ϱ such that $\varphi \in \mathbf{C}^1(\mathbb{R}^2)$. It is interesting to note, though, that we may choose $\varrho = 0$ and thus $\lambda = 0$, and nevertheless have polynomial reproduction up to total degree 3, although the mask h_2 is only supported along the axes, *i.e.*, on the set $(\mathbb{Z} \times \{0\}) \cup (\{0\} \times \mathbb{Z})$. This can be explained by the following.

Observation. (i) In every refinement step, each point of the new grid lies on a unique straight line of the old grid.
(ii) If h_2 is only supported along the axes, the new value at each new point is determined by the old values at the old points on this straight line only.

Part (i) is almost trivial to see, part (ii) is fairly easy to see but complicated to show formally, so we invite the reader to convince him– or herself of this fact. An important consequence of this observation is though that by taking an appropriate interpolating, univariate mask and using its coefficients along the axes in h_2, we can easily generate masks with polynomial reproduction of arbitrarily high degree. This of course is simply due to the fact that the restriction of a bivariate polynomial to any straight line is a univariate polynomial of no higher degree.

But this phenomenon also has its negative sides. As usual, we can generate φ by taking the δ–sequence as input data and running the subdivision scheme. By the above observation, we know that φ vanishes on all lines in the original grid that do not run through the origin. This means that at every point of the original grid different from $(0,0)$, there exist two vanishing directional derivatives, so if $\varphi \in \mathbf{C}^1$, we have $\nabla\varphi = (0,0)$ at all points of the original grid – including the origin, since our assumptions on the symmetry of the mask ensure that φ is symmetric. But this makes it impossible to reproduce linear polynomials, which is a necessary condition for φ to be in $\mathbf{C}^1$. So we have shown the following.

Proposition 4.1. *If $h_1 = \delta$, and h_2 is symmetric and supported only along the axes, then $\varphi \notin \mathbf{C}^1(\mathbb{R}^2)$.*

References

[1] C. Conti, K. Jetter: *A new subdivision method for bivariate splines on the four-directional mesh*, J. Comput. Appl. Math. **119** (2000), 936–952.

[2] C. Conti, G. Zimmermann: *Interpolatory vector subdivision schemes*, submitted.

[3] S. Dekel, N. Dyn: *Poly–scale refinability and subdivision,* Appl. Comput. Harmon. Anal. **13** (2002), 35–62.

[4] S. Dubuc: *Interpolation through an iterative scheme,* J. Math. Anal. Appl. **114** (1986), 185–204.

[5] N. Dyn, J. A. Gregory, D. Levin: *A four–point interpolatory subdivision scheme for curve design,* Comput. Aided Geom. Design **4** (1987), 257–268.

[6] T. Goodman: *Pairs of refinable bivariate splines,* In: *Advanced Topics in Multivariate Approximation, Approximations and Decompositions* Vol. **8**, 125–138, World Scientific, Singapore 1996.

[7] K. Jetter, G. Zimmermann: *Polynomial reproduction in subdivision,* Adv. Comp. Math., to appear (2002).

Address:

Georg Zimmermann
Institut für Angewandte Mathematik und Statistik
Universität Hohenheim (110)
D–70593 Stuttgart
Germany

GPSR Compliance

The European Union's (EU) General Product Safety Regulation (GPSR) is a set of rules that requires consumer products to be safe and our obligations to ensure this.

If you have any concerns about our products, you can contact us on ProductSafety@springernature.com

In case Publisher is established outside the EU, the EU authorized representative is:

Springer Nature Customer Service Center GmbH
Europaplatz 3
69115 Heidelberg, Germany

Batch number: 10577880

Printed by Printforce, the Netherlands